T0340203

Pet-to-Man Travelling Staphylococci

COVER PHOTO LEGENDS

Left picture: *Staphylococcus pseudintermedius* on Tryptic Soy Agar, Liofilchem®, Italy (photo by Edoardo Carretto; published in Savini et al., J Clin Microbiol. 2013;51(5):1636-8)

Right picture: methicillin-resistant *Staphylococcus aureus* on Chromatic™ MRSA, Liofilchem®, Italy (photo by Roberta Marrollo; published in Savini et al., Int J Clin Exp Pathol. 2014;7(5):2670-2. eCollection 2014)

Background picture: Fata, my sweet kitten; Zorro, my tender doggie (photo by Alessia Arvanitis)

Pet-to-Man Travelling Staphylococci

A World in Progress

Edited by

Vincenzo Savini

ACADEMIC PRESS

An imprint of Elsevier

Academic Press is an imprint of Elsevier
125 London Wall, London EC2Y 5AS, United Kingdom
525 B Street, Suite 1800, San Diego, CA 92101-4495, United States
50 Hampshire Street, 5th Floor, Cambridge, MA 02139, United States
The Boulevard, Langford Lane, Kidlington, Oxford OX5 1GB, United Kingdom

Notices
Knowledge and best practice in this field are constantly changing. As new research and
experience broaden our understanding, changes in research methods, professional practices,
or medical treatment may become necessary.

Practitioners and researchers must always rely on their own experience and knowledge in
evaluating and using any information, methods, compounds, or experiments described herein.
In using such information or methods they should be mindful of their own safety and the safety
of others, including parties for whom they have a professional responsibility.

To the fullest extent of the law, neither the Publisher nor the authors, contributors, or editors,
assume any liability for any injury and/or damage to persons or property as a matter of products
liability, negligence or otherwise, or from any use or operation of any methods, products,
instructions, or ideas contained in the material herein.

Library of Congress Cataloging-in-Publication Data
A catalog record for this book is available from the Library of Congress

British Library Cataloguing-in-Publication Data
A catalogue record for this book is available from the British Library

ISBN: 978-0-12-813547-1

For information on all Academic Press publications
visit our website at https://www.elsevier.com/books-and-journals

Working together
to grow libraries in
developing countries

www.elsevier.com • www.bookaid.org

Publisher: John Fedor
Acquisition Editor: Linda Versteeg-Buschman
Editorial Project Manager: Ana Claudia A. Garcia
Production Project Manager: Punithavathy Govindaradjane
Cover Designer: Greg Harris

Typeset by SPi Global, India

Dedication

To Giorgia and Alessia, my life.
To Taziana and Francesco, who have given me life.
To Michela, Daniela, Nicola, my nest.
To Adalgisa, Ester, Enrica, grannies of Light.
To Elda, aunt of my infancy.
To Domenico D'Antonio, charismatic master.
To my Guardian Angel, walking with me.
To angels on Earth, protecting childhood.

Contents

7.   The Staphylococcal Coagulases

Emilia Bonar, Jacek Międzobrodzki, Benedykt Władyka

8.   The Staphylococcal Hemolysins

Eugenio Pontieri

9.   The Staphylococcal Panton-Valentine Leukocidin (PVL)

Paweł Nawrotek, Jolanta Karakulska, Karol Fijałkowski

10.   The Staphylococcal Exfoliative Toxins

Michał Bukowski, Benedykt Władyka,
Adam Dubin, Grzegorz Dubin

11. Extracellular Proteases of *Staphylococcus* spp.

Natalia Stach, Paweł Kaszycki, Benedykt Władyka, Grzegorz Dubin

12. Staphylococcal Lipases

Aldo Lepidi

13. Staphylococcal Bacteriocins

Paweł Mak

17. Methicillin Resistance in *Staphylococcus aureus*

Edoardo Carretto, Rosa Visiello, Paola Nardini

18. In Vivo Resistance Mechanisms: Staphylococcal Biofilms

Barbara Różalska, Beata Sadowska

19. Autovaccines in Individual Therapy of Staphylococcal Infections

Stefania Giedrys-Kalemba, Danuta Czernomysy-Furowicz, Karol Fijałkowski, Joanna Jursa-Kulesza

20. Experimental Animal Models in Evaluation of Staphylococcal Pathogenicity

Jacek Międzobrodzki, Maja Kosecka-Strojek

21. Application of Staphylococci in the Food Industry and Biotechnology

Benedykt Władyka, Emilia Bonar

Contributors

Numbers in Parentheses indicate the pages on which the author's contributions begin.

Anna Białecka (185), Center of Microbiological Research and Autovaccines of Dr. Jan Bóbr Ltd., Krakow, Poland

Joanna Białecka (185), Center of Microbiological Research and Autovaccines of Dr. Jan Bóbr Ltd., Krakow, Poland

Emilia Bonar (95,281), Department of Analytical Biochemistry, Faculty of Biochemistry, Biophysics and Biotechnology, Jagiellonian University, Krakow, Poland

Aneta Buda (11), Department of Microbiology, Faculty of Biochemistry, Biophysics and Biotechnology, Jagiellonian University, Krakow, Poland

Michał Bukowski (127), Department of Analytical Biochemistry, Faculty of Biochemistry, Biophysics and Biotechnology, Jagiellonian University, Krakow, Poland

Edoardo Carretto (225), Clinical Microbiology Laboratory—IRCCS Arcispedale Santa Maria Nuova, Reggio Emilia, Italy

Danuta Czernomysy-Furowicz (253), Department of Immunology, Microbiology and Physiological Chemistry, Faculty of Biotechnology and Animal Husbandry, West Pomeranian University of Technology, Szczecin, Poland

Giovanni Di Bonaventura (1), Department of Experimental and Clinical Sciences, "G. D'Annunzio" University of Chieti-Pescara; Center of Excellence on Aging, "G. D'Annunzio" University Foundation, Chieti, Italy

Cristina E. Di Francesco (43), Faculty of Veterinary Medicine, University of Teramo, Teramo, Italy

Barbara Di Martino (43), Faculty of Veterinary Medicine, University of Teramo, Teramo, Italy

Adam Dubin (127), Department of Analytical Biochemistry, Faculty of Biochemistry, Biophysics and Biotechnology, Jagiellonian University, Krakow, Poland

Grzegorz Dubin (127,135), Malopolska Centre of Biotechnology, Jagiellonian University; Department of Microbiology, Faculty of Biochemistry, Biophysics and Biotechnology, Jagiellonian University, Krakow, Poland

Marco Favaro (25), Department of Experimental Medicine and Surgery, Tor Vergata University of Rome, Rome, Italy

Paolo Fazii (51), Clinical Microbiology and Virology, Civic Hospital of Pescara, Pescara, Italy

Karol Fijałkowski (117,253), Department of Immunology, Microbiology and Physiological Chemistry, Faculty of Biotechnology and Animal Husbandry, West Pomeranian University of Technology, Szczecin, Poland

Carla Fontana (25), Department of Experimental Medicine and Surgery, Tor Vergata University of Rome; Foundation Polyclinic Tor Vergata, Clinical Microbiology Laboratories, Rome, Italy

Giovanni Gherardi (1), Integrated Research Centre (CIR), University Campus Biomedico, Rome, Italy

Stefania Giedrys-Kalemba (253), Department of Microbiology and Immunology, Pomeranian Medical University, Szczecin, Poland

Weronika M. Ilczyszyn (173), Department of Microbiology, Faculty of Biochemistry, Biophysics and Biotechnology, Jagiellonian University, Krakow, Poland

Joanna Jursa-Kulesza (253), Department of Microbiology and Immunology, Pomeranian Medical University, Szczecin, Poland

Jolanta Karakulska (117), Department of Immunology, Microbiology and Physiological Chemistry, Faculty of Biotechnology and Animal Husbandry, West Pomeranian University of Technology, Szczecin, Poland

Andrzej Kasprowicz (185), Center of Microbiological Research and Autovaccines of Dr. Jan Bóbr Ltd., Krakow, Poland

Paweł Kaszycki (135), Biochemistry Unit, Institute of Plant Biology and Biotechnology, Faculty of Biotechnology and Horticulture, University of Agriculture, Krakow, Poland

Maja Kosecka-Strojek (11,173,265), Department of Microbiology, Faculty of Biochemistry, Biophysics and Biotechnology, Jagiellonian University, Krakow, Poland

Beata Krawczyk (199), Department of Molecular Biotechnology and Microbiology, Faculty of Chemistry, Gdańsk University of Technology, Gdańsk, Poland

Józef Kur (199), Department of Molecular Biotechnology and Microbiology, Faculty of Chemistry, Gdańsk University of Technology, Gdańsk, Poland

Aldo Lepidi (147), Department of Life, Health and Environmental Sciences (MeSVA), L'Aquila University, L'Aquila, Italy

Clemencia Chaves López (71), Faculty of Bioscience and Technology for Food, Agriculture and Environment, University of Teramo, Teramo, Italy

Paweł Mak (161), Department of Analytical Biochemistry, Faculty of Biochemistry, Biophysics and Biotechnology, Jagiellonian University, Krakow, Poland

Roberta Marrollo (51), Clinical Microbiology and Virology, Laboratory of Bacteriology and Mycology, Civic Hospital of Pescara, Pescara, Italy

Fulvio Marsilio (43), Faculty of Veterinary Medicine, University of Teramo, Teramo, Italy

Giovanni Mazzarrino (71), Faculty of Bioscience and Technology for Food, Agriculture and Environment, University of Teramo, Teramo, Italy

Jacek Międzobrodzki (11,95,173,265), Department of Microbiology, Faculty of Biochemistry, Biophysics and Biotechnology, Jagiellonian University, Krakow, Poland

Paola Nardini (225), Clinical Microbiology Laboratory—IRCCS Arcispedale Santa Maria Nuova, Reggio Emilia, Italy

Paweł Nawrotek (117), Department of Immunology, Microbiology and Physiological Chemistry, Faculty of Biotechnology and Animal Husbandry, West Pomeranian University of Technology, Szczecin, Poland

Antonello Paparella (71), Faculty of Bioscience and Technology for Food, Agriculture and Environment, University of Teramo, Teramo, Italy

Eugenio Pontieri (103), Department of Biotechnological and Applied Clinical Sciences, University of L'Aquila, L'Aquila, Italy

Chiara Rossi (71), Faculty of Bioscience and Technology for Food, Agriculture and Environment, University of Teramo, Teramo, Italy

Barbara Różalska (237), Department of Immunology and Infectious Biology, Faculty of Biology and Environmental Protection, University of Lodz, Lodz, Poland

Beata Sadowska (237), Department of Immunology and Infectious Biology, Faculty of Biology and Environmental Protection, University of Lodz, Lodz, Poland

Vincenzo Savini (1,51), Clinical Microbiology and Virology, Laboratory of Bacteriology and Mycology, Civic Hospital of Pescara, Pescara, Italy

Annalisa Serio (71), Faculty of Bioscience and Technology for Food, Agriculture and Environment, University of Teramo, Teramo, Italy

Natalia Stach (135), Department of Microbiology, Faculty of Biochemistry, Biophysics and Biotechnology, Jagiellonian University, Krakow, Poland

Rosa Visiello (225), Clinical Microbiology Laboratory—IRCCS Arcispedale Santa Maria Nuova, Reggio Emilia, Italy

Benedykt Władyka (95,127,135,281), Department of Analytical Biochemistry, Faculty of Biochemistry, Biophysics and Biotechnology, Jagiellonian University, Krakow, Poland

Preface

Staphylococci are pathogens with medical and veterinary impacts, causing a wide range of infection states both in humans and animals.

Their pathogenicity relies on virulence factors such as coagulase and PVL (Panton-Valentine leucocidin). Also, they spread easily across a patient population in the hospital environment if hygiene practices are not in place.

Worse still, these bacteria continue to evolve resistance to antibiotic compounds, even including newer agents, such as vancomycin and daptomycin, and antimicrobial resistance is a major implication in treatment failure.

In a world where an increasing population and global travel facilitate circulation of these organisms, they should always be considered as a matter of concern.

This volume would like to shed light on the issue concerning animal-to-man transmission, and vice versa, of staphylococci and mobile genetic elements they may harbor.

Today, in fact, animals, especially pets, have become true family members, so they share domestic environments with their owners, including sofas and beds.

This leads to closer contact between companion animals' owners and vets, due to their pets' health problems, mainly dogs and cats.

Therefore, an exchange of bacteria and microbial genetic material between man and animals does occur, especially in the household, and this book aims to unearth trouble concerning the interspecies travel of staphylococci, as well as ecologic niches they inhabit.

Vincenzo Savini

Chapter 1

Staphylococcal Taxonomy

Giovanni Gherardi*, Giovanni Di Bonaventura[†,‡], Vincenzo Savini[§]
*Integrated Research Centre (CIR), University Campus Biomedico, Rome, Italy, †Department of Experimental and Clinical Sciences, "G. D'Annunzio" University of Chieti-Pescara, Chieti, Italy, ‡Center of Excellence on Aging, "G. D'Annunzio" University Foundation, Chieti, Italy, §Clinical Microbiology and Virology, Laboratory of Bacteriology and Mycology, Civic Hospital of Pescara, Pescara, Italy

1.1 INTRODUCTION

Historically, the bacterial species belonging to the two related genera *Staphylococcus* and *Micrococcus* were considered, along with the species, to belong to the genera *Stomatococcus* and *Planococcus* as part of the family *Micrococcaceae*. Later, molecular analysis and phylogenetic and chemotaxonomic data have revealed that staphylococci and micrococci are not closely related [1]. The *Staphylococcus* genus belongs to the *Bacillus-Lactobacillus-Streptococcus* cluster, which consists of Gram-positive bacteria with a low G/C content in chromosomal DNA. The 2nd edition of *Bergey's Manual of Systematic Bacteriology* [2] updated in 2004 reclassified *Staphylococcus* genus in a new family, named *Staphylococcaceae*, together with the genera *Jeotgalicoccus, Macrococcus, Salinicoccus,* and *Gemella* [3]. The *Staphylococcaceae* family together with *Bacillaceae, Planococcaceae, Listeriaceae,* and other families are part of the order *Bacillales* [3].

In addition, some species previously belonging to the *Micrococcus* genus have been reclassified into the newly established genera *Kocuria, Nesterenkonia, Kytococcus,* and *Dermacoccus.* These genera were reclassified into two related families, the newly redefined *Micrococcaceae* and the newly established *Dermacoccaceae*, typically consisting of species of Gram-positive bacteria with DNA with a high G/C content [4–8]. Both families belong to the suborder *Micrococcineae* [1]. The *Micrococcaceae* family now consists of the genera *Kocuria, Nesterenkonia, Acaricomes, Arthrobacter, Citricoccus, Renibacterium, Rothia;* and *Stomatococcus mucillaginosus*, which is the only species belonging to the former genus *Stomatococcus*, has been reclassified as *Rothia mucilaginosa* [9]. The other family of the *Micrococcineae*, designated *Dermacoccaceae*, contains the genera *Dermacoccus, Demetria, Kytococcus, Luteipulveratus,* and *Yimella*, other than the previously species belonging to *Micrococcus.*

Pet-to-Man Travelling Staphylococci: A World in Progress. https://doi.org/10.1016/B978-0-12-813547-1.00001-7

1

Staphylococci are Gram-positive, nonmotile cocci, that upon microscopic examination, appear as clusters, with a typical cell wall of Gram positive bacteria, containing teichoic acid and peptidoglycan [10]. Staphylococci are facultative anaerobes, with the exception of the anaerobic species *S. saccharolyticus* and *S. aureus* subsp. *anaerobius*. Although staphylococci are usually catalase positive, rare strains that are catalase-negative have been reported [11]. Most staphylococcal species are oxidase negative in the modified oxidase test, with the exception of *S. fleurettii, S. lentus, S. sciuri,* and *S. vitulinus.* Staphylococci are able to grow in the presence of 10% NaCl at a temperature ranging between 18°C and 40°C. They present a metabolism that is typically respiratory and fermentative. Moreover, a common characteristic of all staphylococcal species is that they are susceptible to lysostaphin, with only rare exceptions [6,12]. The percentage of G/C content in chromosomal DNA of staphylococcal species is approximately of 30%–40%. Coagulase-positive staphylococci (CoPS) represent the major pathogenic species within the genus, and possess coagulase, an enzyme able to coagulate rabbit plasma by converting fibrinogen into fibrin. Conversely, those lacking coagulase are classified as coagulase-negative staphylococci (CoNS), and are relatively minor pathogens that generally cause opportunistic infections in compromised hosts.

Staphylococci, including *S. aureus*, generally are opportunistic pathogens or commensals resident on host skin and mucosae in animals and humans. Staphylococci from carriage sites can spread and be transmitted into the environment where they are able to survive for a long time [13,14]. Staphylococci that are commensals may act as pathogens if they succeed in entering the host by several mechanisms, such as skin trauma, inoculation, device implantation, both in immunocompromised patients, and in all those showing an altered microbiota [15–17]. In human beings, more than 80% of hospital-acquired *S. aureus* diseases are endogenous infections that are caused by strains carried in the patients' nose [18]. Taken together, an accurate and reliable species identification of all staphylococci is highly mandatory to permit detailed determination of the host-pathogen relationships [19,20].

1.2 METHODS USED IN STAPHYLOCOCCAL TAXONOMY

In 1925, the first differentiation within the genus *Staphylococcus* consisted of the introduction of two separate groups, that is those of CoPS (originally named "*S. aureus* group") and CoNS, [21]. Later, another classification of CoNS followed, based on their susceptibility or resistance to novobiocin, with novobiocin-susceptible CoNS species belonging to the "*S. epidermidis* group," and novobiocin-resistant belonging to the "*S. saprophyticus* group" [22,23]. Despite limitations, coagulase activity and novobiocin susceptibility represent phenotypic tests that are still used for presumptive identification of staphylococcal isolates. Since 1962, where only three staphylococcal species were identified, an extensive revision of staphylococcal taxonomy has been performed.

Overall, 45 staphylococcal species and 24 subspecies have been described so far in the *Staphylococcus* genus [2,12,24–28]. This has been accomplished through molecular methods. The most clinically significant species in human and veterinary medicine can be identified on the basis of several key characteristics [12], mainly, colony morphology, coagulase production, agglutination assays, and novobiocin and polymixin B susceptibility. Also, the classical fermentation, oxidation, degradation, and hydrolysis (of various substrates) assays have been incorporated into commercial manual and automated biochemical systems [12]. Nevertheless, an accurate characterization of staphylococci to the species level is quite laborious, with phenotypic methods frequently failing in providing correct identification. For this reason, various molecular biology methods have been introduced into microbiology laboratories. These molecular techniques typically require the use of several species-specific PCR primers or hybridization probes, or may necessitate multiple restriction enzymes, although they usually are not able to differentiate all known species simultaneously. Partial 16S rRNA gene sequencing and PCR-restriction fragment length polymorphism (PCR-RFLP) analysis have been described for *Staphylococcus* species identification [29–31], but these methods do not differentiate among some staphylococcal species, that is, between *S. lentus* and *S. sciuri*. The use of PCR-RFLP analysis of the 23S rRNA gene with two restriction enzymes, instead, have been observed to correctly identify *Staphylococcus* species [32], although interpretation of the results is complicated [33]. Recently, amplified fragment length polymorphism fingerprinting has been introduced, and has proved to be highly discriminating, although time-consuming and expensive [34]. Whole-genome DNA-DNA hybridization analysis, again, showed a good performance [35], but it proved to be unsuitable for routine practice. The use of nucleic acid targets provides an alternative option to reach accurate *Staphylococcus* classification, due to their high sensitivity and specificity.

Because a large amount of 16S rRNA sequence data is available in public databases, it is not surprising that the 16S rRNA gene is the most commonly used target for bacterial species identification. Nevertheless, reliability of 16S rRNA gene sequences, although useful in phylogenetic studies at the genus level, is debatable when applied to the species level. In this regard, the 16S rRNA sequence similarity has been shown to be very high for several *Staphylococcus* species [36], such as *S. caprae* and *S. capitis*, that cannot be distinguished by their 16S rRNA gene sequences [34]; and *S. vitulinus*, *S. saccharolyticus*, *S. capitis* subsp. *ureolyticus*, *S. caprae*, the two *S. aureus* subspecies, and the *S. cohnii* subspecies, that have identical 16S rRNA gene sequences in variable regions V1, V3, V7, and V9 [37]. In addition to the 16S rRNA gene [29–31], the 16S-23S rRNA intergenic spacer region [32], and several gene targets, such as the heat shock protein 60 (*hsp60*) gene [38–40], the *fem*A gene [41], the *sod*A gene [42], the *tuf* gene [43], the *rpo*B gene [44,45], and the *gap* gene [46,47] proved to be useful markers for accurate identification. The *tuf* gene-derived data often showed more intraspecies sequence divergence than the 16S rRNA-derived data. Apparently, the 16S rRNA gene is more highly conserved than the *tuf* gene, indicating that the *tuf* gene constitutes a more discriminatory target than the 16S rRNA to differentiate closely related *Staphylococcus* species (Fig. 1.1).

With DNA-DNA reassociation, the *Staphylococcus* species could be divided in eight distinct species groups, represented by *S. epidermidis*, *S. saprophyticus*, *S. simulans*, *S. intermedius*, *S. hyicus*, *S. sciuri*, *S. auricularis*, and *S. aureus* [15,16]. The same groups could be identified by using *hsp60* and

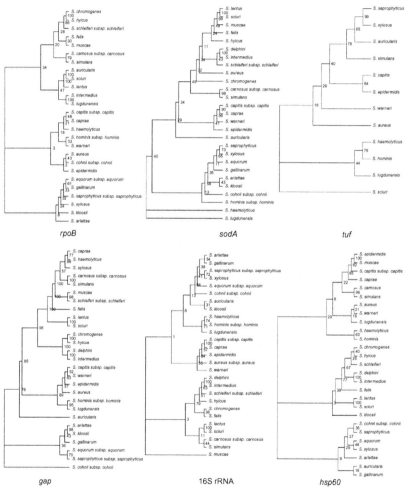

FIG. 1.1 Neighbor-joining tree based on the *gap*, 16S rRNA, *rpoB*, *sodA*, and *tuf* gene sequences showing the phylogenetic relationships among the staphylococcal species [34]. *(From Ghebremedhin B, Layer F, Konig W, Konig, B. Genetic classification and distinguishing of Staphylococcus species based on different partial gap, 16S rRNA, hsp60, rpoB, sodA, and tuf gene sequences. J Clin Microbiol 2008;46:1019–25.)*

the *sod*A gene sequence analysis [36,42]. With *rpo*B-based data, nine clusters were found, including an additional *S. haemolyticus* group. The 16S rRNA sequence analysis allowed researchers to identify 11 genogroups (*S. epidermidis, S. saprophyticus, S. simulans, S. carnosus, S. hyicus/S. intermedius, S. sciuri, S. auricularis, S. warneri, S. haemolyticus, S. lugdunensis,* and *S. aureus*) within 38 taxa [40,48]. With the *gap* sequences, the *Staphylococcus* species were classified into four clusters: the *S. sciuri* group, the *S. hyicus/S. intermedius* group, the *S. haemolyticus/S. simulans* group, and the *S. aureus/S. epidermidis* group. The *gap* sequences analysis proved to be useful for species discrimination, thus representing a valuable approach for interpreting the phylogenetic relationship of staphylococci [37]. The *gap* sequences were more discriminative compared with the abovementioned genes, as shown for *S. caprae* and *S. capitis,* which were clearly distinguished from each other, while they were not by 16S rRNA gene analyses [37].

In detail, with regard to the four groups identified with *gap* sequence analysis, the first clade was represented by the species belonging to the "*S. hyicus/S. intermedius*" group, comprised of *S. hyicus, S. chromogenes, S. delphini, S. intermedius,* and *S. pseudintermedius.* Staphylococcal species closely related to the *S. intermedius* group (including *S. intermedius, S. pseudintermedius* and *S. delphini*), which have importance in the veterinary community, have been shown to present some phenotypic characteristics that do not allow an easy discrimination among them [37,40,49,50]. *S. intermedius* was first described in 1976, and was found to show intermediate biochemical properties between *S. aureus* and *S. epidermidis,* hence the term "*S. intermedius*" [51]. The high phenotypic variability observed with *S. intermedius* isolates [51,52] was associated with a significant genotypic variation [49,53,54]. The first description of *S. pseudintermedius* dates back to 2005 and it was observed that its phenotype was similar to that of *S. intermedius* and *S. delphini* (first reported in 1988 from dolphins) [55,56]. In 2007, studies on phylogenetic analysis of *S. intermedius* collections [49,50] showed that all strains from dogs, cats, and human beings were actually *S. pseudintermedius.* Most feral pigeon-derived strains were *S. intermedius,* instead, and most equine and domestic pigeon-derived isolates belonged to *S. delphini.* Moreover, it has been described that within the *S. intermedius* group, reliable discrimination could be obtained by using a specific multiplex PCR method [57]. Taken together, these findings suggested that isolates with traditional *S. intermedius*-like phenotypic features should be identified as *S. pseudintermedius* when they are from dogs [25,58].

The second clade could be further subdivided into the "*S. sciuri*" and the "*S. haemolyticus/S. simulans*" groups. The first is comprised of the species *S. sciuri* and *S. lentus,* together with *S. vitulinus.* These species are novobiocin-resistant and oxidase positive, and they all share the same characteristic pattern of amino acid substitution in their *hsp60* proteins [36,39]. The "*S. haemolyticus/S. simulans*" group was made of *S. haemolyticus, S. xylosus, S. muscae, S. simulans, S. schleiferi* subsp. *schleiferi, S. carnosus* subsp. *carnosus, S. caprae,* and *S. felis.*

Finally, the third clade representing the fourth group consisted of the "*S. aureus/S. epidermidis*" group and comprised *S. aureus, S. hominis* subsp. *hominis, S. warneri, S. epidermidis, S. capitis* subsp. *capitis,* and *S. lugdunensis.* By 16S rRNA-derived data, the "*S. epidermidis*" group could be divided into five clusters, that is *S. lugdunensis, S. haemolyticus, S. warneri, S. epidermidis,* and *S. aureus* [40,59], with that of *S. epidermidis* being composed of *S. epidermidis, S. capitis, S. caprae,* and *S. saccharolyticus* [36]. Based on *rpo*B sequence data, *S. caprae* and *S. capitis* appeared to be in the *S. haemolyticus* group. Similar to the *S. saprophyticus* group, the *S. epidermidis* group did not form a clearly distinct lineage in the *sod*A-based study and a low percentage of similarity was observed among the species of the group. Similar results were obtained by using *gap*-based sequences. Moreover, based on the *gap* gene data, *S. caprae* showed no close relationship to *S. epidermidis* or *S. capitis.* Indeed, the association of *S. warneri* with the *S. epidermidis* group was inferred from both *gap* and *rpo*B sequence analysis.

By *gap* sequence analysis, *S. auricularis, S. cohnii,* and the heterogeneous *S. saprophyticus* group, (comprised of *S. saprophyticus* subsp. *saprophyticus, S. equorum* subsp. *equorum, S. gallinarum, S. arlettae,* and *S. Kloosii*), were not reliably defined [37]. The *S. saprophyticus* group, as defined by 16S rRNA sequence analysis, includes the novobiocin-resistant and oxidase-negative species *S. saprophyticus* subsp. *saprophyticus, S. arlettae, S. kloosii, S. cohnii, S. gallinarum, S. equorum* subsp. *equorum,* and *S. xylosus.* Interestingly, the *rpo*B-derived data indicated that *S. cohnii* is outside of the group. By *gap* sequence analysis, *S. cohnii* and *S. xylosus* were outside of the *S. saprophyticus* group, too, while *S. cohnii* belonged to the *S. saprophyticus* group, according to the 16S rRNA- and *hsp60*-derived data.

1.3 CONCLUDING REMARKS

An accurate identification of staphylococci to the species level is quite laborious, with phenotypic methods that, in several cases, may fail to correctly identify staphylococcal species. For this reason, various molecular biology methods have been introduced. The determination of the sequences of several genes is an important tool for pathogen identification and phylogenetic studies. Although each gene-derived datum differs from the others, it has been found that groups obtained with two different sequences with a similarity of >90% are stable and reliable.

CONFLICT OF INTEREST

None.

REFERENCES

[1] Stackebrandt E, Rainey FA, Ward-Rainey NL. Proposal for a new hierarchic classification system, *Actinobacteria* classis nov. Int J Syst Bacteriol 1997;47:479–91.

[2] Garrity GM, Johnson KL, Bell J, Searles DB. Taxonomic outline of the procaryotes. Bergey's manual of systematic bacteriology. 2nd ed. New York, NY: Springer-Verlag; 2004.

[3] Ludwig W, Schleifer KH, Whitman WB. Class I. *Bacilli* class nov. In: De Vos P, Garrity GM, Jones D, Krieg NR, Ludwig W, Rainey FA, et al., editors. Bergey's manual of systematic bacteriology: vol. 3: the firmicutes. 2nd ed. New York, NY: Springer; 2009. p. 19–20.

[4] Stackebrandt E, Koch C, Gvozdiak O, Schumann P. Taxonomic dissection of the genus *Micrococcus*: *Kocuria* gen. nov., *Nesterenkonia* gen. nov., *Kytococcus* gen. nov., *Dermacoccus* gen. nov., and *Micrococcus* Cohn 1872 gen. emend. Int J Syst Bacteriol 1995;45:682–92.

[5] Stackebrandt E, Schumann P. Description of *Bogoriellaceae* fam. nov., *Dermacoccaceae* fam. nov., *Rarobacteraceae* fam. nov. and *Sanguibacteraceae* fam. nov. and emendation of some families of the suborder *Micrococcineae*. Int J Syst Evol Microbiol 2000;50:1279–85.

[6] Savini V, Catavitello C, Bianco A, Balbinot A, D'Antonio D. Epidemiology, pathogenicity and emerging resistances in *Staphylococcus pasteuri*: from mammals and lampreys, to man. Recent Pat Antiinfect Drug Discov 2009;4:123–9.

[7] Savini V, Catavitello C, Masciarelli G, Astolfi D, Balbinot A, Bianco A, et al. Drug sensitivity and clinical impact of members of the genus Kocuria. J Med Microbiol 2010;59:1395–402.

[8] Savini V, Catavitello C, Masciarelli G, Astolfi D, Balbinot A, Bianco A, et al. Review of airway illnesses by *Kytococcus* and *Rothia* and a look at inhalatory vancomycin as a treatment support. Recent Pat Antiinfect Drug Discov 2011;6:64–71.

[9] Collins MD, Hutson RA, Baverud V, Falsen E. Characterization of a *Rothia*-like organism from a mouse: description of *Rothia nasimurium* sp. nov. and reclassification of *Stomatococcus mucilaginosus* as *Rothia mucilaginosa* comb. nov. Int J Syst Evol Microbiol 2000;50:1247–51.

[10] Noble WC. Staphylococci on the skin. In: Noble WC, editor. The skin microflora and microbial skin disease. Cambridge: Cambridge University Press; 1992. p. 135–52.

[11] Över U, Tüc Y, Söyletir G. Catalase-negative *Staphylococcus aureus*: a rare isolate of human infection. Clin Microbiol Infect 2000;6:681–2.

[12] Becker K, von Eiff C. *Staphylococcus*, *Micrococcus*, and other catalase-positive cocci. In: Versalovic J, Carroll KC, Funke G, Jorgensen JH, Landry ML, Warnock DW, editors. Manual of clinical microbiology. 10th ed. Washington, DC: ASM Press; 2011. p. 308–30.

[13] Neely AN, Maley MP. Survival of Enterococci and Staphylococci on hospital fabrics and plastic. J Clin Microbiol 2000;38:724–6.

[14] Wagenvoort JHT, Sluijsmans W, Penders RJR. Better environmental survival of outbreak vs. sporadic MRSA isolates. J Hosp Infect 2000;45:231.

[15] Kloos WE, Schleifer KH. Genus IV. *Staphylococcus*. In: Sneath PHA, Mair NS, Sharpe ME, Holt JG, editors. Bergey's manual of systematic bacteriology. vol. 2. Baltimore, MD: Williams & Wilkins; 1986. p. 1013–35.

[16] Kloos WE, George CG. Identification of *Staphylococcus* species and subspecies with the Microscan Pos ID and Rapid Pos ID panel systems. J Clin Microbiol 1991;29:738–44.

[17] Kloos WE, Bannerman TL. Update on clinical significance of coagulase-negative staphylococci. Clin Microbiol Rev 1994;7:117–40.

[18] von Eiff C, Becker K, Machka K, Stammer H, Peters G. Nasal carriage as a source of *Staphylococcus aureus* bacteremia. N Engl J Med 2001;344:11–6.

[19] Gribaldo S, Cookson B, Saunders N, Marples R, Stanley J. Rapid identification by specific PCR of coagulase-negative staphylococcal species important in hospital infection. J Med Microbiol 1997;46:45–53.

[20] Kleeman KT, Bannerman TL, Kloos WE. Species distribution of coagulase-negative staphylococcal isolates at a community hospital and implications for selection of staphylococcal identification procedures. J Clin Microbiol 1993;31:1318–21.

[21] von Darányi J. Qualitative Untersuchungen der Luftbakterien. Arch Hyg (Berlin) 1925;96:182.

[22] Baird-Parker AC. A classification of micrococci and staphylococci based on physiological and chemical tests. J Gen Microbiol 1963;30:409–27.

[23] Mitchell RG, Baird-Parker AC. Novobiocin resistance and the classification of staphylococci and micrococci. J Appl Bacteriol 1967;30:251–4.

[24] Euzéby JP. List of bacterial names with standing in nomenclature: a folder available on the internet. Int J Syst Bacteriol 1997;47:590–2.

[25] Kwok AY, Chow AW. Phylogenetic study of Staphylococcus and Macrococcus species based on partial *hsp60* gene sequences. Int J Syst Evol Microbiol 2003;53:87–92.

[26] Place RB, Hiestand D, Gallmann HR, Teuber M. *Staphylococcus equorum* subsp. *linens*, subsp. nov., a starter culture component for surface ripened semi-hard cheeses. Syst Appl Microbiol 2003;26:30–7.

[27] Spergser J, Wieser M, Taubel M, Rossello-Mora RA, Rosengarten R, Busse HJ. *Staphylococcus nepalensis* sp. nov., isolated from goats of the Himalayan region. Int J Syst Evol Microbiol 2003;53:2007–11.

[28] Bond R, Loeffler A. What's happened to *Staphylococcus intermedius*? Taxonomic revision and emergence of multi-drug resistance. J Small Anim Pract 2012;53:147–54.

[29] Becker K, Harmsen D, Mellmann A, et al. Development and evaluation of a quality-controlled ribosomal sequence database for 16S ribosomal DNA-based identification of *Staphylococcus* species. J Clin Microbiol 2004;42:4988–95.

[30] Bialkowska-Hobrzanska H, Harry HV, Jaskot D, Hammerberg O. Typing of coagulase-negative staphylococci by Southern hybridization of chromosomal DNA fingerprints using a ribosomal RNA probe. Eur J Clin Microbiol Infect Dis 1990;9:588–94.

[31] De Buyser ML, Morvan A, Aubert S, Dilasser F, El Solh N. Evaluation of ribosomal RNA gene probe for the identification of species and sub-species within the genus *Staphylococcus*. J Gen Microbiol 1992;138:889–99.

[32] Maes N, De Gheldre Y, DeRyck R, et al. Rapid and accurate identification of *Staphylococcus* species by tRNA intergenic spacer length polymorphism analysis. J Clin Microbiol 1997;35:2477–81.

[33] Fujita S, Senda Y, Iwagami T, Hashimoto T. Rapid identification of staphylococcal strains from positive-testing blood culture bottles by internal transcribed spacer PCR followed by microchip gel electrophoresis. J Clin Microbiol 2005;43:1149–57.

[34] Taponen S, Simojoki H, Haveri M, Larsen HD, Pyorala S. Clinical characteristics and persistence of bovine mastitis caused by different species of coagulase-negative staphylococci identified with API or AFLP. Vet Microbiol 2006;115:199–207.

[35] Svec P, Vancanneyt M, Sedlacek I, et al. Reclassification of *Staphylococcus pulvereri* Zakrzewska-Czerwinska et al. 1995 as a later synonym of *Staphylococcus vitulinus* Webster et al. 1994. Int J Syst Evol Microbiol 2004;54:2213–5.

[36] Kwok AY, Su SC, Reynolds RP, et al. Species identification and phylogenetic relationships based on partial HSP60 gene sequences within the genus *Staphylococcus*. Int J Syst Bacteriol 1999;49:1181–92.

[37] Ghebremedhin B, Layer F, Konig W, Konig B. Genetic classification and distinguishing of *Staphylococcus* species based on different partial *gap*, 16S rRNA, *hsp60*, *rpoB*, *sodA*, and *tuf* gene sequences. J Clin Microbiol 2008;46:1019–25.

[38] Goh SH, Potter S, Wood JO, Hemmingsen SM, Reynolds RP, Chow AW. HSP60 gene sequences as universal targets for microbial species identification: studies with coagulase-negative staphylococci. J Clin Microbiol 1996;34:818–23.

[39] Goh SH, Santucci Z, Kloos WE, et al. Identification of *Staphylococcus* species and subspecies by the chaperonin 60 gene identification method and reverse checkerboard hybridization. J Clin Microbiol 1997;35:3116–21.

[40] Takahashi T, Satoh I, Kikuchi N. Phylogenetic relationships of 38 taxa of the genus *Staphylococcus* based on 16S rRNA gene sequence analysis. Int J Syst Bacteriol 1999;49:725–8.

[41] Vannuffel P, Heusterspreute M, Bouyer M, Vandercam B, Philippe M, Gala JL. Molecular characterization of *femA* from *Staphylococcus hominis* and *Staphylococcus saprophyticus*, and *femA*-based discrimination of staphylococcal species. Res Microbiol 1999;150:129–41.

[42] Poyart C, Quesne G, Boumaila C, Trieu-Cuot P. Rapid and accurate species-level identification of coagulase-negative staphylococci by using the *sodA* gene as a target. J Clin Microbiol 2001;39:4296–301.

[43] Martineau F, Picard FJ, Ke D, et al. Development of a PCR assay for identification of staphylococci at genus and species level. J Clin Microbiol 2001;39:2541–7.

[44] Drancourt M, Raoult D. *rpoB* gene sequence-based identification of *Staphylococcus* species. J Clin Microbiol 2002;40:1333–8.

[45] Mollet C, Drancourt M, Raoult D. *rpoB* sequence analysis as a novel basis for bacterial identification. Mol Microbiol 1997;26:1005–11.

[46] Yugueros J, Temprano A, Berzal B, et al. Glyceraldehyde-3-phosphate dehydrogenase- encoding gene as a useful taxonomic tool for *Staphylococcus* spp. J Clin Microbiol 2000;38:4351–5.

[47] Yugueros J, Temprano A, Sanchez M, Luengo JM, Naharro G. Identification of *Staphylococcus* spp. by PCR-restriction fragment length polymorphism of *gap* gene. J Clin Microbiol 2001;39:3693–5.

[48] Takahashi T, Kaneto M, Mori Y, Tsuji M, Kikuchi N, Hiramune T. Phylogenetic analyses of *Staphylococcus* based on the 16S rRNA sequence and assignment of clinical isolates from animals. J Vet Med Sci 1997;59:775–83.

[49] Bannoehr J, Ben Zakour NL, Waller AS, et al. Population genetic structure of the *Staphylococcus intermedius* group: insights into *agr* diversification and the emergence of methicillin-resistant strains. J Bacteriol 2007;189:8685–92.

[50] Sasaki T, Kikuchi K, Tanaka Y, Takahashi N, Kamata S, Hiramatsu K. Reclassification of phenotypically identified *Staphylococcus intermedius* strains. J Clin Microbiol 2007;45:2770–8.

[51] Hajek V. *Staphylococcus intermedius*, a new species isolated from animals. Int J Syst Evol Microbiol 1976;26:401–8.

[52] Devriese LA, van de Kerckhove A. A comparison of methods and the validity of deoxyribonuclease tests for the characterization of staphylococci isolated from animals. J Appl Bacteriol 1979;46:385–93.

[53] Meyer SA, Schleifer KH. Deoxyribonucleic acid reassociation in the classification of coagulase-positive staphylococci. Arch Microbiol 1978;117:183–8.

[54] Chesneau O, Morvan A, Aubert S, El Solh N. The value of rRNA gene restriction site polymorphism analysis for delineating taxa in the genus *Staphylococcus*. Int J Syst Evol Microbiol 2000;50:689–97.

[55] Varaldo PE, Kilpper-Balz R, Biavasco F, Satta G, Schleifer KH. *Staphylococcus delphini* sp. *nov.*, a coagulase-positive species isolated from dolphins. Int J Syst Evol Microbiol 1988;38:436–9.

[56] Devriese LA, Vancanneyt M, Baele M, et al. *Staphylococcus pseudintermedius* sp. nov., a coagulase-positive species from animals. Int J Syst Evol Microbiol 2005;55:1569–73.

[57] Sasaki T, Tsubakishita S, Tanaka Y, et al. Multiplex-PCR method for species identification of coagulase-positive staphylococci. J Clin Microbiol 2010;48:765–9.

[58] Hermans K, Devriese LA, Haesebrouck F. Staphylococcus. In: Gyles CL, Prescott JF, Songer JG, Thoen CO, editors. Pathogenesis of bacterial infections in animals. Ames, IA: Wiley-Blackwell; 2010. p. 75–89.

[59] Kloos WE. Taxonomy and systematics of staphylococci indigenous to humans. In: Crossley KB, Archer GL, editors. The staphylococci in humans and disease. New York, NY: Churchill Livingstone; 1997. p. 113–37.

Chapter 2

Staphylococcal Ecology and Epidemiology

Maja Kosecka-Strojek, Aneta Buda, Jacek Międzobrodzki
Department of Microbiology, Faculty of Biochemistry, Biophysics and Biotechnology, Jagiellonian University, Krakow, Poland

2.1 INTRODUCTION

Microbial ecology is a broad discipline encompassing many aspects of the colonization of various environments by microorganisms, as well as their interactions with other nonmicrobial organisms. First, the concept of microbial ecology must be defined. It is a field of research that studies the presence of microorganisms in the environment, in a particular biocenosis, and focuses on the transfer of microorganisms from the environment to higher organisms, as well as between various higher organisms themselves. The different interactions of microorganisms with the environment, higher organisms, or other microorganisms, also belong to the broad area of ecology. Microbial ecology also focuses on the question of how microbial populations assemble to form communities and how these communities interact with each other and with their living and nonliving environments. Staphylococci are widespread in the biosphere, and resist extreme environmental conditions. They can survive at temperatures of 60°C for up to 30 min, and are capable of growing at temperatures ranging from 6.5°C to 45°C, as well as in environments where pH values range from 4.2 to 9.3; hence, they show the ability to colonize many different niches [1–3]. Staphylococci exist in the human body and in animals as well as in soil, water, air, and food [4]. They also share various environments with humans and animals, for example, everyday objects, dust particles, liquids, food products, landfills, and wastewater. Staphylococci are then widespread in nature, but the main ecological niche and the biggest reservoir of them (especially of *Staphylococcus aureus*) are human nares. On average, every third human is colonized with this facultative pathogen [5]; alternatively, it has also been reported that all healthy humans behave as *S. aureus* carriers [6]. However, despite the fact that only a very small part of them ever develop a staphylococcal disease, this species' global burden is remarkable [5]. Farm animals potentially constitute an enormous reservoir of any pathogen, because of their sheer population size. The largest reservoir of *S.*

Pet-to-Man Travelling Staphylococci: A World in Progress. https://doi.org/10.1016/B978-0-12-813547-1.00002-9

aureus is represented by human nares, while the second are cows. The prevalence of *S. aureus* isolated from bovine mastitis is estimated to range from 3% to 5% [7,8]. Given the world's dairy cow population of 1.5 billion [5], up to 75 million cows can be infected worldwide. Although staphylococci are immobile, they can be transmitted on dust particles in the air, with water, or during direct physical contact. Currently, environmental transmission of strains plays the major role, although human-to-animal mutual exchange has been described. The widespread colonization of pigs with the specific methicillin-resistant *S. aureus* (MRSA) lineage ST398 [9,10], and the emerging infections in humans it may cause [11,12], showed that this animal reservoir of *S. aureus* can potentially lead to serious consequences for human health. Typically, human lineages of *S. aureus* are rarely found in animals, suggesting host range barriers [13,14]. However, the rapidly emerging colonization of pig farmers with pig-associated MRSA, known as farm-associated MRSA (FA-MRSA) or livestock-associated MRSA (LA-MRSA) [15], shows that some lineages have a broader host range. Recent evidence implies that the possibility of MRSA transmission to neighboring farms via the contaminated environment or to animals living in the proximity of barns cannot be excluded. Moreover, recontamination of cleaned and disinfected livestock housing from soils or surfaces may occur via the air, either passively, by wind, or actively, by air conditioning systems; otherwise, staff and intruding animals such as rodents (which come into contact with contaminated soil surfaces and subsequently enter animal housing areas) may be responsible for such a phenomenon [16].

2.2 STAPHYLOCOCCAL ECOLOGY

2.2.1 Colonization of Ecosystems and Different Habitats

An ecosystem can be considered a dynamic complex of plants, animals, microbial communities, and their nonliving surroundings, which interact as a functional unit. It may contain many different habitats, or parts of the ecosystem, which are best suited to one or a few populations. In any habitable ecosystem, where resources and growth conditions are suitable, individual microbial cells are always present and grow in the form of populations [17].

Colonization of different environments by staphylococci is associated with the diverse composition and different features of individual biocenosis, the most important of which are represented by humans and animals. Because of optimal biochemical conditions and physical interactions, staphylococci can grow and multiply in such macroorganisms, and there are many reports focusing on human and animal colonization by various staphylococcal species, principally on human skin (i.e., axillae, head, legs, arms) and mucous membranes [6,18]. An individual biocenosis such as soil, water, air, food, or drinking water has different chemical and physical properties than those of humans and animals. The word "soil" refers to the loose outer material of the Earth's surface, a layer distinct from the bedrock that lies underneath. Soils are composed of at least four components, that is, (1) inorganic mineral matter (typically about 40% of the soil volume); (2) organic matter, around 5%; (3) air and water, roughly 50%; (4) microorganisms and macroorganisms, about 5%. Microbial growth is most extensive on the surfaces of soil particles and is highly promoted within the rhizosphere; however, it is not limited to it. The availability of water and nutrients in the soil are two major factors that affect microbial activity. Staphylococci are the microbial components of soil and can even occur in sands, despite unfavorable conditions [17,19].

Another habitat that is colonized by staphylococci is water. Freshwater environments are highly variable in resources and conditions that facilitate microbial growth. Water is an example of an environment where staphylococci may exist because of human or animal presence, or as they are from the other environments; many reports on staphylococci occurrence in swimming pools or water parks have been published [20].

It has also been found that staphylococci colonize air periodically. They cannot reproduce in the air; nonetheless, staphylococci can easily be transferred by it from or to other environments, making them a great threat to higher organisms due to their constant contact with the air. There are several articles reporting the presence of zoonotic staphylococci in the air surrounding farms and barns, and it has been demonstrated that airborne, LA-MRSA is transmitted by and deposited into the air. This is strongly influenced by wind direction and season [16]. Surprisingly, one of the coagulase-negative staphylococci (CoNS) species—*Staphylococcus pasteuri* was isolated from the stratosphere. It is argued that if the microorganisms are able to survive in such an extreme environment, there are several mechanisms explaining their survival, such as clump formation, carbonization of outer layers, or association with stratospheric dust [21].

Staphylococci can also be found in food products and drinking water, causing serious infections; they even can be found in milk, in spite of sanitation procedures [22], as well as in fresh chicken and turkey meat [23], and other food products [24]. There are several publications showing the existence of CoNS in Italian fermented sausages (salsiccia and soppressata lucana) and in dry feedstuffs, raw and processed swine meat products (carcasses, raw unprocessed swine products, raw minced pork, and fresh sausages) after processing in the slaughterhouses. Two other interesting findings about staphylococci food colonization were shown by Italian scientists. They documented *S. pasteuri* appearance in vacuum-packaged charcoal-broiled European river lamprey (*Lampetra fluviatilis*) from three lamprey processing plants and in *Tuber borchii* (commonly named "truffles") [21]. *S. aureus* was also isolated from rural drinking water [25] and drinking water networks [26]. In addition to being the culprits of food poisoning, staphylococci can also be used as starter cultures for surface ripened cheeses; in this context, subspecies of *Staphylococcus equorum* are responsible for the production of cheeses of a high quality [27]. Other CoNS strains have been selected as starter cultures for sausages aiming to guarantee reproducibility of flavor, standardization of product properties, and shortening of ripening time [21].

2.2.2 How Staphylococci Affect the Environment

It is assumed that staphylococci can impact the environment in several ways. Both in human and animal organisms as well as in the environment, the presence of staphylococci is necessary to preserve the ecological balance of a given ecosystem. The enzymes produced catalyze biogeochemical transformations and allow the circulation of chemical elements in the biosphere. Bioremediation is also possible due to the presence of staphylococci in the soil, although they are not the main group of bacteria responsible for this natural process. Studies showed that staphylococci take part in the bioremediation of hydrocarbon and arsenic. It has been reported that arsenic contaminated rhizospheric soils were sampled for arsenic resistant bacteria that had the ability of transforming different arsenic forms; particularly, *Staphylococcus* sp. NBRIEAG-8, identified by *16S rDNA* ribotyping, was capable of growing in several concentrations of arsenite, and was found to posses arsenical resistance (*ars*) genes, which may be associated with high-level tolerance of arsenic-contaminated areas. The same study demonstrated that the mentioned *Staphylococcus* strain is capable of both arsenic uptake and volatilization by expressing *ars* genes and 8 new upregulated proteins, which may have played an important role in reducing bacterial susceptibility to arsenic toxicity and can be used, therefore, in arsenic bioremediation [28–30].

2.2.3 How Are Staphylococci Affected by the Environment?

From a diverse perspective, the environment itself can affect microorganisms. Seasonal fluctuations in temperature, UV radiation, soil composition, water activity, and the composition of microbial communities on soil surfaces are all factors influencing the viability of *S. aureus* in an outdoor environment [16]. Additionally, certain antibiotics can be found in the environment, and exert their activity both on CoPS and CoNS. Nevertheless, some strains show the ability to uptake resistance genes, enabling them to resist antibiotic treatments.

2.2.4 The Interactions of Microorganisms With Other Organisms and Their Outcomes

Many interactions exist among microorganisms, and between micro- and macroorganisms. These interactions can be divided into two groups—indirect or direct—and are either antagonistic or synergistic. The most common antagonistic interactions in microorganism populations are represented by competition for nutritional substrates along with toxin secretion (as a defense mechanism). An example is inhibition of *Corynebacterium diphtheriae* growth by staphylococci [31]. Synergistic interactions produce instead effects that could not otherwise be achieved independently. For instance, when staphylococci and *Salmonella paratyphi* B grow in lactose-enriched media, they are capable of fermenting lactose, thus producing gas. Conversely, when these two organisms grow independently, they do not ferment lactose, nor do they produce a gas [32]. As mentioned before, microorganisms may also interact with macroorganisms, by establishing symbiosis, parasitism, or commensalism; the latter may be absolute or facultative; in this case it also referred to as opportunism. In the light of what has been discussed thus far, staphylococci are considered to be commensals and opportunistic pathogens, both in humans and animals, meaning that they can be harmful for macroorganisms under certain conditions, and they show the potential to cause infections, even serious, for example, in immunocompromised patients.

2.2.5 Interactions Between Microorganisms Causing Changes in Bacterial Genomes

Increasing antibiotic resistance and the emergence of biochemically atypical strains that are hard to identify in laboratory activity are two major issues that diagnostics and novel therapy treatment must deal with. These problems are associated with the high consumption of antibiotics and mobile genetic elements (MGEs) involved in the horizontal gene transfer (HGT) phenomenon. MGEs are typically identified as fragments of DNA that encode a variety of virulence or resistance determinants, as well as the enzymes that mediate their own transfer and integration into new host DNA [33]. Transfer of MGEs between cells is known as lateral or horizontal gene transfer (HGT). Transfer of DNA or HGT, may occur between prokaryote-prokaryote, prokaryote-eukaryote, eukaryote-eukaryote. Staphylococci are capable of obtaining genetic information from other cells or the surrounding environment by (1) free DNA uptake from the environment (transformation), (2) bacteriophage transduction, and (3) direct contact between diverse bacterial cells, a phenomenon labeled as conjugation. Compared with bacteria, such as *Escherichia coli* or *Bacillus subtilis*, *S. aureus* has a limited ability to acquire DNA from the environment (low natural competence). As a consequence, intercellular transfer of plasmids mostly occurs by transduction or conjugation [34]. Transfer of a plasmid can result in the acquirement of genes encoding antibiotic resistance. As a related phenomenon, isolation of both methicilin-resistant *S. aureus* and methicilin-resistant coagulase-negative staphylococci (MR-CoNS) from different sources (such as hospitals, medical devices, ambulatory patients, and healthy humans) is reported [35–37]. Additionally, MR-CoNS carriage by healthy animals such as horses, dogs, cattle, sheep, goats, and pigs is described as well. It is assumed that methicilin-resistance-associated genes had evolved in CoNS and were horizontally transferred, subsequently, within the whole *Staphylococcus* genus [38,39]. More recent studies have shown that *S. aureus* has acquired vancomycin resistance elements from enterococci, resulting in the emergence of vancomycin-resistant *S. aureus* (VRSA) [40]. Another issue of concern is the emergence of vancomycin-resistant coagulase-negative staphylococci (VR-CoNS) among *Staphylococcus capitis* strains [41]. This is a particularly worrisome phenomenon, as vancomycin is among the antibiotics of first choice, or of last resort, as well, against most Gram positive bacteria, and is then often the last chance when other drugs are exhausted [42]. In addition to genes encoding antibiotic resistance and molecules involved in metabolism, staphylococcal plasmids encode resistance to a variety of organic and inorganic ions, such as cadmium, mercury, arsenate, etc., which are highly toxic for living cells [43]. Resistance to disinfectants and antiseptics based on quaternary ammonium compounds (QACs) is widespread in clinical human staphylococcal strains and it is assumed that utilization of QACs in the nosocomial environment and in farm animal operations could be responsible for selection of penicillin-resistant staphylococci and vice versa [21]. The HGT results in either acquiring or releasing important pathogenicity mobile elements. Even environmental bacteria that do not affect man, directly, can in fact harbor and exchange antibiotic resistance genes, through lateral transfer to other, potentially pathogenic organisms that may cause disease cases and outbreaks [44]. The uptake of new genes enriches the bacterial genome and enables microorganisms to express new enzymes and exploit new nutritive substrates; hence, bacteria may get the ability to colonize new ecosystems and grow in their context. Staphylococcal biochemical variations are coded in genetic regulatory systems, especially accessory gene regulation (*agr*) and staphylococcal accessory regulation (*sar*), and are also based on the quorum sensing communication system and the HGT phenomenon [45].

2.3 STAPHYLOCOCCAL EPIDEMIOLOGY

2.3.1 Epidemiology of Staphylococcal Infections

Epidemiology investigates the occurrence and distribution of health-related states or events, including disease, in different populations. The term *epidemiology* is derived from the ancient Greek word "epi" meaning "on," "demos" meaning "people," and "logos" meaning "word and science." This scientific discipline examines the impact of factors that influence health, as well as the application of this knowledge aiming to control health problems. The discipline may concern people, animals, or plants [46] and epidemiology of microbial infections is described by the chain of successive events and processes leading to symptomatic infection [47].

It is known that, among the sources of infections in humans or animals, there are pathogenic or potentially pathogenic (opportunistic) bacteria. The risk of infection increases in patients with underlying predisposing conditions, like in those who undergo surgery, suffer from burn wounds, receive medical device implantation (i.e., catheters, prostheses) or long-term antibiotic and/or antifungal treatments. In hospitals, there are various reservoirs of exogenous or endogenous microorganisms. The first group includes medical staff, co-patients, visitors, inanimate objects, water, and food. To the second one, instead, we may find the patient's physiological microflora.

Outside the hospital, the main reservoirs of microorganisms are animals, along with areas (i.e., rural environments) where particular bacteria circulate and may cause diseases. Nonnosocomial infections involving closed groups of people are referred to as "community acquired infections (CAI)," and mostly occur among team-sports players, military service personnel, prisoners, homeless people, children living in closed institutions, as well as outpatients that belong to what is general labeled as the "community," conventionally including the wide group of nonhospitalized people [48]. Microbial diseases occurring during hospitalization are called instead "nosocomial infections," and are more and more frequently referred to as hospital-acquired infections (HAI) [49]. Several reports deal instead with FAIs (farm-acquired infections), also known as LAIs (livestock-acquired infections), that are animal-related; the etiologic agents of such conditions colonize people who have or had direct or indirect contact with animals and that, therefore, represent a new source of colonization and infection for both hospitalized subjects and outpatients [50–52].

Factors determining the survival and distribution of microorganisms in a given population are their active transmission from infected or colonized hosts, and their ability to periodically colonize macroorganisms, that are more susceptible and vulnerable. Particularly, bacteria may be transmitted by either direct or indirect contact [53].

Staphylococci colonize the skin and mucous membranes of healthy individuals and are also isolated from many animal species [39]. Over the last few decades, they have emerged as major nosocomial pathogens, especially those exerting methicillin and vancomycin resistance [37,54–56].

2.3.2   Epidemiology of Coagulase-Positive Staphylococci

Eight CoPS species have been described, namely, *S. aureus*, *Staphylococcus intermedius*, *Staphylococcus pseudintermedius*, *Staphylococcus delphini*, *Staphylococcus hyicus*, *Staphylococcus schleiferi* subsp. *coagulans* (*Staphylococcus schleiferi* subsp. *schleiferi*, instead, is coagulase-negative), *Staphylococcus lutrae*, and *Staphylococcus agnetis*. *S. intermedius*, *S. pseudintermedius*, and *S. delphini* are classified as members of the *Staphylococcus intermedius* group (SIG). *S. hyicus* and *S. agnetis* are coagulase-variable (*S. agnetis* is coagulase-negative after 4h, while 20%–25% isolates are coagulase-positive after 24 h; *S. hyicus* coagulase expression is strain-dependent [57]. Again, *S. agnetis* has been detected in cow's milk and teat apex and *S. lutrae* is known to inhabit otters [58].

S. aureus is the most frequently encountered CoPS species in the medical setting, and is the main reservoir of it. The organism colonizes skin, and often inhabits the nasal vestibule [59]. Two types of *S. aureus* nasal carriers exist, that is permanent carriers, 20%–30% of the general population, and temporary carriers, representing 50%–60% of subjects. In addition, *S. aureus* often colonizes the nasopharynx, the skin of the anal zone, and that surrounding the hairline across the forehead as well as the jaw area. Increased carriage rates, moreover, are observed in people with type 1 diabetes, those undergoing hemodialysis, peritoneal dialysis and intravenous treatments, and HIV patients. *S. aureus* may cause local or systemic (invasive) infections [60], and some strains or their toxins may be responsible for food poisoning [61].

Local disease instead includes skin and soft tissue infections, such as impetigo, folliculitis, boils, furunculosis, cellulitis, abscesses, and human mastitis [62]. *S. aureus* is the most common cause of bacterial conjunctivitis, as well, and may also be responsible for upper respiratory tract infections (sinusitis, otitis media), meningitis, brain abscess, and prosthetic-valve endocarditis, which are associated with a high mortality rate. *S. aureus* is an important etiologic factor for patients of all age groups (excluding neonates), categorized with acute primary osteoarthritis, chronic osteomyelitis, and sepsis [63,64].

An important CoPS, *S. intermedius,* was first described in 1976 [65]. It inhabits the mouths and surfaces of mostly dogs, birds, minks, cats, foxes, raccoons, gray squirrels, goats, and horses. Moreover, it is a known opportunistic pathogen in animals, as it has been known to cause infections of skin, bones, the respiratory and urogenital tract, and the central nervous system. *S. intermedius* was first identified as a human pathogen in 1989 from dog bite-related cellulitis cases. Nowadays, there is increasing evidence of *S. intermedius* transmission to humans [66].

S. pseudintermedius, additionally, can be mostly found on the skin, ears, and mouth of healthy dogs, and is an easily misrecognized agent of canine infections [67]. Discrimination within the SIG is particularly challenging and the emergence of methicillin-resistant *S. intermedius* (MRSI) and *S. pseudintermedius* (MRSP) strains is of concern [68–70].

S. hyicus causes exudative epidermitis in piglets, and is isolated from the nose, eyes, and skin of healthy pigs, or from the vaginal flora of healthy sows. It is also responsible for causing skin infections in cattle, horses, and goats [71]. *S. hyicus* strains can be misidentified as *S. aureus* or even CoNS [58].

2.3.3 Epidemiology of Coagulase-Negative Staphylococcal Infections

CoNS include more than 44 species [72] and are the most frequently isolated microorganisms in clinical microbiology laboratories [73]. Since the 1970s, the growing number of reported results suggests that CoNS may be true pathogens and cause a variety of infections, mostly among predisposed patients (neonates, patients undergoing catheters or prosthetic implants placement, dialyzed and neutropenic hosts) [74,75]. CoNS disease, moreover, may commonly involve the central nervous system, heart, bones, and joints [76].

CoNS are the most common cause of late-onset neonatal sepsis in neonatal intensive care units (NICU), worldwide, that are related to the CoNS ability to form biofilms, as well as to the still immature immune system of low birth weight newborns (LBWNs) who, moreover, undergo invasive procedures. Notably, most CoNS strains causing LBWN sepsis can be found on the NICU personnel's hands [77]. *S. epidermidis* is the most known member of the CoNS group. Although it is a common saprophyte on mammalian skin, it is also a key-pathogen in hospitalized patients (major clinical picture is catheter-related bacteremia), mostly among immunocompromised subjects [78]. The crucial pathogenic trait for colonization and difficult-to-treat infections of medical devices is *S. epidermidis* formation of biofilms, which may also involve contact lenses and artificial hips. Therapy of biofilm-related infections often requires that the infected surface be removed [79,80].

Again, CoNS, after *S. aureus*, are the most frequent cause of mastitis [81]. *Staphylococcus saprophyticus* is instead a constituent of the normal genito-urinary tract flora and can cause urinary tract infections, especially in young women (it is the second most common etiologic agent of female cystitis, after *Escherichia coli*) [82]. *S. capitis* colonizes the head, face, neck, and ears of humans and can then be easily introduced into the bloodstream through open wounds, implanted medical devices, or venipunctures, thus causing bacteremia, prosthetic valve endocarditis, LBWN late-onset sepsis, and is even promoted by biofilm formation [83,84].

S. sciuri, finally, is widely distributed in nature, and is as a commensal of rodent species, pets in general, and farm animals [85]. Also, it has been associated with human diseases such as endocarditis, urinary tract, and wound infections [86].

2.4 CONCLUSIONS

In conclusion, staphylococci are widespread in the environment, colonizing different niches, but closely related to humans and animals. They do interact both with the environment by contribution in bioremediation and acquisition of various MGEs and with other organisms by antagonistic or synergistic interactions. As opportunistic pathogens, the staphylococci are the most frequently isolated microorganisms in clinical microbiology laboratories both from humans and animals, causing the infections of different localizations, manifestations, and outcomes.

ACKNOWLEDGMENTS

This work was partly financed by funds granted by the National Science Centre (NCN, Poland) on the basis of the projects: no. 2016/21/N/NZ6/00981 (for MKS) and no. N N401 017740 (for JM).

CONFLICT OF INTEREST

None declared.

REFERENCES

[1] Kloos WE. Taxonomy and systematics of staphylococci indigenous to humans. In: Crossley KB, Archer GL, editors. The staphylococci in human disease. New York: Churchill Livingstone; 1997. p. 113–37.

[2] Kloos WE, Schleifer KH. Isolation and characterization of staphylococci from human skin II. Descriptions of four new species: *Staphylococcus warneri, Staphylococcus capitis, Staphylococcus hominis*, and *Staphylococcus simulans*. Int J Syst Bacteriol 1975;25(1):62–79.

[3] Vora P, Senecal A, Schaffner DW. Survival of *Staphylococcus aureus* ATCC 13565 in intermediate moisture foods is highly variable. Risk Anal 2003;23(1):229–36.

[4] Kloos WE, Schleifer KH, Smith RF. Characterization of *Staphylococcus sciuri sp.nov.* and its subspecies. Int J Syst Bacteriol 1976;26(1):22–37.

[5] Sakwinska O, Giddey M, Moreillon M, et al. *Staphylococcus aureus* host range and human-bovine host shift. Appl Environ Microbiol 2011;77(17):5908–15.

[6] Kloos WE, Musselwhite MS. Distribution and persistence of *Staphylococcus* and *Micrococcus* species and other aerobic bacteria on human skin. Appl Microbiol 1975;30(3):381–95.

[7] Moret-Stalder S, Fournier C, Miserez R, et al. Prevalence study of *Staphylococcus aureus* in quarter milk samples of dairy cows in the Canton of Bern, Switzerland. Prev Vet Med 2009;88(1):72–6.

[8] Roesch MG, Doherr M, Schären W, et al. Subclinical mastitis in dairy cows in Swiss organic and conventional production systems. J Dairy Res 2007;74(1):86–92.

[9] de Neeling AJ, van den Broek MJ, Spalburg EC, et al. High prevalence of methicillin resistant *Staphylococcus aureus* in pigs. Vet Microbiol 2007;122(3–4):366–72.

[10] Kosecka-Strojek M, Ilczyszyn WM, Buda A, et al. Multiple-locus variable-number tandem repeat fingerprinting as a method for rapid and cost-effective typing of animal-associated *Staphylococcus aureus* strains from lineages other than sequence type 398. J Med Microbiol 2016;65(12):1494–504.

[11] Golding GR, Bryden L, Levett PN, et al. Livestock-associated methicillin-resistant *Staphylococcus aureus* sequence type 398 in humans, Canada. Emerg Infect Dis 2010;16(4):587–94.

[12] Krziwanek K, Metz-Gercek S, Mittermayer H. Methicillin-resistant *Staphylococcus aureus* ST398 from human patients, upper Austria. Emerg Infect Dis 2009;15(5):766–9.

[13] Smith EM, Green LE, Medley GF, et al. Multilocus sequence typing of intercontinental bovine *Staphylococcus aureus* isolates. J Clin Microbiol 2005;43(9):4737–43.

[14] Sung JM, Lloyd DH, Lindsay JA. *Staphylococcus aureus* host specificity: comparative genomics of human versus animal isolates by multi-strain microarray. Microbiology 2008;154 (Pt 7):1949–59.

[15] van Loo I, Huijsdens X, Tiemersma E, et al. Emergence of methicillin-resistant *Staphylococcus aureus* of animal origin in humans. Emerg Infect Dis 2007;13(12):1834–9.

[16] Schulz J, Friese A, Klees S, et al. Longitudinal study of the contamination of air and of soil surfaces in the vicinity of pig barns by livestock-associated methicillin-resistant *Staphylococcus aureus*. Appl Environ Microbiol 2012;78(16):5666–71.

[17] Madigan MT, Martinko JM, Stahl D, et al. Brock biology of microorganisms. 13th ed. London, United Kingdom: Pearson; 2012.

[18] Kloos WE, Zimmerman RJ, Smith RF. Preliminary studies on the characterization and distribution of *Staphylococcus* and *Micrococcus* species on animal skin. Appl Environ Microbiol 1976;31(1):53.

[19] Yamahara KM, Sassoubre LM, Goodwin KD, et al. Occurrence and persistence of bacterial pathogens and indicator organisms in beach sand along the California coast. Appl Environ Microbiol 2012;78(6):1733–45.

[20] Davis TL, Standridge JH, Degnan AJ. Bacteriological analysis of indoor and outdoor water parks in Wisconsin. J Water Health 2009;7(3):452–63.

[21] Savini V, Catavitello C, Bianco A, et al. Epidemiology, pathogenicity and emerging resistances in *Staphylococcus pasteuri*: from mammals and lampreys, to man. Recent Pat Antiinfect Drug Discov 2009;4(2):123–9.

[22] Cleto S, Matos S, Kluskens L, et al. Characterization of contaminants from a sanitized milk processing plant. PLoS One 2012;7(6):e40189.

[23] Fessler AT, Kadlec K, Hassel M, et al. Characterization of methicillin-resistant *Staphylococcus aureus* isolates from food and food products of poultry origin in Germany. Appl Environ Microbiol 2011;77(20):7151–7.

[24] Crago B, Ferrato C, Drews SJ, et al. Prevalence of *Staphylococcus aureus* and methicillin-resistant *S. aureus* (MRSA) in food samples associated with foodborne illness in Alberta, Canada from 2007 to 2010. Food Microbiol 2012;32(1):202–5.

[25] LeChevallier MW, Seidler RJ. *Staphylococcus aureus* in rural drinking water. Appl Environ Microbiol 1980;39(4):739–42.

[26] Faria C, Vaz-Moreira I, Serapicos E, et al. Antibiotic resistance in coagulase negative staphylococci isolated from wastewater and drinking water. Sci Total Environ 2009;407(12):3876–82.

[27] Place RB, Hiestand D, Gallmann HR, et al. *Staphylococcus equorum* subsp. *linens*, subsp. nov., a starter culture component for surface ripened semi-hard cheeses. Syst Appl Microbiol 2003;26(1):30–7.

[28] Srivastava S, Verma PC, Singh A, et al. Isolation and characterization of *Staphylococcus sp.* strain NBRIEAG-8 from arsenic contaminated site of West Bengal. Appl Microbiol Biotechnol 2012;95(5):1275–91.

[29] Srivastava S, Verma PC, Chaudhry V, et al. Influence of inoculation of arsenic-resistant *Staphylococcus arlettae* on growth and arsenic uptake in Brassica juncea (L.) Czern. Var. R-46. J Hazard Mater 2013;262:1039–47.

[30] Eddouaouda K, Mnif S, Badis A, et al. Characterization of a novel biosurfactant produced by *Staphylococcus sp.* strain 1E with potential application on hydrocarbon bioremediation. J Basic Microbiol 2012;52(4):408–18.

[31] Barrow GI. Microbial antagonism by *Staphylococcus aureus*. Microbiology 1963;31(3):471–81.

[32] Kunicki-Goldfinger WJH. Life of bacteria [Życie bakterii]. Warszawa: PWN; 1998 [in Polish].

[33] Frost LS, Leplae R, Summers AO, et al. Mobile genetic elements: the agents of open source evolution. Nat Rev Microbiol 2005;3(9):722–32.

[34] Malachowa N, DeLeo FR. Mobile genetic elements of *Staphylococcus aureus*. Cell Mol Life Sci 2010;67(18):3057–71.

[35] Dakić I, Morrison D, Vuković D, et al. Isolation and molecular characterization of *Staphylococcus sciuri* in the hospital environment. J Clin Microbiol 2005;43(6):2782–5.

[36] Ruppé E, Barbier F, Mesli Y, et al. Diversity of staphylococcal cassette chromosome mec structures in methicillin-resistant *Staphylococcus epidermidis* and *Staphylococcus haemolyticus* strains among outpatients from four countries. Antimicrob Agents Chemother 2009;53(2):442–9.

[37] Carbon C. MRSA and MRSE: is there an answer? Clin Microbiol Infect 2000;6(Suppl. 2):17–22.

[38] Huber H, Ziegler D, Pflüger V, et al. Prevalence and characteristics of methicillin-resistant coagulase-negative staphylococci from livestock, chicken carcasses, bulk tank milk, minced meat, and contact persons. BMC Vet Res 2011;7:6.

[39] Kasprowicz A, Białecka A, Białecka J, et al. The occurrence and comparative phenotypic characteristics of *Staphylococcus* spp. from healthy and diseased, household and shelter dogs, based on routine biochemical diagnostic methods. Pol J Microbiol 2011;60(1):19–26.

[40] Zhu W, Clark NC, McDougal LK, et al. Vancomycin-resistant *Staphylococcus aureus* isolates associated with Inc18-like vanA plasmids in Michigan. Antimicrob Agents Chemother 2008;52(2):452–7.

[41] D'mello D, Daley AJ, Rahman MS, et al. Vancomycin heteroresistance in bloodstream isolates of *Staphylococcus capitis*. J Clin Microbiol 2008;46(9):3124–6.

[42] Salyers AA, Whitt DD. Microbiology: diversity, disease, and the environment. Bethesda, MD: Fitzgerald Science Press; 2001. p. 28, 40, 95, 226, 228.

[43] Jensen SO, Lyon BR. Genetics of antimicrobial resistance in *Staphylococcus aureus*. Future Microbiol 2009;4(5):565–82.

[44] Uyaguari MI, Fichot EB, Scott GI, et al. Characterization and quantitation of a novel β-lactamase gene found in a wastewater treatment facility and the surrounding coastal ecosystem. Appl Environ Microbiol 2011;77(23):8226–33.

[45] Savini V, Catavitello C, Astolfi D, et al. Bacterial contamination of platelets and septic transfusions: review of the literature and discussion on recent patents about biofilm treatment. Recent Pat Antiinfect Drug Discov 2010;5(2):168–76.

[46] Porta MA. Dictionary of epidemiology. Oxford: Oxford University Press; 2008.

[47] Przondo-Mordarska A. Nosocomial infections. Wrocław: Conlinuo; 1999.

[48] Gelatti LC, Bonamigo RR, Inoue FM, et al. Community-acquired methicillin-resistant *Staphylococcus aureus* carrying SCCmec type IV in southern Brazil. Rev Soc Bras Med Trop 2013;46(1):34–8.

[49] Porto JP, Mantese OC, Arantes A, et al. Nosocomial infections in a pediatric intensive care unit of a developing country: NHSN surveillance. Rev Soc Bras Med Trop 2012;45(4):475–9.

[50] Kadlec K, Fessler AT, Hauschild T, et al. Novel and uncommon antimicrobial resistance genes in livestock-associated methicillin-resistant *Staphylococcus aureus*. Clin Microbiol Infect 2012;18(8):745–55.

[51] Köck R, Harlizius J, Bressan N, et al. Prevalence and molecular characteristics of methicillin-resistant *Staphylococcus aureus* (MRSA) among pigs on German farms and import of livestock-related MRSA into hospitals. Eur J Clin Microbiol Infect Dis 2009;28(11):1375–82.

[52] Savini V, Salutari P, Sborgia M, et al. Brief tale of a bacteraemia by *Rhodococcus equi*, with concomitant lung mass: what came first, the chicken or the Egg? Mediterr J Hematol Infect Dis 2011;3(1):e2011006.

[53] Adaszek Ł, Górna M, Ziętek J, et al. Bacterial nosocomial infections in dogs and cats. Z Wet 2009;84:10.

[54] Schwalbe RS, Stapleton JT, Gilligan PH. Vancomycin-resistant *Staphylococcus*. N Engl J Med 1987;317:766–8.

[55] Stefani S, Varaldo PE. Epidemiology of methicillin-resistant staphylococci in Europe. Clin Microbiol Infect 2003;9(12):1179–86.

[56] Appelbaum PC. The emergence of vancomycin-intermediate and vancomycin-resistant *Staphylococcus aureus*. Clin Microbiol Infect 2006;12(Suppl. 1):16–23.

[57] Savini V, Passeri C, Mancini G, et al. Coagulase-positive staphylococci: my pet's two faces. Res Microbiol 2013;164(5):371–4.

[58] Casanova C, Iselin L, von Steiger N, et al. *Staphylococcus hyicus* bacteremia in a farmer. J Clin Microbiol 2011;49(12):4377–8.

[59] Malachowa N, Kohler PL, Schlievert PM, et al. Characterization of a *Staphylococcus aureus* surface virulence factor that promotes resistance to oxidative killing and infectious endocarditis. Infect Immun 2011;79(1):342–52.

[60] Safdar N, Bradley EA. The risk of infection after nasal colonization with *Staphylococcus aureus*. Am J Med 2008;121(4):310–5.

[61] Pliszka A, Windyga B, Maciejska K, et al. Comparison of methods of detecting staphylococcal enterotoxin in food. Rocz Panstw Zakl Hig 1973;24(5):585–95.

[62] Oliveira M, Bexiga R, Nunes SF, et al. Invasive potential of biofilm-forming Staphylococci bovine subclinical mastitis isolates. J Vet Sci 2011;12(1):95–7.

[63] Cianciara J, Juszczyk J. Infectious and parasitic diseases. Lublin: Czelej; 2012. p. 825–7.

[64] Rieg S, Jonas D, Kaasch AJ, et al. Microarray-based genotyping and clinical outcomes of *Staphylococcus aureus* bloodstream infection: an exploratory study. PLoS One 2013;8(8):e71259.

[65] Hajek V. Staphylococcus intermedius, a new species isolated from animals. Int J Syst Bacteriol 1976;26:401–8.

[66] Hatch S, Sree A, Tirrell S, Torres B, et al. Metastatic complications from *Staphylococcus intermedius*, a zoonotic pathogen. J Clin Microbiol 2012;50(3):1099–101.

[67] Miedzobrodzki J, Kasprowicz A, Białecka A, et al. The first case of a *Staphylococcus pseudintermedius* infection after joint prosthesis implantation in a dog. Pol J Microbiol 2010;59(2):133.

[68] Chrobak D, Kizerwetter-Swida M, Rzewuska M, et al. Antibiotic resistance of canine *Staphylococcus intermedius* group (SIG)—practical implications. Pol J Vet Sci 2011;14(2):213–8.

[69] Dicicco M, Neethirajan S, Singh A, et al. Efficacy of clarithromycin on biofilm formation of methicillin-resistant *Staphylococcus pseudintermedius*. BMC Vet Res 2012;8:225.

[70] Savini V, Kosecka M, Marrollo R, et al. CAMP test detected *Staphylococcus delphini* ATCC 49172 beta-haemolysin production. Pol J Microbiol 2013;62(4):465–6.

[71] Foster AP. Staphylococcal skin disease in livestock. Vet Dermatol 2012;23(4):342–51. e63.

[72] Rahman AZ, Hamzah SH, Hassan SA, et al. The significance of coagulase-negative staphylococci bacteremia in a low resource setting. J Infect Dev Ctries 2013;7(6):448–52.

[73] Hung KH, Yan JJ, Lu YC, et al. Evaluation of discrepancies between oxacillin and cefoxitin susceptibility in coagulase-negative staphylococci. Eur J Clin Microbiol Infect Dis 2011;30(6):785–8.

[74] Dubois D, Leyssene D, Chacornac JP, et al. Identification of a variety of *Staphylococcus* species by matrix-assisted laser desorption ionization-time of flight mass spectrometry. J Clin Microbiol 2010;48(3):941–5.

[75] Huebner J, Goldmann DA. Coagulase-negative staphylococci: role as pathogens. Annu Rev Med 1999;50:223–36.

[76] Casey AL, Lambert PA, Elliott TS. Staphylococci. Int J Antimicrob Agents 2007;29(Suppl. 3):S23–32.

[77] Hira V, Sluijter M, Goessens WHF, et al. Coagulase-negative staphylococcal skin carriage among neonatal intensive care unit personnel: from population to infection. J Clin Microbiol 2010;48(11):3876–81.

[78] Smith PA, Powers ME, Roberts TC, et al. Vitro activities of arylomycin natural-product antibiotics against *Staphylococcus epidermidis* and other coagulase-negative staphylococci. Antimicrob Agents Chemother 2011;55(3):1130–4.

[79] Rumi MV, Huguet MJ, Bentancor AB, et al. The icaA gene in staphylococci from bovine mastitis. J Infect Dev Ctries 2013;7(7):556–60.

[80] Otto M. *Staphylococcus epidermidis*—the 'accidental' pathogen. Nat Rev Microbiol 2009;7(8):555–67.

[81] Biavasco F, Vignaroli C, Varaldo PE. Glycopeptide resistance in coagulase-negative staphylococci. Eur J Clin Microbiol Infect Dis 2000;19:403–17.

[82] Ronald A. The etiology of urinary tract infection: traditional and emerging pathogens. Dis Mon 2003;49(2):71–82.

[83]   Abdalla NM. Evaluation of resistance of commonly used antibiotics on clinical case of *Staphylococcus capitis* from Assir region, Saudi Arabia. Indian J Med Sci 2011;65(12):547–51.

[84]   Rasigade JP, Raulin O, Picaud JC, et al. Methicillin-resistant *Staphylococcus capitis* with reduced vancomycin susceptibility causes late-onset sepsis in intensive care neonates. PLoS One 2012;7(2):e31548.

[85]   Hauschild T, Schwarz S. Differentiation of *Staphylococcus sciuri* strains isolated from free-living rodents and insectivores. J Vet Med B Infect Dis Vet Public Health 2003;50:241–6.

[86]   Coimbra DG, Almeida AG, Jùnior JB, et al. Wound infection by multiresistant *Staphylococcus sciuri* identified by molecular methods. New Microbiol 2011;34(4):425–7.

Chapter 3

Coagulase-Positive and Coagulase-Negative Staphylococci in Human Disease

Carla Fontana*,†, Marco Favaro*
*Department of Experimental Medicine and Surgery, Tor Vergata University of Rome, Rome, Italy
†Foundation Polyclinic Tor Vergata, Clinical Microbiology Laboratories, Rome, Italy

3.1 INTRODUCTION

Staphylococci are a group of Gram-positive cocci (0.5–1 µm in diameter) that occasionally occur singly, in pairs, and tetrads, but much more frequently in "grape-like" clusters. They are nonmotile, nonsporeforming, aerobic, and facultatively anaerobic, capable of prolonged survival on environmental surfaces under varying conditions, and widely distributed in nature [1]. They inhabit skin, skin glands, and mucous membranes, particularly the nasal cavities. Again, they can be found as symbiotic organisms in the mouth, and intestinal, genitourinary, and upper respiratory tracts. Nevertheless, they may adopt a pathogenic behavior as well, and gain access to diverse bodily sites through damaged skin and mucousae and cause opportunistic infections. Those that develop in the colonized site (particularly skin, for *Staphylococcus aureus*) are furuncoles, cellulites, and impetigo. These are acute, purulent infections that, if untreated, may spread to surrounding tissues or disseminate via the blood vessels to various organs, resulting in bacteremia, pneumonia, osteomyelites, encephalitis, meningitis, chorioamnionitis, scalded skin syndrome (SSS), muscle abscesses, intraabdominal, and urogenital tract infections [1]. Among the coagulase-positive staphylococci (CoPS), *S. aureus* represents the most dangerous species, as it may express several pathogenic determinants and/or virulence factors (see respective chapters of this volume) that allow it to adhere to surfaces, colonize, invade, and damage tissues, escape the immune system, and cause harmful toxic effects [2].

Certain coagulase-negative staphylococci (CoNS) are cause for concern, as well, as they may act as human pathogens, mostly *Staphylococcus saprophyticus*, *Staphylococcus epidermidis*, and *Staphylococcus haemolyticus*; although CoNS in general were considered to be relatively innocuous in the past, they

Pet-to-Man Travelling Staphylococci: A World in Progress. https://doi.org/10.1016/B978-0-12-813547-1.00003-0

have gained (*S. epidermidis*, particularly) significant interest in recent years as they have been observed to be potential causes of nosocomial infections, especially in immunocompromised, long-term hospitalized, and critically ill patients [3]. *S. epidermidis*, the most frequently isolated CoNS species, is the leading agent of infections related to medical and surgical devices (such as catheters and prosthetic implants) due to its ability to produce multilayered, highly structured and protective biofilm on artificial surfaces [3,4]. Biofilm-associated organisms (sessile population), moreover, are greatly resistant to antibiotics, mainly due to slime production [5–7] particularly, sessile bacteria are up to 1,000-fold more resistant than their planktonic (free-floating) form (see the chapter *In vivo resistance mechanisms: staphylococcal biofilms* in this volume) [5].

Therefore, it is clear that all staphylococci may be cause for concern both in community- and in nosocomial-acquired infections; and play a prominent role in human pathology that deserves the attention of microbiologists and clinicians.

3.1.1   *Staphylococcus aureus*

S. aureus is a pluripotent and well-known human opportunistic pathogen causing diseases through both toxin-mediated and nontoxin-mediated mechanisms, in the community and in hospital settings. Although it rarely causes severe pictures in healthy carriers, it may be responsible for serious syndromes in vulnerable patients [8]. Infections range from those of the skin and soft tissues to invasive, life-threatening diseases. We might place pathologies it causes into three main groups [2,9], that is:

i. superficial lesions and wound infections,
ii. toxinoses (i.e., the SSS, toxic shock syndrome, and food poisoning), and
iii. systemic infections (i.e., bacteraemia, endocarditis, osteomyelitis, pneumonia, brain abscesses, meningitis).

A colonized carrier is at higher risk of infection, and also represents an *S. aureus* reservoir who potentially disseminates the organism to other subjects [10]. It is estimated that 25%–35% of healthy human individuals carry *S. aureus* on skin or mucosae, meaning that up to two billion individuals in the world are currently carriers [11]. Particularly relevant is methicillin-resistant *S. aureus* (MRSA), which emerged in the 1980s, and is nowadays a critical, difficult-to-treat, nosocomial pathogen. Despite published recommendations for management and control of hospital-acquired MRSA (HA-MRSA) strains, the greatest part of worldwide hospitals (with the exception of some Northern European countries, especially The Netherlands and Scandinavia) are still facing this pathogen. Epidemiological data show the highest HA-MRSA rates (>50%) in North and South America, Asia, and Malta; intermediate rates (25%–50%) in China, Australia, Africa, and some European countries such as Portugal (49%), Greece (40%), Italy (37%), and Romania (34%). In recent years, it has been observed that HA-MRSA prevalence is however decreasing, particularly in some parts of Austria, France, Ireland, the UK, and Greece, and has remained stable elsewhere (ECDC data), with very high rates in East Asia, especially Sri Lanka (86.5%), South Korea (77.6%), Vietnam (74.1%), Taiwan (65.0%), Thailand (57.0%), and Hong Kong (56.8%) [11]. Hence, MRSA is a cause for global concern [12–14]. Strains have acquired variants of staphylococcal cassette chromosome *mec* (SCC*mec*) elements carrying the methicillin resistance determinant (the *mecA* gene). However, the vast majority of isolates worldwide belongs to a limited number of clones, some of which are associated with global epidemics. For instance, MRSA-15 (EMRSA-15) has proven to be transmitted rapidly within and among hospitals as well as to different countries, and has spread across continents. It carries a type IV SCC*mec* element and belongs to multilocus sequence type ST22 [14]. Along with HA-MRSA, community-associated methicillin-resistant *S. aureus* (CA-MRSA) clones are of concern, and mainly infect children, young and middle-aged adults; a new member, the livestock-associated MRSA (LA-MRSA), has recently been added to the family [15,16]. It possesses a novel divergent *mecA* gene, called *mecA*$_{LGA251}$, which is 70% homologous to *S. aureus mecA* [17]. However, LA-MRSA is prevalent in certain high-risk groups of workers only, that is, those who have direct contact with animals [11].

S. aureus genome is plastic, so the organism can keep or lose part of itself according to necessity and growth conditions (for instance, competition for particular ecological niches and antibiotic selective pressure). In fact, most virulence factors are located on mobile genetic elements (MGEs) that can constitute 20% of the whole genome [18]. MGEs are transmitted among *S. aureus* strains by horizontal gene transfer either during cell replication, or by integration into a new host [11]. All among bacteriophages, pathogenicity islands and genomic islands, plasmids, transposons, and SCCmec cassettes are classes of MGEs that include the *mecA* gene, the bacteriophage-encoded Panton-Valentine leukocidin (PVL), and many resistance determinants [11,19,20]. It is noteworthy that multiple virulence factors may have the same function, so it is not surprising that strains with less MGEs are equally or much more pathogenic than those possessing all virulence determinants [21].

To establish infection, *S. aureus* employs the surface proteins called "microbial surface components recognizing adhesive matrix molecules" (MSCRAMMs), commonly referred to as adherence factors or adhesines. These mediate adherence to host tissues by binding collagen, fibronectin, and fibrinogen. MSCRAMMs play a fundamental role in starting endovascular, bone, joint, as well as prosthetic-device infections [2]. The adhesines comprise proteins covalently anchored to cell peptidoglycan (PT) (via the threonine residue in the signal motif at their C-terminus) that recognize the most prominent plasma components or the extracellular matrix. Typical members MSCRAMMs are the protein A (SpA), the fibronectin-binding proteins A and B (FnbpA and FnbpB), the collagen-binding protein, the clumping factors A and B (ClfA and ClfB), [22]. Once *S. aureus* has adhered it may convert host tissues into nutrients through production of exotoxins and enzymes such as nucleases, proteases, lipases, hyaluronidase, collagenase and elastase, that enable *S. aureus* to invade and metastasize to distant sites. Exotoxins possess cytolytic activity, meaning that they form β-barrel pores in the plasma membrane and cause leakage of the cell's content and bacterial lysis [2]. Members of this group are α-hemolysin, β-hemolysin, γ-hemolysin, leukocidins, and PVL. Further exotoxins are the toxic shock syndrome toxin-1 (TSST-1), the staphylococcal enterotoxins (SEA, SEB, SECn, SED, SEE, SEG, SEH, and SEI) and the exfoliative toxins A and B (ETA and ETB). Besides being pyrogenic, particularly, TSST-1 and enterotoxins are superantigens, as they directly stimulate proliferation of T-lymphocytes and initiate the "cytokines storm," resulting in a sepsis-like toxic syndrome and food poisoning, respectively. Although ETA and ETB (causing the SSS) have long been recognized for mitogenic activity toward T lymphocytes, it is not yet clear whether they also behave as superantigens [2]. Cell wall components as well (such as PT, lipoteichoic acid, and α-toxin) may support the septic shock by interacting with and activating the host immune system and coagulation pathways [23]. ETs disrupt desmosomes, triggering exfoliation. The result is a split in the epidermis at the granular level, which is responsible for the typical superficial desquamation [23]. *S. aureus* not only produces toxins, but it also has a strong impact on the immune system response (both innate and adaptive) due to staphylococcal complement inhibitor (SCIN, which blocks phagocytosis by human neutrophils by stopping the C3 convertase and preventing the C3b formation on the bacterium surface); the chemotaxis inhibitory protein (CHIPS) and the formyl peptide receptor-like-1 inhibitory protein (FLIPr), which block neutrophil receptors for chemoattractants; staphylokinase (SAK), which, by binding to α-defensins, inhibit their bactericidal properties; the extracellular fibrinogen binding protein (Efb, which is similar to adhesins) and the extracellular adherence protein (Eap, which inhibits complement activation) 2008 [23]. It is therefore clear that *S. aureus* is an extremely versatile organism capable (via a fine regulation/expression of its virulence factors) of causing moderate to severe infections. Either by switching on/off some regulator genes (*agr*, *mgr*, ArlR/ArlS, SaeRS, and Rot) or by using quorum-sensing systems, *S. aureus* may adapt to metabolic pathways to different environmental conditions in order to reduce undue metabolic demands, and to express some MGEs, only when required [24–27]. Recurrences of *S. aureus* infections are common, mainly because this pathogen can survive and persist in a quiescent state in various tissues, causing recrudescences when suitable conditions arise [23].

3.1.2 Skin and Soft Tissue Infections

Because *S. aureus* is a common component of the saprophyte human skin flora, it is consequently one of the major agents of skin and soft-tissue infections (SSTIs). SSTIs are common both in inpatients and outpatients and have been classified as complicated or uncomplicated by the US Food and Drug

Administration (FDA) [28,29]. Complicated SSTIs range from deep decubitus ulcers to diabetic foot infections (namely: infected burn, deep-tissue infection, major abscess, infected ulcer, perirectal abscess). Impetigo, cellulitis, erysipelas, furuncle, and simple abscesses are instead generally considered as uncomplicated pictures. Development of *S. aureus* uncomplicated SSTIs (folliculitis, furuncle, carbuncle, cellulitis, impetigo, mastitis, surgical wound infections, hidradenitis suppurativa) is supported by predisposing factors, including skin diseases, skin damage (e.g., insect bites, minor trauma), injections (e.g., piercing, drug use), and poor personal hygiene. Superficial infections are all characterized by formation of pus-containing blisters and their spread to the surrounding tissues.

Folliculitis, in particular, is a superficial infection involving the hair follicle, with a central area of purulence surrounded by induration and erythema.

Furuncles (boils), on the other hand, are painful lesions that tend to occur in moist bodily regions (such as follicle pores) and may evolve into true abscesses with a purulent central area.

Carbuncles are most often located in the lower neck and they result from the coalescence of other lesions that extend to a deeper layer of the subcutaneous tissue [23].

Cellulites are depicted by redness, swelling, and pain at the site of infection. It involves the underlying skin layers and the deeper dermis as far down as the subcutaneous fat. It usually starts from a scrape or cut in the skin (or following trauma, surgery, underlying skin lesions, lymphatic stasis, diabetes) that allows bacteria to enter, but there may be no apparent injury, as well.

Erysipelas involves the upper dermis (this feature distinguishes it from cellulites) and may be accompanied by fever, chills, and malaise, followed the formation of a warm, shiny bright red plaque. The lesions are well demarcated and are often present in the lower extremities [30].

Impetigo is highly contagious and appears as an acute infection of the epidermis superficial layers, occurring commonly in children, especially those who live in hot and humid areas. The name comes from the Latin *impetere* (to assail). Two forms, bullous and nonbullous, are acknowledged. The nonbullous is the most frequent form, it is specific to children and tends to affect skin on the face or extremities; and underlying, predisposing injuries are usually recognized, such as bites, cuts, abrasions, or any other trauma. Lesions start as vesicles that quickly rupture, leaving honey-colored crusts. *S. aureus*, together with group A *Streptococcus*, are the main causative agents, and differential diagnosis is therefore crucial for a successful antibiotic treatment. Geographical localization may be helpful for diagnosis, as well, as streptococcal impetigo is more common in warm, humid environments and tropical or subtropical climates, while the staphylococcal form is prevalent in temperate climates, particularly during summer [31]. Bullous impetigo, instead, is caused exclusively by *S. aureus* (even if it frequently replaces an initial streptococcal infection), and if it affects intact skin and is toxin-mediated (exfoliative toxins, encoded by *eta* and *etb* genes), then it may be considered a localized SSSS, where large skin areas are damaged and lost [32,33]. Lesions are represented by vesicles that rapidly progress to bullae containing clear or yellow fluid. Bullae rupture after 2–3 days, forming light brown crusts at the borders of erythematous erosions. Sometimes impetigo evolves to ecthyma, which is an ulcerated infection often accompanied by lymphadenitis [33].

Mastitis is an infection of the breast (generally unilateral) that usually occurs in breastfeeding women only. As *S. aureus* is normally present on the skin, and in fact, any crack on nipples can act as a portal of entry. The infected area becomes swollen, red, hard, and painful, but other signs or symptoms such as fever, chills, and tachycardia may be present [31].

Hidradenitis suppurrativa is a chronic skin condition appearing as a subcutaneous pea-sized lump. It develops in areas rich in apocrine, sweat, or sebaceous glands, mostly where skin rub occurs, such as armpits, the groin, between buttocks, and under breasts (the disease is otherwise known as acne inversa). The exact cause of hidradenitis suppurativa remains unclear, but a predisposing condition is the follicular occlusion. This begins with follicular plugging that obstructs the apocrine gland ducts along with perifolliculitis around the ducts; the follicular epithelium then ruptures, with a following bacterial infection and formation of subcutaneous sinus tracts between abscesses. Hidradenitis suppurativa-associated lumps are painful and may break open and drain foul-smelling pus. In many cases, tunnels connecting lumps will form under the skin. The disease is usually diagnosed after puberty, and it persists for years and may worsen over time [31]. Both complicated and uncomplicated SSTIs are mainly due to MRSA [34]. The CDC encourages us to consider MRSA (either HA or CA-MRSA) in the differential diagnosis of SSTIs, particularly in patients with the complaint of a "spider bite," and every time the lesion is purulent, fluctuant, or a palpable fluid-filled cavity is observed, with a yellow or white center, and a central zone draining pus; possibly, pus should be aspirated with a needle or syringe to allow examination (http://www.cdc.gov/mrsa/diagnosis/ [35]). CA-MRSA SSTIs usually affect children, young adults, and middle-aged adults (20–47 years) without any known risk factors for MRSA. Moreover, a history of spider bites is common in these subjects [36]. Transmission of CA-MRSA usually relies on direct contact with infected/colonized patients, or a contaminated and/or crowded environment where direct contact and turf abrasion (for example, contact among athletes in sports activities as well as environments such as in the military or a closed community) may behave as risk factors. CA-MRSA, furthermore, may carry PVL genes, while HA-MRSA does not usually harbor such pathogenic elements [36]. As PVL may also cause dermonecrosis (see the chapter *The staphylococcal Panton-Valentine toxin*), this toxin is consistently associated with SSTIs, particularly those evolving to necrosis and suppuration [36].

Finally, TSS presents with diffuse macular erythema, fever, and hypotension, as well as with rapid involvement of three or more organs. The disease is prevalent in young people, in menstruating females who use tampons as well as nonmenstruating women, and it is also reported in men and healthy individuals in general. The causative *S. aureus* strains secrete the TSST-1 toxin (see the preceding description), a member of the superantigen family that is able to directly activate T-cells, and then cytokine storm induction. Particularly, inflammatory mediators including interferon, TNF-α and -β, interleukin-1 and -6, are released and their cascade is responsible for a shock syndrome mimicking that caused by endotoxins [37].

3.1.3   Bacteremia and Bacteremic Infections

When *S. aureus* invades the bloodstream (bacteremia) it may be responsible for subsequent systemic and life-threatening bacteremic conditions such as endocarditis, osteomyelitis, pneumonia, brain abscesses, and meningitis [38]. The mortality associated with *S. aureus* bacteremia (SAB) ranges from 0% to 83%, and this variability is likely to be attributable both to different patients' conditions (a worse outcome is expected in critically ill patients as well as those with comorbidities) and to a genetic assortment of *S. aureus* isolates [39,40].

SAB prevalence is high, being 26.0% in North America, 21.6% in Latin America. *S. aureus* is the second most common agent of nosocomial bacteremia in Europe (prevalence of 19.5%), and the major cause of bacteremia in early-onset inpatients [39,41].

Mechanisms used by *S. aureus* to invade the bloodstream are consistently similar to those behind other infections by this pathogen: adhesion and colonization, invasion and evasion, biofilm production, secretion of several enzymes (such as proteases, lipases, and elastases, which allow the invasion and destruction of the host's tissues and therefore the metastatization) and, of course, antibiotic resistance. In establishing SAB, the colonization could be a specific risk [39]. Von Eiff et al. demonstrated that blood isolates were identical to nasal isolates in 82% of colonized patients. The elimination of nasal carriage by locally applied or systemic antibiotics is a useful prevention strategy [42]. Nose-colonizing *S. aureus* is able to adhere to nasoepithelial cells, and at the same time, to cope with the host's defense and compete with resident microorganisms. MSCRAMMs (which include protein A) are the most important components in the adhesion and attachment to nasal epithelial cells, but the clumping factor B and the wall-associated teichoic acids may play a relevant role, too [8]. Slime production and the ability to escape host defenses can have a large part in establishing a SAB, as well. These are different sides of the same coin; in fact, the biofilm (either produced on prosthetic devices or on host's tissues) allows *S. aureus* to adhere, survive, and hide its presence to the immune system. Moreover, a biofilm matrix may restrict the penetration of some antibiotics, thus indirectly contributing to antimicrobial resistance [43]. Evasion of the host's defense is, however, a multifactorial process, in which many components take part including: (i) secretion of chemotaxis inhibitory proteins;

(ii) production of extracellular adherence protein (which specifically interfere with neutrophil extravasation and chemotaxis towards the infection site); (iii) production of leukocidins (such as PVL); (iv) presence of antiphagocytic microcapsule (zwitterionic capsule), charged positively and negatively, and able to both mediate protection from the immune system and induce abscess formation); (v) prevention of opsonization (mediated by protein A, which binds the Fc portion of immunoglobulins). Next to PVL, enterotoxins and TSS are able to subvert the normal immune response by acting as superantigens, then inducing a potent polyclonal stimulation and expansion of T-cells (by direct interaction with Vb-specific receptor), followed by the deletion and/or suppression of T-cells to an anergic state [39]. The SAB infective framework can be further complicated due to the interaction of *S. aureus* with coagulation pathways of the host. Many components produced by *S. aureus* are able to act as procoagulative factors: these are clumping factors, coagulase (CoA), von Willebrand factor-binding protein (vWbp), protein A, fibronectin-binding proteins, cell wall PT [44]. (See the chapter *The staphylococcal coagulases* in this volume.) CflB also plays a prevalent role in adhesion to nose squamous epythelial cells, while CflA not only hampers phagocytosis by interacting with both macrophages and neutrophils (in neutrophils this is due to both a fibrinogen-dependent and a fibrinogen-independent mechanism) but it also activates platelets through the interaction with GPIIb/IIIa and FcγRIIa platelet receptors, mediating fibrinogen-platelet interaction, fibrin cross-linking, and thrombus formation [45]. When the activation of platelets becomes systemic [46], SAB can be accompanied by thrombocytopenia (as a result of platelets consumption) and by disseminated intravascular coagulation (DIC). Thrombus formation as well as the fibrin-clots may contribute to the colonization of endothelium and/or to damage heart valves, then to promote infective endocarditis as well as endovascular infections [46]. It is also postulated that coagulation, explicitly fibrin clots on the inner heart surfaces, and together with biofilm, can be the basis for the formation of endocardial vegetations [47]. Also, the PT is able to activate the extrinsic pathway of coagulation by inducing the expression of tissue factor (TF) in the host's cells. TF (that is a single-chain transmembrane protein composed of 263 amino acids residues recognized as the major physiological initiator of blood coagulation) is expressed either by monocytes or endothelial cells only after induction (i.e., by *S. aureus* PT). TF binds plasma factor VII, forming a potent procoagulant complex, which can rapidly activate factor IX and X, resulting in thrombin generation [48]. On the contrary, protein A may contribute to the coagulative cascade by interacting with vWF and by interacting with the gC1qR on platelets and endothelial cells [44]. The thrombus formation increases the expression of gC1qR, then *S. aureus* adhesion, and this potentially increases platelet aggregation. Staphylococci trapped in a fibrin mesh, containing activated leukocytes and platelets, could bind to platelets and endothelial surfaces and increase thrombin formation. Thrombin further activates endothelial cells, which increase the adhesion of leukocytes and platelets. In this complicated picture, toxins may have a role. PLV can lyse leukocytes and release inflammatory mediators including amino-acid derived vasodilators (histamine and serotonin), acid hydrolases-like β-glucoronidase and lysozyme, proteases (elastase and cathepsin G), chemotactic agents (such as leukotriene B4 and IL-8) which are all chemical mediators contributing to thrombus formation and therefore, consequently, deep vein thrombosis and septic shock, the latter relying on the vasodilators' release [44]. Catheters and synthetic devices in general are potential portals of entry for *S. aureus* into the bloodstream, and then sources for catheter-related bloodstream infections (CRBSIs) and subsequent hematogenous dissemination. Further, thrombophlebitis may develop at the catheter insertion site, causing local fever, pain, and occasionally erythema [49]. Via blood circulation, *S. aureus* may reach bones, thus being responsible for osteomyelitis. It is interesting that, in such a disease, particularly when chronic, *S. aureus* having the ability to form small-colony variants (SCVs) is crucial. SCVs, in fact, support survival in a metabolically inactive state under harsh conditions and may therefore contribute to persistent and recurrent infections [8]. In those of bone, moreover, the SpA role is relevant, as well, as the protein has been observed, in vitro, to directly bind osteoblasts, preventing their proliferation and mineralization and inducing, conversely, cell apoptosis. Infected osteoblasts also increase the expression of RANKL, a key protein involved in initiating bone resorption [50].

S. aureus is also recognized as an important cause of pneumonia in both the adult and pediatric populations. In hospital settings, the ventilator-associated pneumonia (VAP) is particularly aggressive in compromised and critical patients, as well as in the elderly. Mainly, high morbidity and mortality are due to HA-MRSA. Multiple virulence factors are implicated in *S. aureus* VAP, including lipotheicoic acid (LTA), PT, MSCRAMMs, in particular SpA and α-toxin. *S. aureus* is able to adhere to the respiratory epithelium, damage the alveolo capillary barrier, and attract leukocytes. The rapid and excessive recruitment of neutrophils alone causes the intense host inflammatory response that is the basis for lung injury and pneumonia progression. When the disease turns into a necrotizing pneumonia, the action of SpA, α-toxin, and β-toxin (which cause cell damage and play a role in inflammation and necrosis of airway epithelium) can be implied. Conversely, the role of PVL in necrotizing pneumonia is still controversial [2,51].

Not less important is community-acquired pneumonia (CAP) due to CA-MRSA strains, which occurs in otherwise healthy individuals. As a necrotizing infection; it frequently follows influenza, and the rapid and progressive course along with the extensive lung involvement may lead to acute respiratory distress with pleural effusion, hemoptysis, and leucopenia [52]. Moreover, it should be taken into account that *S. aureus* pneumonia may seriously complicate cystic fibrosis, too, and patients receiving immunosuppressive treatments [53].

The bloodstream can bring organisms, finally, to the central nervous system, hence *S. aureus* meningitis and cerebritis can develop. Albeit spread worldwide, that by *S. aureus* represents less than 6% of all cases of bacterial meningitis [54]. It could manifest as a serious postoperative complication as well as a consequence of a microorganism metastatic dissemination or contiguous infections (from epidural abscesses, endocarditis, skin and soft tissue infections, pneumonia, sinusitis, and related cavernous sinus thrombosis). Hematogenous and postoperative-meningitis are, however, distinct clinical entities; the former is usually community-acquired and mainly affects old patients with comorbidities such as cardiovascular and chronic renal impairment (hemodialized patients particularly), diabetes, and intravenous drug use. Again, it presents with a more serious picture than postoperative forms, as patients suffer from altered mental status, fever, and septic shock. On the contrary, the postoperative disease occurs particularly after a postsurgery leakage of cerebrospinal fluid, or due to placement of intraventricular shunts and external ventricular drainage; finally, it may be accompanied by brain abscesses and cerebritis [55,56].

To conclude, along with *S. aureus* ssp. *aureus*, *S. aureus* ssp. *anaerobius* has been reported as the agent of a bacteremia episode (with septic arthritis and multiple lung abscesses); interestingly, this occurred in a man who used to prepare strawberry soil with sheep manure, without wearing gloves, thus causing hand injuries acting as a portal of entry to the bloodstream. This case emphasized, particularly, the animal origin of the pathogen, along with the human pathogenic potential of this strictly anaerobic *S. aureus* subspecies [57].

3.1.4 Enteritis

The presence of *S. aureus* in foods potentially represents a public health hazard because many strains produce enterotoxins [1]. Staphylococcal enterotoxins (SEs) are heat-stable proteins and are produced by other CoPS as well, such as certain *Staphylococcus intermedius* and *Staphylococcus hyicus* strains. Although SEs form a group of serologically different proteins, namely A, B, C_1, C_2, C_3, D, F, G, H, I, J, K, L, M, N, and O, they all share important features as they have similar structures, they all cause emesis and gastroenteritis, they show superantigenicity properties, and are resistant to heat and pepsin digestion [58]. The most common presentation of staphylococcal food poisoning includes nausea, vomiting, abdominal pain, diarrhea, chills, with or without fever, sometimes dizziness, and myalgias. Symptoms develop 2–4 hours after the ingestion of toxin-contaminated products and the illness may vary from quite mild pictures (lasting a few hours only and usually self-limiting) to severe conditions, even fatal, that require hospitalization (mostly children and the elderly). The exact mechanism on which staphylococcal food poisoning relies is not yet completely understood. Nevertheless, it is known that meat (particularly if minced), salads, cream, cream-filled bakery products, and dairy foods are the main vehicles of enterotoxic *S. aureus* isolates [59].

3.2 COAGULASE-POSITIVE STAPHYLOCOCCI OTHER THAN *S. AUREUS*

Together with *S. aureus* (both subspecies), other CoPS including *S. intermedius*, *Staphylococcus pseudintermedius*, *Staphylococcus schleiferi* subsp. *coagulans*, and *S. hyicus* have been involved in human disease [60], although they are primarily animal pathogens [61]. *S. hyicus* has been responsible for a septic episode in a farmer who was in close contact with piglets, as well as a wound infection after a donkey bite; it is also able to produce an exfoliative toxin (although the human receptor, desmoglein-1, is likely to be resistant to toxin activity) [62]. *S. intermedius* is only rarely found as a human nose colonizer and in man, it may be responsible for piodermitis, bacteremia, pneumonia, brain abscesses, and infections of wounds, (mostly canine-inflicted). *S. pseudintermedius* is recognized as an occasional human pathogen (a case of rhinosinus-itis, a catheter-related bacteremia, an implantable cardioverter-defibrillator infection, and endocarditis and a case of graft versus host disease-related wound infection were reported) [63]. The ability to form biofilm is variable among bacterial species; in this context, *S. pseudintermedius* has been classified as either a strong or a moderate biofilm producer [64]. *S. schleiferi* subspecies *coagulans* expresses potential human pathogenicity, as well, and a case of a finger wound infection and an episode of endocarditis in a dog owner have been described [65]. Finally, *S. intermedius* and *S. hyicus* (which produce A to E toxins) are potentially able to cause enterotoxin-related enteritis [66]. Particularly, an outbreak due to *S. intermedius* and related to a butter blend and margarine was described in the United States [67].

3.3 COAGULASE-NEGATIVE STAPHYLOCOCCI

CoNS include about 40 species, and are the most prevalent organisms among the routinely identified clinical isolates. They are commensals on skin and mucosae; for instance, *S. capitis* mostly colonize the head, *Staphylococcus auricolaris* preferably inhabits the ear canal, *S. saprophyticus* is present in the inguinal area, *S. haemolyticus* in the axilla, while *S. epidermidis* and *Staphylococcus hominis* may be collected from all body sites; in fact, they do not show any tropism for specific niches but can be readily man-to-man transferred through contact or sloughing of skin [68,69]. Although considered for a long time to be as nonpathogens, being commensals of skin and mucosa, they evoked interest when they became frequently isolated pathogens in hospital-acquired infections and, particularly, in patients receiving foreign implants (catheters as well as any type of prosthetic device) along with those with surgical wounds [70]. Infections due to these pathogens start from any iatrogenic procedures that move the colonizing flora from skin or mucosae to usually inaccessible sites, or devices, that consequently become contaminated. Patients who are at high risk of CoNS infection are immunocompromised, immunosuppressed, oncologic, and onco-hematologic people, transplanted subjects, as well as those who are long-term hospital patients, or intravenous drug abusers. The most important feature of their pathogenicity is the ability to produce slime (the sugar matrix that, together with sessile cells, form biofilms) either on foreign bodies or on the host's tissue. Biofilms have been known to confer reduced antibiotic suscep-tibility and defense against the host immune system [71]. Also, this bacterial organization allows long-term survival of microorganisms and, therefore, long-lasting, chronic infections [72].

If compared with CoPS, CoNS express fewer virulence factors; in fact, with the exception of native valve endocarditis, they cannot cause purulent infec-tions, maybe with the only exception of *Staphylococcus lugdunensis*.

The main CoNS infections are endocarditis and sepsis, infection of joint prostheses, endophthamitis (due to lens), peritonitis, ventriculitis (shunt-related), and urinary tract infections (in patients with permanent catheters or otherwise healthy people with no underlying disorders) [70]. In general, the largest amounts of staphylococci are found in sweat glands and on mucous mem-branes surrounding body openings. *S. lugdunensis* tends, however, to be more aggressive than other CoNS as it may cause purulent infections [69]. It is in fact responsible of endocarditis, sepsis, osteomyelitis, and brain abscesses [69].

Several factors may play a role in colonizing human skin and mucosae. In *S. epidermidis,* it has been evidenced as a protein that promotes aggregation. Particularly, the accumulation-associated protein Aap (which needs zinc ions and proteolytic processing to display aggregating activity) forms large fibrils on the bacterial surface [69,70]. Moreover, *S. epidermidis* forms biofilms based on a two-step process: first, bacteria adhere to the surface (initial attachment phase); second, they form cell-cell aggregates and a multilayered architecture (accumulative phase). In the first phase the autolysin AtlE is involved in the bacterial attachment to medical device surfaces, while in the accumulative phase, the abovementioned Aap may play a role together with a polysaccha-ride intercellular adhesin (PIA) and a linear poly-Nacetyl-1,6-β-glucosamine (PNAG; a component mediating intercellular adhesion) [73]. But colonization is also supported by the production of several substances facing the host barriers such as fatty acid-modifying enzymes (which detoxify the host fatty acids), li-pases (which guarantee the survival of CoNS in fatty secretions), and proteasea (which cause tissue damage).

The ability to produce substances forming bonds with the host cells and protecting the microorganism is also described for *S. lugdunensis, S. haemo-lyticus* and with different gradualness of expression, in *S. simulans, S. capitis, S. hominis* (both subspecies), *S. schleiferii* (both subspecies), *S. cohnii* (both subspecies) and *S. caprae*. On the contrary, the urease production is specific to *S. capitis* subsp. *ureolyticus, S. simulans, S. hominis, S. warneri, S. caprae*; the lipase, conversely, is typical in *S. simulans, S. caprae, S. hominis, S. schleiferii* subsp coagulans, and *S. cohnii* subsp urealyticus. *S. saprophyticus* not only produces urease, instead, which causes tissue damage and invasion, but also exhibits a predilection for the uropoietic tract (particularly in young women) by secreting Ass autolysin that mediates hemoagglutination and adhesion to uroepithelial cells fibronectin (this is the reason why this CoNS species has been known to be potentially relevant when isolated from urinary infections) [74]. The Ssp protein, finally, mediates *S. saprophyticus* aggregation and formation of multilayered colonies [69,70]. To conclude, aiming to contrast the host defenses, CoNS produce phenol-soluble modulins (PSMs) and PSM-like compounds; these are short, amphipathic, and α-helical peptides that may have (some of them) strong cytolytic activity toward human neutrophils and other cell types. Some others exhibit instead a bactericidal effect, making CoNS gain a competitive advantage toward other resident organisms; in this context, for instance, *S. haemolyticus* produces a PSM-like factor that inhibits gonococcal growth [69].

3.3.1   CoNS and Their Role in Device Infections

Following their placement, medical implants are usually rapidly coated with a host-derived, protein-based film, that contains receptors allowing bacterial attachment (to form, them biofilms). Infecting CoNS can originate from several sources, such as skin at the insertion site, the medical device itself (if colonized before implant), air, and healthcare workers. Bacterial colonization often causes device dysfunction and, furthermore, infectious agents can disseminate from the original site of colonization, thus causing disease in other suitable niches. This is the case of CRBSIs that result in significantly increased hospital costs, duration of hospitalization, and the patient's morbidity. Diagnosis of CRBSIs is well approached by the guidelines of the American Society of Infectious Diseases [75]. It can represent a true challenge because clinical findings are unreliable and mostly include fever, chills, hypotension (having poor specificity), inflammation and/or purulent discharge around the device entrance and bloodstream infection (showing greater specificity but poor sensitivity). Therefore, to achieve an accurate diagnosis, both microbiological criteria and a clinical picture should be taken into account [75]. Once a CRBSI is documented and the etiologic agent identified, the catheter should be removed, with this decision being based on illness severity, evidence of device infection, underlying complications such as endocarditis, septic thrombosis, tunnel infection, or metastatic seeding. Sometimes, and particularly when a long-term catheter has been implanted (in oncologic patients for instance the device is used for chemiotherapic treatment), removal is impossible, hence antibiotic lock therapy may be administered [75]. Together with the intravenously placed catheter, all artificial devices, once implanted, may be potentially colonized by CoNS: this is the case of infection of pacemakers (presenting as a simple pocket infection, septicemia, or endocarditis), orthopedic prostheses (resulting in local pain, swelling, increased inflammatory markers, alterations at bone-cement interface, periprosthetic osteolysis, localized periosteal new bone formation), ventriculoperitoneal or cerebrospinal fluid shunts (mostly implanted to reduce hydrocephalus; shunt infection may present with intracranial hypertension due to device malfunction, with or without headache, nausea, vomiting, and mental state impairment; the disease is often complicated by ventriculitis and associated with high morbidity and mortality) [76].

Bacterial endophthalmitis, to conclude, involves the vitreous and/or aqueous humor. Most cases are exogenous, and organisms are introduced via trauma (blunt trauma injuries, retained intraocular foreign bodies, lens-related), surgery, or an infected cornea. Conversely, endogenous endophthalmitis follows bloodstream dissemination. Staphylococci, particularly CoNS, are the causative agents in 70% of cases, and up to 80% in postcataract endophthalmitis. Systemic antibiotics are not effective and intravitreal administration is required, but, due to drug resistant biofilm communities, treatment can be challenging [77]. CoNS are also involved in conjunctivitis, especially in newborns during the first month of life, who may suffer hyperemia and ocular discharge. Wadhwani, Prentice, and Mohile showed that the most common organisms in newborns' conjunctivitis are CoNS (up to 60% of cases) and that history of midwife interference and premature rupture of amniotic membranes as well as the presence of the same CoNS in the mother's vaginal mucosa, which may represent risk factors promoting the development of babies' conjunctivitis [78–80].

REFERENCES

[1] Bannerman TL. *Staphylococcus*, *Micrococcus*, and other catalase-positive cocci that grow aerobically. [Chapter 28]. In: Murray P, Baron EJ, Jorgensen JH, Pfaller MA, Yolken RH, editors. Manual of clinical microbiology. 8th ed. Washington, DC: ASM Press; 2004. p. 384–404.

[2] Bien J, Sokolova O, Bozko P. Characterization of virulence factors of *Staphylococcus aureus*: novel function of known virulence factors that are implicated in activation of airway epithelial proinflammatory response. J Pathogens 2011;601905. https://doi.org/10.4061/2011/601905.

[3] McCann MT, Gilmore BF, Gorman SP. *Staphylococcus epidermidis* device-related infections: pathogenesis and clinical management. J Pharm Pharmacol 2008;60:1551–71.

[4] Von Eiff C, Arciola CR, Montanaro L, Becker K, Campoccia D. Emerging *Staphylococcus* species as new pathogens in implant infections. Int J Artific Organs 2006;29:360–7.

[5] Gilbert P, Das J, Foley I. Biofilm susceptibility to antimicrobials. Adv Dental Res 1997;11:160–7.

[6] Costerton JW, Stewart PS, Greenberg EP. Bacterial biofilms; a common cause of persistent infections. Science 1999;284:1318–22.

[7] Natoli S, Fontana C, Favaro M, et al. Characterization of coagulase-negative staphylococcal isolates from blood with reduced susceptibility to glycopeptides and therapeutic options. BMC Infect Dis 2009;9:83. https://doi.org/10.1186/1471-2334-9-83.

[8] Liu GY. Molecular pathogenesis of *Staphylococcus aureus* infection. Pediatr Res 2009; 65(5 Pt 2).

[9] Aires De Sousa M, De Lencastre H. Bridges from hospitals to the laboratory: genetic portraits of methicillin-resistant *Staphylococcus aureus* clones. FEMS Immunol Med Microbiol 2004;40(2):101–11.

[10] Chambers HF, DeLeo FR. Waves of resistance: *Staphylococcus aureus* in the antibiotic era. Nat Rev Microbiol 2009;7(9):629–41.

[11] Stefani S, Chung DR, Jodi A, et al. Meticillin-resistant *Staphylococcus aureus* MRSA: global epidemiology and harmonisation of typing methods. Int J Antimicrob Agents 2012;39:273–82.

[12] Boyce JM. Should we vigorously try to contain and control methicillin-ressitant *Staphylococcus aureus*? Infect Control Hosp Epidemiol 1991;12:46–54.

[13] Woodford N, Livermore DM. Infections caused by Gram-positive bacteria: a review of the global challenge. J Infect 2009;59(Suppl 1):S4–16. https://doi.org/10.1016/S0163-4453(09)60003-7.

[14] Holden MT, Hsu LY, Kurt K, et al. A genomic portrait of the emergence, evolution, and global spread of a methicillin-resistant *Staphylococcus aureus* pandemic. Genome Res 2013;23(4):653–64. https://doi.org/10.1101/gr.147710.112.

[15] Diep BA, Otto M. The role of virulence determinants in community-associated MRSA pathogenesis. Trends Microbiol 2008;16:361–9.

[16] Monaco M, Pedroni P, Sanchini A, Bonomini A, Indelicato A, Pantosti A. Livestock-associated methicillin-resistant *Staphylococcus aureus* responsible for human colonization and infection in an area of Italy with high density of pig farming. BMC Infect Dis 2013;13:258. https://doi.org/10.1186/1471-2334-13-258.

[17] García-Álvarez L, Holden MT, Lindsay H, Webb CR, Brown DF, Curran MD, et al. Meticillin-resistant *Staphylococcus aureus* with a novel mecA homologue in human and bovine populations in the UK and Denmark: a descriptive study. Lancet Infect Dis 2011;11:595–603.

[18] Malachowa N, DeLeo FF. Mobile genetic elements of *Staphylococcus aureus*. Cell Mol Life Sci 2010;67(18):3057–71.

[19] Firth N, Skurray RA. Genetics accessory elements and genetic exchange. In: Fischetti VA, Novick RP, Ferretti JJ, Portnoy DA, Rood JI, editors. Gram-positive pathogens. 2nd ed. Washington, DC: ASM Press; 2006. p. 413–26.

[20] Novick RP. Staphylococcal plasmids and their replication. Annu Rev Microbiol 1989;43:537–65.

[21] McCarthy AJ, Breathnach AS, Linsday AJ. Detection of mobile genetic element variations between colonizing and infecting hospital associated methicillin resistant *Sthaphylococcus aureus* isolates. J Clin Microbiol 2012;50:1073–5.

[22] Foster TJ, Hook M. Surface protein adhesins of *Staphylococcus aureus*. Trends Microbiol 1998;6(12):484–8.

[23] Gordon RJ, Lowy FD. Pathogenesis of methicillin-resistant *Staphylococcus aureus* infection. Clin Infect Dis 2008;46(Suppl 5):S350–9.

[24] Cheung AL, Koomey JM, Butler CA, Projan SJ, Fischetti VA. Regulation of exoprotein expression in *Staphylococcus aureus* by a locus (sar) distinct from agr. Proc Natl Acad Sci U S A 1992;89:6462–6.

[25] Novick RP. Autoinduction and signal transduction in the regulation of staphylococcal virulence. Mol Microbiol 2003;48:1429–49.

[26] Said-Salim B, Dunman PM, McAleese FM, et al. Global regulation of *Staphylococcus aureus* genes by Rot. J Bacteriol 2003;185:610–9.

[27] Liang X, Yu C, Sun J, et al. Inactivation of a two-component signal transduction system, SaeRS, eliminates adherence and attenuates virulence of *Staphylococcus aureus*. Infect Immun 2006;74:4655–65.

[28] US Department of Health and Human Services, Food and Drug Administration, Center for Drug Evaluation and Research (CDER). Guidance for industry: uncomplicated and complicated skin and skin structure infections—developing antimicrobial drugs for treatment (draft guidance); 1998.

[29] US Department of Health and Human Services, Food and Drug Administration, Center for Drug Evaluation and Research (CDER). Guidance for industry: acute bacterial skin and skin structure infections: developing drugs for treatment (draft guidance); 2010.

[30] Denis O, Deplano A, De Beenhouwer H, Hallin M, Huysmans G, Garrino MG, et al. Polyclonal emergence and importation of community-acquired methicillin-resistant *Staphylococcus aureus* strains harbouring Panton-Valentine leucocidin genes in Belgium. J Antimicrob Chemother 2005;56:1103–6.

[31] Stevens DL, Bisno AL, Chambers HF, et al. Practice guidelines for the diagnosis and management of skin and soft-tissue infections. Clin Infect Dis 2005;41:1373–406.

[32] Amagai M, Matsuyoshi N, Wang ZH, Andl C, Stanley JR. Toxin in bullous impetigo and staphylococcal scalded-skin syndrome targets desmoglein 1. Nat Med 2000;6(11):1275–7.

[33] Cole C, Gazewood J. Diagnosis and treatment of impetigo. Am Farm Physician 2007;75(6):859–64.

[34] Sabitha R. Skin and soft-tissue infections: classifying and treating a spectrum. Cleveland J Med 2012;79(1):57–66.

[35] http://www.cdc.gov/mrsa/diagnosis/.

[36] Stryjewski ME, Chambers HF. Skin and soft-tissue infections caused by community-acquired methicillin-resistant *Staphylococcus aureus*. Clin Infect Dis 2008;46:S368–77.

[37] Schleivert PM, Shands KN, Dan BB, Schmid GP, Nishimura RD. Identification and characterization of an exotoxin from *Staphylococcus aureus* associated with toxic shock syndrome. J Infect Dis 1981;143:509–16.

[38] Corey RG. *Staphylococcus aureus* bloodstream infections: definitions and treatment. Clin Infect Dis 2009;48:S254–9.

[39] Naber CK. *Staphyloccus aureus* bacteremia: epidemiology, pathophysiology, and management strategies. Clin Infect Dis 2009;48:231–7.

[40] del Rio A, Cervera C, Moreno A, Moreillon P, Miró JM. Patients at risk of complications of staphylococcus aureus bloodstream infection. Clin Infect Dis 2009;48:S246–53.

[41] Laupland KB, Lyytikäinen O, Søgaard M, et al. International Bacteremia Surveillance Collaborative. The changing epidemiology of *Staphylococcus aureus* bloodstream infection: a multinational population-based surveillance study. Clin Microbiol Infect 2013;19(5):465–71.

[42] Von Eiff C, Becker K, Machka K, Stammer H, Peters G. Nasal carriage as a source of *Staphylococcus aureus* bacteremia. N Engl J Med 2001;344:11–6.

[43] Patel R. Biofilms and antimicrobial resistance. Clin Orthop Relat Res 2005;41–7.

[44] Martin E, Cevik C, Nugent K. The role of hypervirulent *Staphylococcus aureus* infections in the development of deep vein thrombosis. Thromb Res 2012;130:302–8. https://doi.org/10.1111/j.1469-0691.2012.03903.x.

[45] Miajlovic H, Loughman A, Brennan M, Cox D, Foster TJ. Both complement- and fibrinogen-dependent mechanisms contribute to platelet aggregation mediated by Staphylococcus aureus clumping factor B. Infect Immun 2007;75(7):3335.

[46] Gafter-Gvili A, Mansur N, Bivas A, et al. Thrombocytopenia in *Staphylococcus aureus* bacteremia: risk factors and prognostic importance. Mayo Clin Proc 2011;86(5):389–96.

[47] Drake TA, Pang M. *Staphylococcus aureus* induces tissue factor expression in human cardiac valve endothelial cells. J Infect Dis 1988;157:749–56.

[48] Mattsson E, Herwald H, Björck L, Egesten A. Peptidoglycan from *Staphylococcus aureus* induces tissue factor expression and procoagulant activity in human monocytes. Infect Immun 2002;70:3033–9.

[49] Fowler Jr. VG, Justice A, Moore C, et al. Risk factors for hematogenous complications of intravascular catheter-associated *Staphylococcus aureus* bacteremia. Clin Infect Dis 2005;40(5):695–703.

[50] Claro T, Widaa A, O'Seaghdha M, et al. *Staphylococcus aureus* protein A binds to osteoblasts and triggers signals that weaken bone in osteomyelitis. PLoS One 2011;6(4):e18748. https://doi.org/10.1371/journal.pone.0018748.

[51] González C, Rubio M, Romero-Vivas J, González M, Picazo JJ. Bacteremic pneumonia due to *Staphylococcus aureus*: a comparison of disease caused by methicillin-resistant and methicillin-susceptible organisms. Clin Infect Dis 1999;29(5):1171–7.

[52] Rubinstein E, Kollef MH, Nathwani D. Pneumonia caused by methicillin-resistant *Staphylococcus aureus*. Clin Infect Dis 2008;46:S378–85.

[53] Conway SP, Brownlee KG, Denton M, Peckham DG. Antibiotic treatment of multidrug-resistant organisms in cystic fibrosis organisms in cystic fibrosis. Am J Respir Med 2003;2(4):321–32.

[54] Schlesinger LS, Ross SC, Schaberg DR. *Staphylococcus aureus* meningitis: a broad-based epidemiologic study. Medicine (Baltimore) 1987;66(2):148–56.

[55] Pistella E, Campanile F, Bongiorno D, Stefani S. Successful treatment of disseminated cerebritis complicating methicillin-resistant *Staphylococcus aureus* endocarditis unresponsive to vancomycin therapy with linezolid. Scand J Infect Dis 2004;36(3):222–5.

[56] Aguilar J, Urday-Cornejo V, Donabedian S, Perri M, Tibbetts R, Zervos M. *Staphylococcus aureus* meningitis: case series and literature review. Medicine (Baltimore) 2010;89(2):117–25. https://doi.org/10.1097/MD.0b013e3181d5453d.

[57] Peake SL, Peter JV, Chan L, Wise RP, Butcher AR, Grove DI. First report of septicemia caused by an obligately anaerobic *Staphylococcus aureus* infection in a human. J Clin Microbiol 2006;44(6):2311–3.

[58] Casman EP, Bergdoll MS, Robinson J. Designation of staphylococcal enterotoxin. J Bacteriol 1963;85:715–6.

[59] Veras JF, do Carmo LS, Tong LC, et al. A study of the enterotoxigenicity of coagulase-negative and coagulase-positive staphylococcal isolates from food poisoning outbreaks in Minas Gerais, Brazil. Int J Infect Dis 2008;12(4):410–5. https://doi.org/10.1016/j.ijid.2007.09.018.

[60] Savini V, Barbarini D, Polakowska K, et al. Methicillin-resistant *Staphylococcus pseudintermedius* infection in a bone marrow transplant recipient. J Clin Microbiol 2013;51(5):1636–8. https://doi.org/10.1128/JCM.03310-12.

[61] Hanselman BA, Kruth SA, Rousseau J, Scott Weese J. Coagulase positive staphylococcal colonization of humans and their household pets. Can Vet J 2009;50:954–8.

[62] Casanova C, Iselin L, von Steiger N, Droz S, Sendi P. *Staphylococcus hyicus* bacteremia in a farmer. J Clin Microbiol 2011;49:4377–8.

[63] Savini V, Passeri C, Mancini G, et al. yi: my pet's two faces. Res Microbiol 2013;164(5):371–4.

[64] Singh A, Walker M, Rousseau J, Weese JS. Characterization of the biofilm forming ability of *Staphylococcus pseudintermedius* from dogs. BMC Vet Res 2013;9:93.

[65] Kumar D, Cawley JJ, Irizarry-Alvarado JM, Alvarez A, Alvarez S. Case of *Staphylococcus schleiferi* subspecies coagulans endocarditis and metastatic infection in an immune compromised host. Transpl Infect Dis 2007;9(4):336–8.

[66] Adesiyun AA, Tatini SR, Hoover DG. Production of enterotoxins by *Staphylococcus hyicus*. Vet Microbiol 1984;9:487–95.

[67] Khambaty FM, Bennett RW, Shah DB. Application of pulse field gel electrophoresis to the epidemiological characterisation of *Staphylococcus intermedius* implicated in a food-related outbreak. Epidemiol Infect 1994;113:75–81.

[68] Fey PF, Olson ME. Current concepts in biofilm formation of *Staphylococcus epidermidis*. Future Microbiol 2010;5(6):917–33. https://doi.org/10.2217/fmb.10.56.

[69] Otto M. *Staphylococcus* colonization of the skin and antimicrobial peptides. Expert Rev Dermatol 2010;5(2):183–95. https://doi.org/10.1586/edm.10.6.

[70] Longaureova A. Coagulase negative staphylococci and their participation in pathogenesis of human infections. Bratisl Lek Listy 2006;107(11-12):448–52.

[71] Yao Y, Sturdevant DE, Otto M. Genomewide analysis of gene expression in *Staphylococcus epidermidis* biofilms: insights into the pathophysiology of *S. epidermidis* biofilms and the role of phenol-soluble modulins in formation of biofilms. J Infect Dis 2005;191:289–98.

[72] Bjarnsholt T. The role of bacterial biofilms in chronic infections. APMIS Suppl 2013;136:1–51. https://doi.org/10.1111/apm.12099.

[73] Qin Z, Yang X, Yang L, et al. Formation and properties of in vitro biofilms of icanegative *Staphylococcus epidermidis* clinical isolates. J Med Microbiol 2007;56:83–93.

[74] Hovelius B, Mårdh PA. *Staphylococcus saprophyticus* as a common cause of urinary tract infections. Rev Infect Dis 1984;6(3):328–37.

[75] Mermel LA, Allon M, Bouza E, et al. Clinical practice guidelines for the diagnosis and management of intravascular catheter-related infection: update by the infectious diseases. Clin Infect Dis 2009;49(1):1–45.

[76] Conen A, Walti LN, Merlo A, Fluckiger U, Battegay M, Trampuz A. Characteristics and treatment outcome of cerebrospinal fluid shunt-associated infections in adults: a retrospective analysis over an 11-year period. Clin Infect Dis 2008;47:73–82.

[77] Durand ML. Endophthalmitis. Clin Microbiol Infect 2013;19:227–34.

[78] Prentice MJ, Hutchinson GR, Taylor-Robinson D. A microbiological study of neonatal conjunctivae and conjunctivitis. Br J Ophthalmol 1977;61:601–7.

[79] Mohile M, Deorari KA, Satpathy G, Sharma A, Singh M. Microbiological study of neonatal conjunctivitis with special reference to chlamydia trachomatis. Indian J Ophthalmol 2002;50:295–9.

[80] Wadhwani M, D'souza P, Jain R, et al. Conjunctivitis in the newborn: a comparative study. Indian J Pathol Microbiol 2011;54:254–7.

FURTHER READING

www.bacterio.net (http://www.bacterio.net/allnamessz.html).

Chapter 4

Coagulase-Positive and Coagulase-Negative Staphylococci Animal Diseases

Fulvio Marsilio, Cristina E. Di Francesco, Barbara Di Martino
Faculty of Veterinary Medicine, University of Teramo, Teramo, Italy

4.1 INTRODUCTION

Staphylococci are among the major groups of bacterial commensals isolated from skin, skin glands, and mucous membranes of mammals. Although staphylococci may colonize inner and/or external surfaces of healthy individuals, they may also behave as opportunistic pathogens as well as leading causes of community-associated and hospital-acquired disease in humans and animals worldwide. Coagulase positive staphylococci (CoPS) (i.e., *Staphylococcus aureus* in humans, *S. aureus* and *Staphylococcus (pseudo)intermedius* in animals) are most commonly implicated in pathologic processes. Coagulase-negative staphylococci (CoNS—mostly *Staphylococcus epidermidis* in humans and dogs), instead, are considered to be less common causes of disease in animals. Although their role as nosocomial pathogens in humans is becoming increasingly important, their zoonotic potential and importance in veterinary medicine is still unclear [1].

4.2 COAGULASE-POSITIVE STAPHYLOCOCCI

S. aureus, along with *Escherichia coli* and *Streptococcus uberis*, are considered the three most important pathogens of the udder, and sometimes their high prevalence in a dairy herd is looked at as a threat to the economic sustainability of the livestock. Mastitis is the leading cause of economic losses in dairy cattle herds, because of poor yields in the infected udder, need for veterinary treatments, the amount of milk to be discarded (due to contamination with pathogens and/or antibiotic residues), and anticipated culling. Particularly if leukocytes counts in bulk milk are higher than 400,000 cells/mL, the product is considered unfit for human consumption (EEC directive 94/71) [2].

Pet-to-Man Travelling Staphylococci: A World in Progress. https://doi.org/10.1016/B978-0-12-813547-1.00004-2

S. aureus remains the most important agent of bovine mastitis. The prevalence of the staphylococcal infection of mammary glands in the herds may differ among countries. It varies from 5% to 30% of clinical mastitis and from 5% to 10% of subclinical diseases [3]. *S. aureus* may be a potential problem from the public health point of view, because some strains are able to produce enterotoxins. The shedding from the infected udder is low, but contamination of bulk milk is pivotal for staphylococcal poisoning in fermented raw milk food [4].

Cows may be silent carriers of *S. aureus* in their udder, teat skin, nasal cavity, and rectum. However, the infected udder is the most important staphylococcal reservoir. Even if *S. aureus* needs to infect animals in order to multiply and survive, it may inhabit the environment for a long time, then the most important route of transmission "udder to udder" are the milking machine or the farmer's hands [2].

S. aureus is able to cross the teat canal and may establish infection in the lumen of the mammary gland where it can multiply very rapidly, provided that oxygen is present (as it occurs in healthy milk). Furthermore, *S. aureus* adheres at the mammary epithelial cells at cisterns and duct levels, mainly in the presence of microdamage of the epithelium that exposes tissues and the extracellular matrix protein for which *S. aureus* has specific adhesins [2].

The next step is the internalization of the pathogens in the mammary epithelial cells that support chronicization of mastitis via immune system escape. The inflammatory response induced by the invasion of the mammary gland attracts a large number of neutrophils, together with monocytes and lymphocytes, that are able to produce specific mediators that are in turn responsible for the mastitis clinical picture [2].

Several upshots of staphylococcal mastitis are described that can range from subclinical to gangrenous processes, depending on the virulence of the strain involved [5]. Differences in virulence potential correlate with gene content and levels of cytolytic toxin expression [6]. α-Hemolysin (or α-toxin) plays an important role in the severity of infection [7] and producing strains are associated with gangrenous mastitis [2].

Some studies have shown that strains isolated from clinical mastitis may also differ from mild isolates with regard to toxins overproduction and proteases involved in blood clot formation. Furthermore, some strains that appear to be associated with severe mastitis show propensity for iron acquisition [5].

In order to avoid mastitis-related economic losses in dairy herds, it is necessary to introduce specific control measures such as milking procedures, post-milking teat disinfection, reform of infected animals, and preventing the introduction of infected animals during the herd turnover [2]. Furthermore, in order to prevent the infection spread within the herd, the *S. aureus* infected quarters have to be diagnosed as soon as possible to permit treatment and to allow infected animals to be rapidly isolated or reformed [2]. Staphylococcal mastitis is very difficult to treat. Antibiotics are administered via the teat canal, sometimes in combination with the parenteral therapy. The relevant frequency of treatment failure, however, may be explained by the intracellular localization of *S. aureus* or the formation of microabscesses and intramammary biofilms [2].

In order to reduce the impact of the extensive use of antibiotics in livestock, and then to prevent the emergence of resistant strains, several alternative methods for mastitis control are under investigation. The vaccination would be particularly useful, in this context, but currently available vaccines show only limited efficacy when used in the field [8].

The use of lactic acid bacteria as a mammary probiotic is a promising alternative strategy. For example, intrammamary infusions with *Lactococcus lactis* in cows with chronic subclinical or clinical mastitis were shown to be as effective as antibiotics in preventing the onset of mastitis [9]. Furthermore, it was demonstrated that *Lactobacillus casei* is able to inhibit *S. aureus* adhesion to and invasion of bovine mammary epithelial cells, in vitro [10]. These strategies appear promising, but the protective effect still has to be proven experimentally.

The genetic selection of resistance to mastitis in ruminants, based on low somatic cell counts and low mastitis incidence, may be considered as a possible strategy for large and small ruminants [11]. Another direct strategy is to breed transgenic cows that are able to produce the lysostaphin in their milk. As a consequence, milk is able to lyse *S. aureus* and, in a context of experimental mastitis, the recombinant cows were fully protected against different *S. aureus* strains [12]. However, the ethical acceptability of using transgenic animals has to be considered.

Since description of *S. intermedius* and, later, *S. pseudintermedius*, [13,14], more than one isolate previously identified as the former were reclassified as belonging to the second, based on nucleotide sequence analysis of the *sod*A and *hsp*60 genes [13,14]. Nowadays, these two species, together with *S. delphini* [15], are considered to form the so called *S. intermedius* Group (SIG) [16,17]. Based on findings by Authors [15,18] that investigated SIG strains from different countries by multilocus sequence typing, it has been proposed currently, that all canine strains be reported as *S. pseudintermedius*, unless genomic analyses prove they belong to another species [19].

S. pseudintermedius has been recognized as an opportunistic pathogen in several animal species, especially dogs and cats. It is mostly associated with canine pyoderma, otitis externa, wound and urinary tract infections, as well as other kinds of infections in pets [20]. To date, in birds, *S. pseudintermedius* has been isolated from healthy pigeons [15]. Some recent reports indicated that it could occasionally cause human colonization and, sporadically, infection suggesting that *S. pseudintermedius* may be a zoonotic pathogen of public health concern [21].

Unlike *S. aureus*, *S. pseudintermedius* colonization is unusual in humans, even among individuals with frequent contact with animals. When studying 144 healthy veterinary college staff members, the authors found one only person to harbor the organism [22]. Again, among 3397 CoPS isolates from hospitalized patients, only two *S. pseudintermedius* isolates were identified [23]. However, by reanalyzing 14 isolates from human dog-bite wounds that were originally identified as *S. aureus*, three strains (22%) could be reclassified as *S. pseudintermedius* [24], highlighting that the two species may be misidentified as each other, and the real incidence in humans might be consequently underestimated.

A study on *S. pseudintermedius* prevalence in 13 dogs affected by deep pyoderma, their owners, and 13 individuals without daily dog contact [25] showed that the occurrence in owners of ill dogs was significantly higher (6/13) than in the control group (1/13) and that owners often carried the same *S. pseudintermedius* strains as their pets. Interestingly, all individuals were sampled a second time and were found not to be carriers when the dogs no longer had purulent lesions, so it may be argued that contact with animal lesions (rather than healthy surfaces) is the true risk factor for transmission to humans [25].

S. pseudintermedius is a potentially invasive pathogen in the case of dog bite-related wounds in humans, and was identified in 18% of such infected lesions in Great Britain [26]. In another case, *S. pseudintermedius* was cultured from the ear fluid of a patient with otitis externa as well as from her pet dog [27]. A methicillin-resistant *S. pseudintermedius* (MSP) was isolated from 17 dogs and one staff member at a veterinary teaching hospital in Japan. The human isolate shared susceptibility and a genotype profile with some of the dogs, suggesting animal-to-man transmission [28].

Nosocomial spread of MRSP was also described in a veterinary clinic in the Netherlands [29]. The strain was isolated from infected surgical wounds of five dogs and one cat, and all patients had undergone surgery at the same facility. Again, four of 22 environmental samples and four of seven persons working at the clinic were found to be MRSP carriers. The genomic profiles of the isolates were indistinguishable, and it was concluded the isolates were genotypically related, and that a nosocomial epidemic had occurred. As dogs and cats had had no contact with each other, it seemed likely that the veterinary operators (surgeons, nurses) were the indirect source for the infections. Good practice and careful hygiene are therefore necessary to prevent multidrug-resistant isolates' spread in veterinary facilities.

4.3 COAGULASE NEGATIVE STAPHYLOCOCCI

CoNS have become the most common bacterial pathogens isolated from milk samples in many countries and are responsible for bovine intramammary infections; therefore, they might be considered as emerging agents of mastitis [30]. They are opportunistic bacteria able to adhere to metal devices, thus producing a protective biofilm. This ability enables CoNS to persist on milking equipment as well as on the milker's hands, which may be a major source of staphylococcal spread. CoNS have traditionally been considered to be part of the normal skin microbiota; in this context, as opportunistic bacteria, they may adopt a pathogenic behavior and cause mastitis. In dairy farms that have successfully controlled mammary gland infections *S. aureus*, CoNS have become frequently encountered causes of bovine mastitis [31].

To date, several CoNS species have been described. *Staphylococcus chromogenes* belong to the bovine skin microbiota and is the most commonly isolated CoNS species found in bovine milk, especially in heifers around calving and in first lactation. *Staphylococcus simulans*, a commonly isolated organism in mastitic milk samples, has been reported to show propensity to cause a stronger inflammatory reaction than other CoNS species. *Staphylococcus agnetis*, again, is a recently described bovine-associated coagulase-variable species; it is closely related to and was initially classified as *Staphylococcus hyicus*, coaulase expression of which is strain-dependent [16], and that has been reported to cause a particularly strong inflammatory reaction in bovine mammary glands [30]. Furthermore, some CoNS colonizing animal skin and mucous membrane as harmless inhabitants are now recognized to be implicated in human skin and soft tissue infections as well as bacteremia. Among them, *Staphylococcus haemolyticus*, *Staphylococcus capitis*, and *Staphylococcus xylosus* are commensals of farm animals, but can cause opportunistic infections in man [32]. CoNS pathogenicity has long been underestimated as they were associated with chronic or subacute infections that were milder when compared with those by CoPS species. Nonetheless, the etiological role of CoNS in prosthesis and foreign body infections is increasingly being recognized in human medicine [33].

In pets, the pathogenic potential of these microorganisms has not yet been clearly defined, although there have been some reports of infections related to methicillin-resistant CoNS in these hosts. To date, CoNS strains in dogs and cats have been neglected, although the recent development of new molecular techniques has allowed accurate CoNS identification [34], leading to a better understanding of them. Further knowledge of CoNS carriage in animals will be of benefit as these bacteria might represent a reservoir of antibiotic resistance traits that might be transmitted to CoPS, especially the *mec*A element. This assumption was mainly based on studies reporting antibiotic resistance in clinical CoPS isolates from dogs and humans living in close contact with pets [35,36]. However, a clear picture of CoPS and CoNS distribution, diversity, and multidrug resistance in pets as well as the role of dogs and cats as drug resistance reservoirs is still lacking nowadays.

4.4   CONCLUDING REMARKS

Staphylococci are pathogens whose versatility in terms of infections and hosts makes them a serious threat to animal and human health, food security, and consequently, public health. Furthermore, they have a notable ability to develop resistance to antimicrobial agents, and the issue of antibiotic resistance in this group of microorganisms has to be addressed by the existing antibiotics, as well as the development of novel molecules. The genomic analysis of staphylococci isolated from animals (from cows and poultry, mainly) may be used to demonstrate the evolutionary origin of some clones from human beings and explain the molecular mechanisms of host adaptation [2]. These approaches have to be extended to other livestock isolates, and population genomic analysis may contribute to provide a better understanding as to how and why some clones spread so efficiently in a given host population. Despite host adaptation and specialization, some strains seem instead to lack specific host tropism and can easily be transmitted from animals to humans and vice versa [36]. This raises the question of whether staphylococcal infections should be considered zoonoses or humanoses [2]; nevertheless, this also means that staphylococci may be used for studies in the framework of the *One Health concept*, involving cooperation among experts in animal, human, and public health sciences.

CONFLICT OF INTEREST

None.

REFERENCES

[1] Weese JS, van Duijkeren E. Methicillin-resistant *Staphylococcus aureus* and *Staphylococcus pseudintermedius* in veterinary medicine. Vet Microbiol 2010;140(3–4):418–29.

[2] Peton V, Le Loir Y. *Staphylococcus aureus* in veterinary medicine. Infect Genet Evol 2014;21:602–15.

[3] Bergonier D, de Crémoux R, Rupp R, Lagriffoul G, Berthelot X. Mastitis of dairy small ruminants. Vet Res 2003;34(5):689–716.

[4] Le Loir Y, Baron F, Gautier M. *Staphylococcus aureus* and food poisoning. Genet Mol Res 2003;2(1):63–76.

[5] Le Maréchal C, Seyffert N, Jardin J, et al. Molecular basis of virulence in *Staphylococcus aureus* mastitis. PLoS ONE 2011;6(11):e27354.

[6] Guinane CM, Sturdevant DE, Herron-Olson L, et al. Pathogenomic analysis of the common bovine *Staphylococcus aureus* clone (ET3): emergence of a virulent subtype with potential risk to public health. J Infect Dis 2008;197(2):205–13.

[7] Zhao X, Lacasse P. Mammary tissue damage during bovine mastitis: causes and control. J Anim Sci 2008;86(Suppl 13):57–65.

[8] Pereira UP, Oliveira DG, Mesquita LR, Costa GM, Pereira LJ. Efficacy of *Staphylococcus aureus* vaccines for bovine mastitis: A systematic review. Vet Microbiol 2011;148(2-4):117–24.

[9] Klostermann K, Crispie F, Flynn J, Ross RP, Hill C, Meaney W. Intramammary infusion of a live culture of *Lactococcus lactis* for treatment of bovine mastitis: comparison with antibiotic treatment in field trials. J Dairy Res 2008;75(3):365–73.

[10] Bouchard DS, Rault L, Berkova N, Le Loir Y, Evan S. Inhibition of *Staphylococcus aureus* invasion into bovine mammary epithelial cells by contact with live *Lactobacillus casei*. Appl Environ Microbiol 2013;79(3):877–85.

[11] Rupp R, Bergonier D, Dion S, et al. Response to somatic cell count-based selection for mastitis resistance in a divergent selection experiment in sheep. J Dairy Sci 2009;92(3):1203–19.

[12] Wall RJ, Powell AM, Paape MJ, et al. Genetically enhanced cows resist intramammary *Staphylococcus aureus* infection. Nat Biotechnol 2005;23(4):445–51.

[13] Hajek V. *Staphylococcus intermedius*, a new species isolated from animals. Int J Syst Bacteriol 1976;26(4):401–8.

[14] Devriese LA, Vancanneyt M, Baele M, et al. *Staphylococcus pseudintermedius* sp. nov., a coagulase-positive species from animals. Int J Syst Evol Microbiol 2005;55(4):1569–73.

[15] Sasaki T, Kikuchi K, Tanaka Y, Takahashi N, Kamata S, Hiramatsu K. Reclassification of phenotypically identified *Staphylococcus intermedius* strains. J Clin Microbiol 2007;45(9):2770–8.

[16] Savini V, Passeri C, Mancini G, et al. Coagulase-positive staphylococci: my pet's two faces. Res Microbiol 2013;164(5):371–4.

[17] Savini V, Barbarini D, Polakowska K, et al. Methicillin-resistant *Staphylococcus pseudintermedius* infection in a bone marrow transplant recipient. J Clin Microbiol 2013;51(5):1636–8.

[18] Bannoehr J, Ben Zakour NL, Waller AS, et al. Population genetic structure of the *Staphylococcus intermedius* group: insights into agr diversification and the emergence of methicillin-resistant strains. J Bacteriol 2007;189(23):8685–92.

[19] Devriese LA, Hermans K, Baele M, Haesebrouck F. *Staphylococcus pseudintermedius* versus *Staphylococcus intermedius*. Vet Microbiol 2009;133(1–2):206–7.

[20] Ruscher C, Lübke-Becker A, Wleklinski CG, Soba A, Wieler LH, Walther B. Prevalence of methicillin-resistant *Staphylococcus pseudintermedius* isolated from clinical samples of companion animals and equidaes. Vet Microbiol 2009;136(1-2):197–201.

[21] van Duijkeren E, Kamphuis M, van der Mije IC, et al. Transmission of methicillin-resistant *Staphylococcus pseudintermedius* between infected dogs and cats and contact pets, humans and the environment in households and veterinary clinics. Vet Microbiol 2011;150(3-4):338–43.

[22] Talan DA, Staatz D, Staatz A, Overturf GD. Frequency of *Staphylococcus intermedius* as human nasopharyngeal flora. J Clin Microbiol 1989;27(10):2393.

[23] Mahoudeau I, Delabranche X, Prevost G, Monteil H, Piemont Y. Frequency of isolation of *Staphylococcus intermedius* from humans. J Clin Microbiol 1997;35(8):2153–4.

[24] Talan DA, Goldstein EJ, Staatz D, Overturf GD. *Staphylococcus intermedius*: clinical presentation of a new human dog bite pathogen. Ann Emerg Med 1989;18(4):410–3.

[25] Guardabassi L, Loeber ME, Jacobson A. Transmission of multiple antimicrobial-resistant *Staphylococcus intermedius* between dogs affected by deep pyoderma and their owners. Vet Microbiol 2004;98(1):23–7.

[26] Lee J. *Staphylococcus intermedius* isolated from dog-bite wounds. J Infect 1994;29(1):105.

[27] Tanner MA, Everett CL, Youvan DC. Molecular phylogenetic evidence for noninvasive zoonotic transmission of *Staphylococcus intermedius* from a canine pet to a human. J Clin Microbiol 2000;38(4):1628–31.

[28] Sasaki T, Kikuchi K, Tanaka Y, Takahashi N, Kamata S, Hiramatsu K. Methicillin-resistant *Staphylococcus pseudintermedius* in a veterinary teaching hospital. J Clin Microbiol 2007;45:1118–25.

[29] van Duijkeren E, Houwers DJ, Schoormans A, et al. Transmission of methicillin-resistant *Staphylococcus intermedius* between humans and animals. Vet Microbiol 2008;128:213–5.

[30] Simojoki H, Orro T, Taponen S, Pyörälä S. Host response in bovine mastitis experimentally induced with *Staphylococcus chromogenes*. Vet Microbiol 2009;134:95–9.

[31] Ruegg PL. The quest for the perfect test: phenotypic versus genotypic identification of coagulase-negative staphylococci associated with bovine mastitis. Vet Microbiol 2009;134:15–9.

[32] Åvall-Jääskeläinen S, Koort J, Simojoki H, Taponen S. Bovine-associated CNS species resist phagocytosis differently. BMC Vet Res 2013;9:227.

[33] Piette A, Verschraegen G. Role of coagulase-negative staphylococci in human disease. Vet Microbiol 2009;134:45–54.

[34] Gandolfi-Decristophoris P, Regula G, Petrini O, Zinsstag J, Schelling E. Prevalence and risk factors for carriage of multi-drug resistant staphylococci in healthy cats and dogs. J Vet Sci 2013;14:449–56.

[35] Weese JS, Dick H, Willey BM, et al. Suspected transmission of methicillin-resistant *Staphylococcus aureus* between domestic pets and humans in veterinary clinics and in the household. Vet Microbiol 2006;115:148–55.

[36] Somavaji R, Privantha MA, Rubin JE, et al. Human infections due to *Staphylococcus pseudintermedius*, an emerging zoonosis of canine origin: report of 24 cases. Diagn Microbiol Infect Dis 2016;85:471–6.

Chapter 5

Transfer of Staphylococci and Related Genetic Elements

Vincenzo Savini*, Roberta Marrollo*, Paolo Fazii[†]
*Clinical Microbiology and Virology, Laboratory of Bacteriology and Mycology, Civic Hospital of Pescara, Pescara, Italy, [†]Clinical Microbiology and Virology, Civic Hospital of Pescara, Pescara, Italy

5.1 BACKGROUND

The socio-economic relationship between dogs and their owners has changed dramatically during the past several decades in many countries [1,2]. Particularly, instead of behaving as working dogs (such as watchdogs or sheepdogs) that usually live in stables or kennels, most pets nowadays share the household with their owners, almost as family members [1,2]. Also, persons who feel disconnected from society have been frequently observed to substitute social contact by companion animals, and assign them supportive anthropomorphic traits (the so-called "humanization" of nonhuman beings) [1–3]. Many dogs have therefore gained a status that is nearly equal to that of a person, in more than one household, as animals are given privileges previously exclusively meant for humans [1–3].

Consequently, the transfer of microorganisms from humans to their pets and vice versa may be enhanced, and especially involve those residing on the dog's skin or mucosae, which might be easily transmitted to owners either by direct contact or indirectly through the household [4–6].

Among such commensals, *Staphylococcus aureus* and *Staphylococcus pseudintermedius* represent the major species. While the former is a common colonizer in humans and widely distributed among mammals in general, the second seems to be mostly associated with small animals (such as dogs and cats) and is only occasionally found in clinical materials of human origin [4–6].

S. pseudintermedius is believed to be more common among dogs and cats than *S. aureus*, and certain *S. pseudintermedius* strains have become a major matter of concern in veterinary medicine due to the frequency of methicillin and multidrug resistance [6–10].

Concomitant human and animal colonization by indistinguishable strains of these two species has been reported in the literature, meaning a man-to-dog-to-man

Pet-to-Man Travelling Staphylococci: A World in Progress. https://doi.org/10.1016/B978-0-12-813547-1.00005-4

transfer, that reasonably occurs bidirectionally, and the question has arisen therefore how changes in the human-to-dog relationship may influence such an interspecies transmission [1–10].

Opportunistic pathogens like coagulase-positive staphylococci (CoPS) are of special interest, as reports dealing with their transferability between animals and humans have increased during recent years, including methicillin-resistant *S. aureus* (MRSA) and methicillin-resistant *S. pseudintermedius* (MRSP) [1–10].

MRSA, particularly, has become one of the major nosocomial pathogens in healthcare settings, as it is responsible for an increased mortality among affected patients. In the field of veterinary medicine, MRSA nosocomial outbreaks have recently gained attention, for instance, in equine and small animal clinics. Nasal colonization of veterinary personnel seems to play a crucial role in rising infection rates in animal patients, just like healthcare operators in hospitals [11–20].

Households where pets live have been known to behave as a potential source not only for MRSA, but even for other staphylococci whose concomitant human and veterinary importance has been acknowledged, including MRSP, *Staphylococcus schleiferi* ssp. *coagulans*, and coagulase-negative staphylococci (CoNS) such as *Staphylococcus schleiferi* ssp. *schleiferi* [11–27].

Also, reports of isolation of MRSA from pets (the organism is mostly found on dogs and cats) have increased over the past years and some researchers have speculated that the human epidemic is indeed driving the veterinary one; really, staphylococci transmission can occur in both directions, that is from people to several species of pets and food animals and vice versa [11–50].

The ability of staphylococci to survive on environmental surfaces enhances the likelihood of strains of the genus establishing communities that persist in the household [11–50].

Moreover, and notably, staphylococci other than *S. aureus* might play a role in the horizontal transfer of genes (including those for antimicrobial resistance) within household bacterial populations [11–50].

5.2 STAPHYLOCOCCI IN THE ENVIRONMENT

Staphylococci are hardy organisms that may remain viable in dry environments for at least a week, up to 3 months, or even longer. Survival seems to depend on several factors, including dust composition, temperature and humidity, the kind of surface material where the organisms reside, along with specific strains' features, that all together may affect staphylococcal longevity in households [11–50].

In general, it has been observed that, among bacteria, Gram-positive organisms are predominant in indoor dust, and those found in households are usually of human origin. Colonized or infected people, as well as, consequently, animals, can shed bacteria into the environment by virtue of direct contact with household surfaces, along with skin cells (with adherent bacteria) shedding, aerosol discharge, sneezing, and gastrointestinal routes. In light of this, it is clear that although *S. aureus* and other staphylococci are commonly and directly transmitted by person-to-person contact, indirect transmission may also occur due to environmental exposure, particularly through aerosols, fomites, and settled dust [11–50].

Notably, although hospital and public community have been known to be characterized by transient contact by diverse persons, the household environment is characterized by persistent, high-intensity contact between the same individuals. Consequently, transmission dynamics in this ambit might differ from those of public settings, and require different control measures [11–50].

Additionally, home environments are beyond administrative authority, so persons who inhabit the household are responsible for ensuring that the indoor environment is, in their opinion, hygienic [51–56].

Commonly touched sites (such as telephones, television remote controls, door knobs, pillows, bedding, toys, hand towels, taps) show high rates of contamination and may play an important role in the abovementioned indirect household transmission.

Bedroom surfaces (i.e., pillows and bedding), particularly, can promote indirect movement of *S. aureus* and other staphylococci from people to pets and vice versa, and this route of transmission might partly rely on exfoliation, then on deposition of skin cells contaminated with staphylococci [11–56].

5.3   PEOPLE AND HOUSEHOLD PETS

Of course, duration of household members' colonization may be important for transmission, as a long colonization time could increase the possibility for a transfer event [11–56].

It is important to note that 20% of people are estimated to be persistent carriers who harbor MRSA for months or years, and, even after decolonization, these subjects can be recolonized, preferentially with the previously persistent strain (if exposed to several diverse ones). This effect makes home decontamination crucial for strategies aiming at decolonizing such persistent carriers. The potential, however, exists for colonization loss both by treatment or natural clearance (both in man and in animals) [57–62].

The choice of anatomical sampling sites can affect surveillance estimates of transmission frequency. In fact, the highest household transmission rates are reported in studies testing multiple sites on tested persons (that is, all among nostrils, throat, skin, perineum, as well as clinical lesions), if compared to surveys where only nasal swabbing was performed. In this ambit, it is important to note that nonnasal colonization may be more common than what was previously believed, particularly with community-acquired MRSA (CA-MRSA), and may affect the dynamics of staphylococcal transmission within households [11–70].

Notably, in fact, colonization of sites outside the nose, such as the pharynx, has been involved in transmission to household contacts, while further risk factors for such a transfer include increased duration of exposure (meaning time spent in the household), number of household contacts, presence of eczema or a history of previous cutaneous infections. Again, colonization of the groin or vagina could cause sexual transmission, or vertical transmission to infants during delivery [11–70].

Children who are 0–17 years old are colonized with *S. aureus* more commonly and for longer periods than adults. Infants, also, can develop *S. aureus* enteric colonization shortly after birth and can stay *S. aureus*-positive for months. They also show high rates of perineal infection compared with other age groups, and both of such two conditions may act as sources of home environmental contamination. In households, infants who change zones can get contaminated with *S. aureus*, as children usually interact closely with their environments, which potentially leads to the establishment of a cycle of contamination and reexposure to staphylococci, and related consequent implications for colonization risk or duration [71–76].

Pets are known to play a crucial role in human colonization in the household, as well; it is estimated that 39% of US households have at least one dog, and 33% have at least one cat. Companion animals become colonized or infected with *S. aureus*, including hospital-acquired MRSA (HA-MRSA) and CA-MRSA strains. Interestingly, this finding has led researchers to maintain that household animals have the so-called "humanosis" of MRSA and are not true *S. aureus* reservoirs [71–80].

Differentiation of true colonization from transient carriage is a veterinary challenge. However, in general, a greater proportion of dogs than cats seem to harbor *S. aureus*, and transmission between people and pets is likely to occur bidirectionally [71–80].

Like people, pets can get colonized or infected due to nosocomial transmission, either in veterinary or in human healthcare settings where, for example, therapy pets visit or reside with patients. Repeated, intermittent carriage of particular MRSA strains observed in companion animals might depend on poor sensitivity of testing protocols or rely on the fact that pets are recolonized from reservoirs, such as humans or environmental surfaces, after decolonization or natural clearance [71–80].

Infection of or transient carriage by pets may provide a means for bacterial transfer to people, and genotypical relatedness between staphylococcal isolates from people and companion animals within households tends to be similar to that existing between isolates from human household members [71–80].

Although collection of related strains from both persons and animals may suggest that transmission has occurred, this does not necessarily show the direction of such a movement. Also, together with contact with the veterinary setting and use of antimicrobials, risk factors for pet colonization with *S. aureus* may include contact with children and licking behaviors. Interactions between pets and children within households may involve, in fact, direct face-to-face contact through licking and biting, or indirect contact through shared household areas.

Whether the concomitant presence of pets and children confers a greater risk for household contamination is unclear [81–88].

As observed in humans, anatomical sites other than the nose (i.e., mouth, perineum, and inguinal skin) are frequently colonized by *S. aureus* both in dogs and in cats and, particularly, a positive dorsal fur site is probably a consequence of contamination from human hands or mouth contact; such a colonization may be important for transmission, but may not be indicative of pet colonization status. Moreover, differentiation of colonization from transient contamination in animals is hard to do [81–88].

Pets other than dogs and cats can also play a relevant role in bacterial transmission, and clinical *S. aureus* strains, including MRSA, have been identified in parrots and other birds, bats, rats, rabbits, hamsters, guinea pigs, small ruminants, iguanas, and turtles. Reptiles seem to be resistant to colonization by staphylococci, possibly because they are ectothermic; nonetheless, they usually have less frequent contact with humans than do furry pets, which could lead to a reduced risk of transmission [89–100].

5.4   VETERINARY STAPHYLOCOCCI

Although *S. aureus* is the major organism within the genus that infects people and can affect, to a lesser extent, pets, *S. pseudintermedius* has been known to do the reverse, then predominantly cause disease in animals, and occasionally in man.

CoPS in general, and particularly MRSP, are leading causes of bacterial cutaneous infections both in dogs and in cats. Although infection by MRSP is rare in humans, pet owners and veterinarians can be colonized, often transiently. As observed with *S. aureus*, when people are colonized with MRSP, they usually share the same strains as their companion animals [1,6,8,9,17,18].

Domestic contamination due to *S. pseudintermedius* has been described for households where MRSP-infected pets live, and it has been assumed that dust particles play a role in such a condition [1,6,8,9,17,18].

Furthermore, it was possible that certain MRSP lineages (e.g., ST71) may have a greater ability to adapt to the human host than methicillin-susceptible *S. pseudintermedius* (MSSP), although the interspecies transferability does not seem to be linked with the methicillin resistance trait at all [1,6,8,9,17,18].

Prevalence in persons exposed to MRSP is estimated to be 4%–12% in household studies, and occupational cohorts and evidence seems to indicate that people can become truly colonized, not only transiently, with MRSP (although it remains unknown whether *S. pseudintermedius* colonization in humans is indeed transient or permanent) [22,43,45]. Moreover, it has been observed that in the event of concurrent environmental contamination, human colonization, and animal colonization in household settings, some animals are persistently colonized with the same MRSP strain, whereas many are intermittently colonized or later become negative, and household environments may be positive, even in the absence of concurrent human or animal colonization. The household may be therefore crucial for the recolonization of people or pets after natural clearance or due to treatment [22,43,45].

5.5 HORIZONTAL GENETIC TRANSFER IN VETERINARY STAPHYLOCOCCI

As widely reported in the literature, MRSA, MRSP, and methicillin-resistant staphylococci in general possess the so-called staphylococcal chromosomal cassette *mec* (SCC*mec*), a large mobile genetic element that contains the *mecA* gene; the latter is known to confer resistance to β-lactam antibiotics by virtue of synthesis of an altered penicillin-binding protein (PBP), named as PBP-2′ or PBP-2a. Actually, *mecA*-harboring CoNS (not only CoPS, then) have been isolated from several domesticated and healthy animals and the "birth" of MRSA may be traced back to the transfer of an SCC*mec* element from *Staphylococcus epidermidis* to methicillin-susceptible *S. aureus;* again, the origin of the *mecA* gene seems to be *Staphylococcus fleurettii*, an animal-associated CoNS species [101–120]. In this context, it is interesting to note that the prevalence of methicilin resistance is greater among staphylococci from racing horses than nonracing ones, and this observation probably partly relies on the frequent antibiotic administration and the close contact with humans in this horse population. Moreover, horses in general often are in close contact with their owners and farm staff members as well as with other animals of the same species, so the risk of bacteria or mobile genetic element transfer between these animals and humans (or vice versa) has to be taken into consideration and, particularly, a putative risk of MRSA cross-transmission between persons and horses has been described in the literature [121–132].

SCC*mec* elements in MRSP are of more diverse types than those harbored by MRSA, or are combinations of elements initially belonging to MRSA and other staphylococcal species. Detailed information on methicillin resistance is reported in a dedicated chapter of this volume. Nonetheless, here it is important to emphasize that when multiple species of staphylococci are present concomitantly horizontal transfer may occur within household bacterial communities [133–137]. However, differences have been found between SCC*mec* types in MRSP from pets and methicillin-resistant CoNS from persons sharing the same household, thus the existence of no common SCC*mec* source for household staphylococci is highlighted [133–159].

Previous receipt of antibiotics, especially cephalosporins and fluoroquinolones, seem to be involved in following colonization or infection with methicillin-resistant staphylococci, both in humans and in animals. Clinical reports of susceptible *S. pseudintermedius* that became resistant after antibiotic therapy in the same animal may suggest either that use of drugs selected the resistant strains, or that *S. pseudintermedius* is receptive to gene transfer within bacterial communities in household or veterinary clinics [160–165].

Also, clinical infection of pets with MRSP or MRSA may make it necessary to administer antibiotics that are more commonly used in humans, unfortunately, thus exposing the household microbiome to such drugs. Consequently, the household microbiome becomes a potential source of resistance-related genetic elements that may undergo transfer among diverse species of staphylococci [160–165].

5.6 CONCLUDING REMARKS

It has been ascertained that pets (mostly dogs) living in close contact with humans who are MRSA carriers can become colonized with this organism. Also, interspecies transfer of staphylococci may involve CoNS as well and, for instance, *Staphylococcus lentus* equine and human isolates from Italian farms have been shown to share the same pulsed-field gel electrophoresis (PFGE) patterns, suggesting a common origin and a possible horse-to-man transmission or vice versa [166–180].

Therefore, it is likely that the intensive daily contact between farm or companion animals and humans (i.e., their owners) may enhance the likelihood of getting colonized through interspecies-transmission for both sides. Failure to detect and treat such colonized pets may result in recurrent MRSA colonization or infection in their owners. Therefore, it is crucial to identify the source for an MRSA colonization among pets, which might support an unexplained carriage or disease relapse in humans. Actually, antimicrobial therapy of healthcare workers and, concomitantly, of MRSA carriers and any family members or pets that have been close to infected subjects can remove recurrent MRSA infections [166–180].

Very little is known about methicillin-resistant CoNS transmission between animals and man and, moreover, although the impact of *S. aureus* nasal carriage in humans is clear, that occurring in animals still plays an uncertain clinical role. Finally, the importance of nasal carriage of methicillin-resistant CoNS in both humans and animals needs further and deeper elucidation.

Although reports indicate that heavy household contamination may contribute to human colonization and recolonization, few decontamination strategies have been studied for this kind of environment. Gaseous ozone, as well as the replacement of carpets and mattress, commercial steam cleaning, and general disinfection of hard surfaces, have been used in households. Again, low-temperature home laundering is effective for clothing [79,102]. However, laundered clothing may newly become contaminated within hours if exposed in a healthcare setting. Therefore, laundering should be done frequently during decolonization of household members [79,102].

Notably, *S. aureus* and other bacteria, although susceptible to detergents used in laboratories, might not be killed by household use of dishwashing liquid in the presence of food residue. Nevertheless, 2.1% sodium hypochlorite with a degreaser has been observed to effectively reduce *S. aureus* amounts on kitchen items [79,102]. In addition, the combination of chlorine and quaternary ammonium-based compounds for disinfection of multiple sites in kitchens, bathrooms, and pet-associated zones may decrease *S. aureus* presence by 99.99% (certain *S. aureus* strains may, however, harbor mutations that confer resistance to household disinfectants). Also, cleansing procedures targeting staphylococci could be effective against other bacteria, too, that spread within households, such as *Escherichia coli* [79,102].

As several contaminated sites in the household are frequently touched, proper hand hygiene practices can help to decrease contamination of such areas. Particularly, household cleaning and laundering of both human and animal bedding should be required when persons or pets are diagnosed with a staphylococcal disease or a mere colonization. Additionally, in 2010, the Clinical and Laboratory Standards Institute (CLSI) recommended screening and decolonization for household people and pets, in the event of recurrent human infection within the household. Concerning human nose decolonization, it has been shown that short-term treatment with topical mupirocin can temporarily eradicate nasal MRSA carriage, although the long-term effect of this treatment procedure is not fully elucidated and recolonization of treated patients may actually occur [58,84,99].

Few interventions have been established for animals. Systemic rifampin has been used to achieve decolonization of people and animals, but the development of rifampin resistance in staphylococci (frequently observed after treatment) and the compound's potential for causing side-effects restrict its use, so this drug should not be routinely used for pet decolonization. Conversely, chlorhexidine scrub or shampoo at either 2% or 4% has been known to be effective against canine pyoderma and could also be appropriate for decolonization of pets [58,79,84,99,102].

If antibiotics are prescribed (by a veterinarian, of course), prescription of both topical and systemic compounds should rely on antibiotic susceptibility testing, and, notably, β-lactam drugs should be avoided. In the absence of conclusive studies on pet decolonization, however, researchers recommend an initial topical therapy, for instance with chlorhexidine and tris-EDTA, to prevent overuse or misuse of oral drugs. Undoubtedly, infected animals should be always treated case-by-case and under veterinary supervision [58,79,84,99,102].

Contact precautions, social distancing, and good hygiene should be recommended for households where MRSA-positive animals live and, if warranted, pets should be included in treatment regimens targeting the whole household [58,79,84,99,102]. Also, when dogs and cats get persistently infected with *S. aureus*, animal treatment has to be combined with people decolonization, as persons are probably *S. aureus* sources for companion animals. In fact, pets may often clear colonization without any drug treatment, but reduction of person and environmental contact is important through temporary contact isolation (during the decolonization phase); for instance, animals to be decolonized should sleep on a surface that can be easily disinfected or laundered rather than in a bed with their owners [58,79,84,99,102].

It is clear that owners and veterinarians may prefer to treat animals rather than remove them permanently from households, as many owners regard pets as part of the family, and the presence of pets notoriously has both physical and mental health benefits, so it is important to shed more light on procedures targeting companion animals' decolonization, when this is necessary [58,79,84,99,102].

Understanding transmission dynamics of staphylococci within households needs data from people, animals, and environment. In fact, variable durations of colonization in humans and animals, and variable survival rates of bacteria on household surfaces, may lead to a dynamic cycle of transmission. Also, sampling several anatomical locations in animals and their owners is crucial to more deeply investigate the existence and persistence of a colonization condition. Nevertheless, one should keep in mind that not only staphylococcal strains may spread bidirectionally between humans and animals, but even mobile genetic elements, which may be given and received among diverse species of staphylococci.

REFERENCES

[1] Abraham JL, Morris DO, Griffeth GC, Shofer FS, Rankin SC. Surveillance of healthy cats and cats with inflammatory skin disease for colonization of the skin by methicillin-resistant coagulase-positive staphylococci and *Staphylococcus schleiferi* ssp. *schleiferi*. Vet Dermatol 2007;18:252–9.

[2] Allen KD, Anson JJ, Parsons LA, Frost NG. Staff carriage of methicillin-resistant *Staphylococcus aureus* (EMRSA 15) and the home environment: a case report. J Hosp Infect 1997;35:307–11.

[3] Amir NH, Rossney AS, Veale J, O'Connor M, Fitzpatrick F, Humphreys H. Spread of community-acquired meticillin-resistant *Staphylococcus aureus* skin and soft-tissue infection within a family: implications for antibiotic therapy and prevention. J Med Microbiol 2010;59:489–92.

[4] Ammerlaan HS, Kluytmans JA, Wertheim HF, Nouwen JL, Bonten MJ. Eradication of methicillin-resistant *Staphylococcus aureus* carriage: a systematic review. Clin Infect Dis 2009;48:922–30.

[5] Archer GL, Climo MW. Antimicrobial susceptibility of coagulase-negative staphylococci. Antimicrob Agents Chemother 1994;38:2231–7.

[6] Baptiste KE, Williams K, Willams NJ, et al. Methicillin-resistant staphylococci in companion animals. Emerg Infect Dis 2005;11:1942–4.

[7] Beard-Pegler MA, Stubbs E, Vickery AM. Observations on the resistance to drying of staphylococcal strains. J Med Microbiol 1988;26:251–5.

[8] Bender JB, Waters KC, Nerby J, Olsen KE, Jawahir S. Methicillin-resistant *Staphylococcus aureus* (MRSA) isolated from pets living in households with MRSA-infected children. Clin Infect Dis 2012;54:449–50.

[9] Black CC, Solyman SM, Eberlein LC, Bemis DA, Woron AM, et al. Identification of a predominant multilocus sequence type, pulsed-field gel electrophoresis cluster, and novel staphylococcal chromosomal cassette in clinical isolates of *mecA*-containing, methicillin-resistant *Staphylococcus pseudintermedius*. Vet Microbiol 2009;139:333–8.

[10] Bloemendaal ALA, Brouwer EC, Fluit AC. Methicillin resistance transfer from *Staphylocccus epidermidis* to methicillin-susceptible *Staphylococcus aureus* in a patient during antibiotic therapy. PLoS One 2010;5:e11841.

[11] Boyce JM, Havill NL, Otter JA, Adams NMT. Widespread environmental contamination associated with patients with diarrhea and methicillin-resistant *Staphylococcus aureus* colonization of the gastrointestinal tract. Infect Control Hosp Epidemiol 2007;28:1142–7.

[12] Boyce JM. Environmental contamination makes an important contribution to hospital infection. J Hosp Infect 2007;65(Suppl. 2):50–4.

[13] Bramble M, Morris D, Tolomeo P, Lautenbach E. Potential role of pet animals in household transmission of methicillin-resistant *Staphylococcus aureus*: a narrative review. Vector Borne Zoonotic Dis 2011;11:617–20.

[14] Briscoe JA, Morris DO, Rankin SC, Hendrick MJ, Rosenthal KL. Methicillin-resistant *Staphylococcus aureus*-associated dermatitis in a Congo African grey parrot (*Psittacus erithacus*). J Avian Med Surg 2008;22:336–43.

[15] Broens EM, Cleef B, Graat E, Kluytmans J. Transmission of methicillin-resistant *Staphylococcus aureus* from food production animals to humans: a review. CAB Rev 2008;3:1–12.

[16] Burden M, Cervantes L, Weed D, Keniston A, Price CS, Albert RK. Newly cleaned physician uniforms and infrequently washed white coats have similar rates of bacterial contamination after an 8-hour workday: a randomized controlled trial. J Hosp Med 2011;6:177–82.

[17] Busscher JF, van Duijkeren E, Sloet van Oldruitenborgh-Oosterbaan MM. The prevalence of methicillin-resistant staphylococci in healthy horses in the Netherlands. Vet Microbiol 2006;113:131–6.

[18] Cain CL, Morris DO, Rankin SC. Clinical characterization of *Staphylococcus schleiferi* infections and identification of risk factors for acquisition of oxacillin-resistant strains in dogs: 225 cases (2003–2009). J Am Vet Med Assoc 2011;239:1566–73.

[19] Calfee DP, Durbin LJ, Germanson TP, Toney DM, Smith EB, Farr BM. Spread of methicillin-resistant *Staphylococcus aureus* (MRSA) among household contacts of individuals with nosocomially acquired MRSA. Infect Control Hosp Epidemiol 2003;24:422–6.

[20] Chemello RML, Giugliani ERJ, Bonamigo RR, Bauer VS, Cecconi MCP, Zubaran GM. Breastfeeding and mucosal and cutaneous colonization by *Staphylococcus aureus* in atopic children. An Bras Dermatol 2011;86:435–9.

[21] Chomel BB, Sun B. Zoonoses in the bedroom. Emerg Infect Dis 2011;17:167–72.

[22] Chuang CY, Yang YL, Hsueh PR, Lee PI. Catheter-related bacteremia caused by *Staphylococcus pseudintermedius* refractory to antibiotic-lock therapy in a hemophilic child with dog exposure. J Clin Microbiol 2010;48:1497–8.

[23] Cimolai N. MRSA and the environment: implications for comprehensive control measures. Eur J Clin Microbiol Infect Dis 2008;27:481–93.

[24] Clinical and Laboratory Standards Institute. Surveillance for methicillin-resistant *Staphylococcus aureus*: principles, practices, and challenges; a report. Wayne: Clinical and Laboratory Standards Institute; 2010.

[25] Cohn LA, Middleton JR. A veterinary perspective on methicillin-resistant staphylococci. J Vet Emerg Crit Care (San Antonio) 2010;20:31–45.

[26] Coughlan K, Olsen KE, Boxrud D, Bender JB. Methicillin-resistant *Staphylococcus aureus* in resident animals of a long-term care facility. Zoonoses Public Health 2010;57:220–6.

[27] Cuny C, Friedrich A, Kozytska S, et al. Emergence of methicillin-resistant *Staphylococcus aureus* (MRSA) in different animal species. Int J Med Microbiol 2010;300:109–17.

[28] Currie K, Cuthbertson L, Price L, Reilly J. Cross-sectional survey of meticillin-resistant *Staphylococcus aureus* home-based decolonization practices in Scotland. J Hosp Infect 2012;80:140–3.

[29] Dancer SJ. Importance of the environment in meticillin-resistant *Staphylococcus aureus* acquisition: the case for hospital cleaning. Lancet Infect Dis 2008;8:101–13.

[30] Davis AO, O'Leary JO, Muthaiyan A, et al. Characterization of *Staphylococcus aureus* mutants expressing reduced susceptibility to common house-cleaners. J Appl Microbiol 2005;98:364–72.

[31] Davis MF, Baron P, Price LB, et al. Dry collection and culture methods for recovery of methicillin-susceptible and methicillin-resistant *Staphylococcus aureus* strains from indoor home environments. Appl Environ Microbiol 2012;78:2474–6.

[32] de Boer HEL, van Elzelingen-Dekker CM, van Rheenen-Verberg CMF, Spanjaard L. Use of gaseous ozone for eradication of methicillin-resistant *Staphylococcus aureus* from the home environment of a colonized hospital employee. Infect Control Hosp Epidemiol 2006;27:1120–2.

[33] Dietze B, Rath A, Wendt C, Martiny H. Survival of MRSA on sterile goods packaging. J Hosp Infect 2001;49:255–61.

[34] Dowling PM. Miscellaneous antimicrobials: ionophores, nitrofurans, nitroimidazoles, rifamycins, oxazolidinones, and others. In: Giguere S, Prescott JF, Baggot JD, Walker RD, Dowling PM, editors. Antimicrobial therapy in veterinary medicine. 4th ed. Ames: Blackwell Publishing; 2006. p. 285–300.

[35] Engemann JJ, Carmeli Y, Cosgrove SE, Fowler VG, Bronstein MZ, et al. Adverse clinical and economic outcomes attributable to methicillin resistance among patients with *Staphylococcus aureus* surgical site infection. Clin Infect Dis 2003;36:592–8.

[36] Enoch DA, Karas JA, Slater JD, Emery MM, Kearns AM, Farrington M. MRSA carriage in a pet therapy dog. J Hosp Infect 2005;60:186–8.

[37] Epley N, Waytz A, Akalis S, Cacioppo JT. When we need a human: motivational determinants of anthropomorphism. Soc Cogn 2008;26:143–55.

[38] Espinosa-Gongora C, Chrobak D, Moodley A, Bertelsen MF, Guardabassi L. Occurrence and distribution of *Staphylococcus aureus* lineages among zoo animals. Vet Microbiol 2012;158:228–31.

[39] Eveillard M, Martin Y, Hidri N, Boussougant Y, Joly-Guillou ML. Carriage of methicillin-resistant *Staphylococcus aureus* among hospital employees: prevalence, duration, and transmission to households. Infect Control Hosp Epidemiol 2004;25:114–20.

[40] Faires MC, Tater KC, Weese JS. An investigation of methicillin-resistant *Staphylococcus aureus* colonization in people and pets in the same household with an infected person or infected pet. J Am Vet Med Assoc 2009;235:540–3.

[41] Ferreira JP, Anderson KL, Correa MT, et al. Transmission of MRSA between companion animals and infected human patients presenting to outpatient medical care facilities. PLoS One 2011;6:e26978.

[42] Ferreira JP, Fowler Jr. VG, Correa MT, Lyman R, Ruffin F, Anderson KL. Transmission of methicillin-resistant *Staphylococcus aureus* between human and hamster. J Clin Microbiol 2011;49:1679–80.

[43] Fitzgerald JR. The *Staphylococcus intermedius* group of bacterial pathogens: species reclassification, pathogenesis and the emergence of meticillin resistance. Vet Dermatol 2009;20:490–5.

[44] Foster TJ. The *Staphylococcus aureus* "superbug". J Clin Invest 2004;114:1693–6.

[45] Frank LA, Kania SA, Kirzeder EM, Eberlein LC, Bemis DA. Risk of colonization or gene transfer to owners of dogs with meticillin-resistant *Staphylococcus pseudintermedius*. Vet Dermatol 2009;20:496–501.

[46] Freitas EAF, Harris RM, Blake RK, Salgado CD. Prevalence of USA300 strain type of methicillin-resistant *Staphylococcus aureus* among patients with nasal colonization identified with active surveillance. Infect Control Hosp Epidemiol 2010;31:469–75.

[47] Fritz SA, Hogan PG, Hayek G, et al. Household versus individual approaches to eradication of community-associated *Staphylococcus aureus* in children: a randomized trial. Clin Infect Dis 2012;54:743–51.

[48] Fritz SA, Hogan PG, Hayek G, et al. *Staphylococcus aureus* colonization in children with community-associated *Staphylococcus aureus* skin infections and their household contacts. Arch Pediatr Adolesc Med 2012;166:551–7.

[49] Fuda C, Suvorov M, Shi Q, et al. Shared functional attributes between the *mecA* gene product of *Staphylococcus sciuri* and penicillin-binding protein 2a of methicillin-resistant *Staphylococcus aureus*. Biochemistry 2007;46:8050–7.

[50] Gandara A, Mota LC, Flores C, Perez HR, Green CF, Gibbs SG. Isolation of *Staphylococcus aureus* and antibiotic-resistant *Staphylococcus aureus* from residential indoor bioaerosols. Environ Health Perspect 2006;114:1859–64.

[51] Gorwitz RJ, Kruszon-Moran D, McAllister SK, et al. Changes in the prevalence of nasal colonization with *Staphylococcus aureus* in the United States, 2001–2004. J Infect Dis 2008;197:1226–34.

[52] Graffunder EM, Venezia RA. Risk factors associated with nosocomial methicillin-resistant *Staphylococcus aureus* (MRSA) infection including previous use of antimicrobials. J Antimicrob Chemother 2002;49:999–1005.

[53] Griffeth GC, Morris DO, Abraham JL, Shofer FS, Rankin SC. Screening for skin carriage of methicillin-resistant coagulase-positive staphylococci and *Staphylococcus schleiferi* in dogs with healthy and inflamed skin. Vet Dermatol 2008;19:142–9.

[54] Guardabassi L, Schwarz S, Lloyd DH. Pet animals as reservoirs of antimicrobial-resistant bacteria. J Antimicrob Chemother 2004;54:321–32.

[55] Halablab MA, Hijazi SM, Fawzi MA, Araj GF. *Staphylococcus aureus* nasal carriage rate and associated risk factors in individuals in the community. Epidemiol Infect 2010;138:702–6.

[56] Hanselman BA, Kruth SA, Rousseau J, Weese JS. Coagulase positive staphylococcal colonization of humans and their household pets. Can Vet J 2009;50:954–8.

[57] Hardy KJ, Gossain S, Henderson N, et al. Rapid recontamination with MRSA of the environment of an intensive care unit after decontamination with hydrogen peroxide vapour. J Hosp Infect 2007;66:360–8.

[58] Hebert C, Robicsek A. Decolonization therapy in infection control. Curr Opin Infect Dis 2010;23:340–5.

[59] Heller J, Innocent GT, Denwood M, Reid SWJ, Kelly L, Mellor DJ. Assessing the probability of acquisition of meticillin-resistant *Staphylococcus aureus* (MRSA) in a dog using a nested stochastic simulation model and logistic regression sensitivity analysis. Prev Vet Med 2011;99:211–24.

[60] Hewlett AL, Falk PS, Hughes KS, Mayhall CG. Epidemiology of methicillin-resistant *Staphylococcus aureus* in a university medical center day care facility. Infect Control Hosp Epidemiol 2009;30:985–92.

[61] Huang YC, Chao AS, Chang SD, et al. Association of *Staphylococcus aureus* colonization in parturient mothers and their babies. Pediatr Infect Dis J 2009;28:742–4.

[62] Huang YC, Ho CF, Chen CJ, Su LH, Lin TY. Nasal carriage of methicillin-resistant *Staphylococcus aureus* in household contacts of children with community-acquired diseases in Taiwan. Pediatr Infect Dis J 2007;26:1066–8.

[63] Huijsdens XW, van Santen-Verheuvel MG, Spalburg E, et al. Multiple cases of familial transmission of community-acquired methicillin-resistant *Staphylococcus aureus*. J Clin Microbiol 2006;44:2994–6.

[64] Ishihara K, Shimokubo N, Sakagami A, Ueno H, Muramatsu Y, et al. Occurrence and molecular characteristics of methicillin-resistant *Staphylococcus aureus* and methicillin-resistant *Staphylococcus pseudintermedius* in an academic veterinary hospital. Appl Environ Microbiol 2010;76:5165–74.

[65] Johansson PJ, Gustafsson EB, Ringberg H. High prevalence of MRSA in household contacts. Scand J Infect Dis 2007;39:764–8.

[66] Jones TF, Creech CB, Erwin P, Baird SG, Woron AM, Schaffner W. Family outbreaks of invasive community-associated methicillin-resistant *Staphylococcus aureus* infection. Clin Infect Dis 2006;42:e76–78.

[67] Kadlec K, van Duijkeren E, Wagenaar JA, Schwarz S. Molecular basis of rifampicin resistance in methicillin-resistant *Staphylococcus pseudintermedius* isolates from dogs. J Antimicrob Chemother 2011;66:1236–42.

[68] Kassem II, Sigler V, Esseili MA. Public computer surfaces are reservoirs for methicillin-resistant staphylococci. ISME J 2007;1:265–8.

[69] Kawano J, Shimizu A, Saitoh Y, et al. Isolation of methicillin-resistant coagulase-negative staphylococci from chickens. J Clin Microbiol 1996;34:2072–7.

[70] Kempker R, Mangalat D, Kongphet-Tran T, Eaton M. Beware of the pet dog: a case of *Staphylococcus intermedius* infection. Am J Med Sci 2009;338:425–7.

[71] Klein E, Smith DL, Laxminarayan R. Community-associated methicillin-resistant *Staphylococcus aureus* in outpatients, United States, 1999–2006. Emerg Infect Dis 2009;15:1925–30.

[72] Kluytmans JAJW, Wertheim HFL. Nasal carriage of *Staphylococcus aureus* and prevention of nosocomial infections. Infection 2005;33:3–8.

[73] Kniehl E, Becker A, Forster DH. Bed, bath and beyond: pitfalls in prompt eradication of methicillin-resistant *Staphylococcus aureus* carrier status in healthcare workers. J Hosp Infect 2005;59:180–7.

[74] Kock R, Mellmann A, Schaumburg F, Friedrich AW, Kipp F, Becker K. The epidemiology of methicillin-resistant *Staphylococcus aureus* (MRSA) in Germany. Dtsch Arztebl Int 2011;108:761–7.

[75] Kottler S, Middleton JR, Perry J, Weese JS, Cohn LA. Prevalence of *Staphylococcus aureus* and methicillin-resistant *Staphylococcus aureus* carriage in three populations. J Vet Intern Med 2010;24:132–9.

[76] Kuijper EJ, Soonawala D, Vermont C, van Dissel JT. Household transmission of haemolytic uraemic syndrome associated with *Escherichia coli* O104:H4 in the Netherlands, May 2011. Euro Surveill 2011;16(25), pii=19897.

[77] Kusumaningrum HD, van Putten MM, Rombouts FM, Beumer RR. Effects of antibacterial dishwashing liquid on foodborne pathogens and competitive microorganisms in kitchen sponges. J Food Prot 2002;65:61–5.

[78] Laarhoven LM, de Heus P, van Luijn J, Duim B, Wagenaar JA, van Duijkeren E. Longitudinal study on methicillin-resistant *Staphylococcus pseudintermedius* in households. PLoS One 2011;6:e27788.

[79] Lakdawala N, Pham J, Shah M, Holton J. Effectiveness of low-temperature domestic laundry on the decontamination of healthcare workers' uniforms. Infect Control Hosp Epidemiol 2011;32:1103–8.

[80] Larson EL, Gomez-Duarte C, Lee LV, Della-Latta P, Kain DJ, Keswick BH. Microbial flora of hands of homemakers. Am J Infect Control 2003;31:72–9.

[81] Larsson AK, Gustafsson E, Nilsson AC, Odenholt I, Ringberg H, Melander E. Duration of methicillin-resistant *Staphylococcus aureus* colonization after diagnosis: a four-year experience from southern Sweden. Scand J Infect Dis 2011;43:456–62.

[82] Lauderdale TLY, Wang JT, Lee WS, et al. Carriage rates of methicillin-resistant *Staphylococcus aureus* (MRSA) depend on anatomic location, the number of sites cultured, culture methods, and the distribution of clonotypes. Eur J Clin Microbiol Infect Dis 2010;29:1553–9.

[83] Lautenbach E, Tolomeo P, Nachamkin I, Hu B, Zaoutis TE. The impact of household transmission on duration of outpatient colonization with methicillin-resistant *Staphylococcus aureus*. Epidemiol Infect 2010;138:683–5.

[84] Layton MC, Perez M, Heald P, Patterson JE. An outbreak of mupirocin-resistant *Staphylococcus aureus* on a dermatology ward associated with an environmental reservoir. Infect Control Hosp Epidemiol 1993;14:369–75.

[85] Leech JA, Nelson WC, Burnett RT, Aaron S, Raizenne ME. It's about time: a comparison of Canadian and American time-activity patterns. J Expo Anal Environ Epidemiol 2002;12:427–32.

[86] Lefebvre SL, Reid-Smith RJ, Waltner-Toews D, Weese JS. Incidence of acquisition of methicillin-resistant *Staphylococcus aureus, Clostridium difficile*, and other health-care-associated pathogens by dogs that participate in animal-assisted interventions. J Am Vet Med Assoc 2009;234:1404–17.

[87] Lefebvre SL, Weese JS. Contamination of pet therapy dogs with MRSA and *Clostridium difficile*. J Hosp Infect 2009;72:268–9.

[88] Levy SB. Antibacterial household products: cause for concern. Emerg Infect Dis 2001;7 (Suppl):512–5.

[89] Lidwell OM, Lowbury EJ. The survival of bacteria in dust: II. The effect of atmospheric humidity on the survival of bacteria in dust. J Hyg (Lond) 1950;48:21–7.

[90] Lin Y, Barker E, Kislow J, et al. Evidence of multiple virulence subtypes in nosocomial and community-associated MRSA genotypes in companion animals from the upper midwestern and northeastern United States. Clin Med Res 2011;9:7–16.

[91] Lindberg E, Adlerberth I, Matricardi P, et al. Effect of lifestyle factors on *Staphylococcus aureus* gut colonization in Swedish and Italian infants. Clin Microbiol Infect 2011;17:1209–15.

[92] Lis DO, Pacha JZ, Idzik D. Methicillin resistance of airborne coagulase-negative staphylococci in homes of persons having contact with a hospital environment. Am J Infect Control 2009;37:177–82.

[93] Lloyd DH. Reservoirs of antimicrobial resistance in pet animals. Clin Infect Dis 2007;45 (Suppl. 2):S148–52.

[94] Loeffler A, Lloyd DH. Companion animals: a reservoir for methicillin-resistant *Staphylococcus aureus* in the community? Epidemiol Infect 2010;138:595–605.

[95] Loeffler A, Boag AK, Sung J, et al. Prevalence of methicillin-resistant *Staphylococcus aureus* among staff and pets in a small animal referral hospital in the UK. J Antimicrob Chemother 2005;56:692–7.

[96] Loeffler A, Pfeiffer DU, Lloyd DH, Smith H, Soares-Magalhaes R, Lindsay JA. Meticillin-resistant *Staphylococcus aureus* carriage in UK veterinary staff and owners of infected pets: new risk groups. J Hosp Infect 2010;74:282–8.

[97] Lucet JC, Paoletti X, Demontpion C, et al, the *Staphylococcus aureus* Resistant a la Meticilline en Hospitalisation A Domicile (SARM HAD) Study Group. Carriage of methicillin-resistant *Staphylococcus aureus* in home care settings: prevalence, duration, and transmission to household members. Arch Intern Med 2009;169:1372–8.

[98] Mahoudeau I, Delabranche X, Prevost G, Monteil H, Piemont Y. Frequency of isolation of *Staphylococcus intermedius* from humans. J Clin Microbiol 1997;35:2153–4.

[99] Manian FA. Asymptomatic nasal carriage of mupirocin-resistant, methicillin-resistant *Staphylococcus aureus* (MRSA) in a pet dog associated with MRSA infection in household contacts. Clin Infect Dis 2003;36:e26–28.

[100] Marschall J, Muhlemann K. Duration of methicillin-resistant *Staphylococcus aureus* carriage, according to risk factors for acquisition. Infect Control Hosp Epidemiol 2006;27:1206–12.

[101] McCullough AC, Seifried M, Zhao X, et al. Higher incidence of perineal community acquired MRSA infections among toddlers. BMC Pediatr 2011;11:96.

[102] Medrano-Felix A, Martinez C, Castro-del Campo N, et al. Impact of prescribed cleaning and disinfectant use on microbial contamination in the home. J Appl Microbiol 2011;110:463–71.

[103] Miller LG, Diep BA. Clinical practice: colonization, fomites, and virulence: rethinking the pathogenesis of community-associated methicillin-resistant *Staphylococcus aureus* infection. Clin Infect Dis 2008;46:752–60.

[104] Miller LG, Eells SJ, Taylor AR, et al. *Staphylococcus aureus* colonization among household contacts of patients with skin infections: risk factors, strain discordance, and complex ecology. Clin Infect Dis 2012;54:1523–35.

[105] Mollema FPN, Richardus JH, Behrendt M, et al. Transmission of methicillin-resistant *Staphylococcus aureus* to household contacts. J Clin Microbiol 2010;48:202–7.

[106] Monnet DL. Methicillin-resistant *Staphylococcus aureus* and its relationship to antimicrobial use: possible implications for control. Infect Control Hosp Epidemiol 1998;19:552–9.

[107] Moodley A, Stegger M, Bagcigil AF, et al. *spa* typing of methicillin-resistant *Staphylococcus aureus* isolated from domestic animals and veterinary staff in the UK and Ireland. J Antimicrob Chemother 2006;58:1118–23.

[108] Morel AS, Wu F, Della-Latta P, Cronquist A, Rubenstein D, Saiman L. Nosocomial transmission of methicillin-resistant *Staphylococcus aureus* from a mother to her preterm quadruplet infants. Am J Infect Control 2002;30:170–3.

[109] Morgan M. Methicillin-resistant *Staphylococcus aureus* and animals: zoonosis or humanosis? J Antimicrob Chemother 2008;62:1181–7.

[110] Morris DO, Boston RC, O'Shea K, Rankin SC. The prevalence of carriage of meticillin-resistant staphylococci by veterinary dermatology practice staff and their respective pets. Vet Dermatol 2010;21:400–7.

[111] Murayama N, Nagata M, Terada Y, Shibata S, Fukata T. Efficacy of a surgical scrub including 2% chlorhexidine acetate for canine superficial pyoderma. Vet Dermatol 2010;21:586–92.

[112] Nerby JM, Gorwitz R, Lesher L, et al. Risk factors for household transmission of community-associated methicillin-resistant *Staphylococcus aureus*. Pediatr Infect Dis J 2011;30:927–32.

[113] Nienhoff U, Kadlec K, Chaberny IF, Verspohl J, Gerlach GF, et al. Transmission of methicillin-resistant *Staphylococcus aureus* strains between humans and dogs: two case reports. J Antimicrob Chemother 2009;64:660–2.

[114] Noble WC. Dispersal of skin microorganisms. Br J Dermatol 1975;93:477–85.

[115] Normand EH, Gibson NR, Reid SW. Antimicrobial resistance trends in bacterial isolates from companion-animal community practice in the UK. Prev Vet Med 2000;46:267–78.

[116] Oehler RL, Velez AP, Mizrachi M, Lamarche J, Gompf S. Bite-related and septic syndromes caused by cats and dogs. Lancet Infect Dis 2009;9:439–47.

[117] Oie S, Hosokawa I, Kamiya A. Contamination of room door handles by methicillin-sensitive/methicillin-resistant *Staphylococcus aureus*. J Hosp Infect 2002;51:140–3.

[118] Oie S, Kamiya A. Survival of methicillin-resistant *Staphylococcus aureus* (MRSA) on naturally contaminated dry mops. J Hosp Infect 1996;34:145–9.

[119] Oie S, Suenaga S, Sawa A, Kamiya A. Association between isolation sites of methicillin-resistant *Staphylococcus aureus* (MRSA) in patients with MRSA-positive body sites and MRSA contamination in their surrounding environmental surfaces. Jpn J Infect Dis 2007;60:367–9.

[120] Ojima M, Toshima Y, Koya E, Ara K, Kawai S, Ueda N. Bacterial contamination of Japanese households and related concern about sanitation. Int J Environ Health Res 2002;12:41–52.

[121] Ojima M, Toshima Y, Koya E, et al. Hygiene measures considering actual distributions of microorganisms in Japanese households. J Appl Microbiol 2002;93:800–9.

[122] Oller AR, Mitchell A. *Staphylococcus aureus* recovery from cotton towels. J Infect Dev Ctries 2009;3:224–8.

[123] Otter JA, Yezli S, French GL. The role played by contaminated surfaces in the transmission of nosocomial pathogens. Infect Control Hosp Epidemiol 2011;32:687–99.

[124] Pancholi P, Healy M, Bittner T, et al. Molecular characterization of hand flora and environmental isolates in a community setting. J Clin Microbiol 2005;43:5202–7.

[125] Pano-Pardo JR, Garcia A, Aracil F, Martinez A. Community-associated methicillin-resistant *Staphylococcus aureus* disease in two members of a household in Spain. Enferm Infecc Microbiol Clin 2010;28:472–3.

[126] Paul NC, Moodley A, Ghibaudo G, Guardabassi L. Carriage of methicillin-resistant *Staphylococcus pseudintermedius* in small animal veterinarians: indirect evidence of zoonotic transmission. Zoonoses Public Health 2011;58:533–9.

[127] Peacock SJ, de Silva I, Lowy FD. What determines nasal carriage of *Staphylococcus aureus?*. Trends Microbiol 2001;9:605–10.

[128] Perez HR, Johnson R, Gurian PL, Gibbs SG, Taylor J, Burstyn I. Isolation of airborne oxacillin-resistant *Staphylococcus aureus* from culturable air samples of urban residences. J Occup Environ Hyg 2011;8:80–5.

[129] Perreten V, Kadlec K, Schwarz S, et al. Clonal spread of methicillin-resistant *Staphylococcus pseudintermedius* in Europe and North America: an international multicentre study. J Antimicrob Chemother 2010;65:1145–54.

[130] Popovich KJ, Weinstein RA, Aroutcheva A, Rice T, Hota B. Community-associated methicillin-resistant *Staphylococcus aureus* and HIV: intersecting epidemics. Clin Infect Dis 2010;50:979–87.

[131] Porrero MC, Hasman H, Vela AI, Fernandez-Garayzabal JF, Dominguez L, Aarestrup FM. Clonal diversity of *Staphylococcus aureus* originating from the small ruminants goats and sheep. Vet Microbiol 2012;156:157–61.

[132] Pozzi Langhi SA, Robinson JO, Pearson JC, Christiansen KJ, Coombs GW, Murray RJ. Intrafamilial transmission of methicillin-resistant *Staphylococcus aureus*. Emerg Infect Dis 2009;15:1687–9.

[133] Rich M, Roberts L. Methicillin-resistant *Staphylococcus aureus* isolates from companion animals. Vet Rec 2004;154:310.

[134] Rintala H, Pitkaranta M, Toivola M, Paulin L, Nevalainen A. Diversity and seasonal dynamics of bacterial community in indoor environment. BMC Microbiol 2008;8:56.

[135] Roberts MC, Soge OO, No D, Helgeson SE, Meschke JS. Characterization of methicillin-resistant *Staphylococcus aureus* isolated from public surfaces on a university campus, student homes and local community. J Appl Microbiol 2011;110:1531–7.

[136] Rubin JE, Ball KR, Chirino-Trejo M. Antimicrobial susceptibility of *Staphylococcus aureus* and *Staphylococcus pseudintermedius* isolated from various animals. Can Vet J 2011;52:153–7.

[137] Ruscher C, Lübke-Becker A, Semmler T, Wleklinski CG, Paasch A, et al. Widespread rapid emergence of a distinct methicillin- and multidrug-resistant *Staphylococcus pseudintermedius* (MRSP) genetic lineage in Europe. Vet Microbiol 2010;144:340–6.

[138] Ruscher C, Lübke-Becker A, Wleklinski CG, Soba A, Wieler LH, et al. Prevalence of methicillin-resistant *Staphylococcus pseudintermedius* isolated from clinical samples of companion animals and equidaes. Vet Microbiol 2009;136:197–201.

[139] Rusin P, Maxwell S, Gerba C. Comparative surface-to-hand and fingertip-to-mouth transfer efficiency of gram-positive bacteria, gram-negative bacteria, and phage. J Appl Microbiol 2002;93:585–92.

[140] Salgado CD, Farr BM, Calfee DP. Community-acquired methicillin-resistant *Staphylococcus aureus*: a meta-analysis of prevalence and risk factors. Clin Infect Dis 2003;36:131–9.

[141] Scott E, Duty S, Callahan M. A pilot study to isolate *Staphylococcus aureus* and methicillin-resistant *S. aureus* from environmental surfaces in the home. Am J Infect Control 2008;36:458–60.

[142] Scott E, Duty S, McCue K. A critical evaluation of methicillin-resistant *Staphylococcus aureus* and other bacteria of medical interest on commonly touched household surfaces in relation to household demographics. Am J Infect Control 2009;37:447–53.

[143] Seguin JC, Walker RD, Caron JP, et al. Methicillin resistant *Staphylococcus aureus* outbreak in a veterinary teaching hospital: potential human-to-animal transmission. J Clin Microbiol 1999;37:1459–63.

[144] Sexton T, Clarke P, O'Neill E, Dillane T, Humphreys H. Environmental reservoirs of methicillin-resistant *Staphylococcus aureus* in isolation rooms: correlation with patient isolates and implications for hospital hygiene. J Hosp Infect 2006;62:187–94.

[145] Sherertz RJ, Reagan DR, Hampton KD, et al. A cloud adult: the *Staphylococcus aureus*-virus interaction revisited. Ann Intern Med 1996;124:539–47.

[146] Smith TC, Moritz ED, Leedom Larson KR, Ferguson DD. The environment as a factor in methicillin-resistant *Staphylococcus aureus* transmission. Rev Environ Health 2010;25:121–34.

[147] Soge OO, Meschke JS, No DB, Roberts MC. Characterization of methicillin-resistant *Staphylococcus aureus* and methicillin-resistant coagulase-negative *Staphylococcus* spp. isolated from US West Coast public marine beaches. J Antimicrob Chemother 2009;64:1148–55.

[148] Strommenger B, Kehrenberg C, Kettlitz C, et al. Molecular characterization of methicillin-resistant *Staphylococcus aureus* strains from pet animals and their relationship to human isolates. J Antimicrob Chemother 2006;57:461–5.

[149] Taubel M, Rintala H, Pitkaranta M, et al. The occupant as a source of house dust bacteria. J Allergy Clin Immunol 2009;124:834–40.

[150] Tegnell A, Grabowska K, Jacobsson A, Andersson M, Giesecke J, Ohman L. Study of developed resistance due to antibiotic treatment of coagulase-negative staphylococci. Microb Drug Resist 2003;9:1–6.

[151] Tenover FC, Goering RV. Methicillin-resistant *Staphylococcus aureus* strain USA300: origin and epidemiology. J Antimicrob Chemother 2009;64:441–6.

[152] Tomlin J, Pead MJ, Lloyd DH, et al. Methicillin resistant *Staphylococcus aureus* infections in 11 dogs. Vet Rec 1999;144:60–4.

[153] Tsubakishita S, Kuwahara-Arai K, Sasaki T, Hiramatsu K. Origin and molecular evolution of the determinant of methicillin resistance in staphylococci. Antimicrob Agents Chemother 2010;54:4352–9 Erratum in Antimicrob Agents Chemother 2011;55: 946.

[154] Uhlemann AC, Knox J, Miller M, et al. The environment as an unrecognized reservoir for community-associated methicillin resistant *Staphylococcus aureus* USA300: a case-control study. PLoS One 2011;6:e22407.

[155] van Belkum A, Verkaik NJ, de Vogel CP, et al. Reclassification of *Staphylococcus aureus* nasal carriage types. J Infect Dis 2009;199:1820–6.

[156] Van den Eede A, Martens A, Lipinska U. High occurrence of methicillin-resistant *Staphylococcus aureus* ST398 in equine nasal samples. Vet Microbiol 2009;133:138–44.

[157] van Duijkeren E, Catry B, Greko C, et al The Scientific Advisory Group on Antimicrobials (SAGAM). Review on methicillin-resistant *Staphylococcus pseudintermedius*. J Antimicrob Chemother 2011;66:2705–14.

[158] van Duijkeren E, Kamphuis M, van der Mije IC, et al. Transmission of methicillin-resistant *Staphylococcus pseudintermedius* between infected dogs and cats and contact pets, humans and the environment in households and veterinary clinics. Vet Microbiol 2011;150:338–43.

[159] van Duijkeren E, Wolfhagen MJM, Box ATA, et al. Human-to-dog transmission of methicillin-resistant *Staphylococcus aureus*. Emerg Infect Dis 2004;10:2235–7.

[160] Vancraeynest D, Haesebrouck F, Deplano A, et al. International dissemination of a high virulence rabbit *Staphylococcus aureus* clone. J Vet Med B Infect Dis Vet Public Health 2006;53:418–22.

[161] Vengust M, Anderson MEC, Rousseau J, Weese JS. Methicillin-resistant staphylococcal colonization in clinically normal dogs and horses in the community. Lett Appl Microbiol 2006;43:602–6.

[162] Vincze S, Paasch A, Walther B, Ruscher C, Lübke-Becker A, et al. Multidrug- and methicillin resistant *Staphylococcus pseudintermedius* as a cause of canine pyoderma: a case report. Berl Munch Tierarztl Wochenschr 2010;123:335–58.

[163] Wagenvoort JH, Sluijsmans W, Penders RJ. Better environmental survival of outbreak vs. sporadic MRSA isolates. J Hosp Infect 2000;45:231–4.

[164] Wagenvoort JHT, De Brauwer EI, Sijstermans ML, Toenbreker HM. Risk of re-introduction of methicillin-resistant *Staphylococcus aureus* into the hospital by intrafamilial spread from and to healthcare workers. J Hosp Infect 2005;59:67–8.

[165] Walther B, Hermes J, Cuny C, et al. Sharing more than friendship-nasal colonization with coagulase-positive staphylococci (CPS) and co-habitation aspects of dogs and their owners. PLoS One 2012;7:e35197.

[166] Walther B, Wieler LH, Friedrich AW, et al. Methicillin-resistant *Staphylococcus aureus* (MRSA) isolated from small and exotic animals at a university hospital during routine microbiological examinations. Vet Microbiol 2008;127:171–8.

[167] Walther B, Wieler LH, Friedrich AW, Kohn B, Brunnberg L, Lubke-Becker A. *Staphylococcus aureus* and MRSA colonization rates among personnel and dogs in a small animal hospital: association with nosocomial infections. Berl Munch Tierarztl Wochenschr 2009;122:178–85.

[168] Weese JS, Da Costa T, Button L, et al. Isolation of methicillin-resistant *Staphylococcus aureus* from the environment in a veterinary teaching hospital. J Vet Intern Med 2004;18:468–70.

[169] Weese JS, Dick H, Willey BM, et al. Suspected transmission of methicillin-resistant *Staphylococcus aureus* between domestic pets and humans in veterinary clinics and in the household. Vet Microbiol 2006;115:148–55.

[170] Weese JS, Faires MC, Frank LA, Reynolds LM, Battisti A. Factors associated with methicillin-resistant versus methicillin-susceptible *Staphylococcus pseudintermedius* infection in dogs. J Am Vet Med Assoc 2012;240:1450–5.

[171] Weese JS, Rousseau J, Willey BM, Archambault M, McGeer A, et al. Methicillin-resistant *Staphylococcus aureus* in horses at a veterinary teaching hospital: frequency, characterization, and association with clinical disease. J Vet Intern Med 2006;20:182–6.

[172] Weese JS, Rousseau J. Attempted eradication of methicillin-resistant *Staphylococcus aureus* colonization in horses on two farms. Equine Vet J 2005;37:510–4.

[173] Weese JS, van Duijkeren E. Methicillin-resistant *Staphylococcus aureus* and *Staphylococcus pseudintermedius* in veterinary medicine. Vet Microbiol 2010;140:418–29.

[174] Wertheim HFL, Melles DC, Vos MC, et al. The role of nasal carriage in *Staphylococcus aureus* infections. Lancet Infect Dis 2005;5:751–62.

[175] White LF, Dancer SJ, Robertson C. A microbiological evaluation of hospital cleaning methods. Int J Environ Health Res 2007;17:285–95.

[176] Wieler LH, Ewers C, Guenther S, Walther B, Lübke-Becker A. Methicillin-resistant staphylococci (MRS) and extended-spectrum beta-lactamases (ESBL)-producing *Enterobacteriaceae* in companion animals: nosocomial infections as one reason for the rising prevalence of these potential zoonotic pathogens in clinical samples. Int J Med Microbiol 2011;301:635–41.

[177] Yamamoto T, Takano T, Yabe S, et al. Super-sticky familial infections caused by Panton-Valentine leukocidin-positive ST22 community-acquired methicillin-resistant *Staphylococcus aureus* in Japan. J Infect Chemother 2012;18:187–98.

[178] Yasuda R, Kawano J, Humiaka O, et al. Methicillin resistant coagulase negative staphylococci isolated from healthy horses in Japan. Am J Vet Res 2000;61:1451–5.

[179] Yasuda R, Kawano J, Matsuo E, et al. Distribution of *mecA*-harboring staphylococci in healthy mares. J Vet Med Sci 2002;64:821–7.

[180] Zafar U, Johnson LB, Hanna M, et al. Prevalence of nasal colonization among patients with community-associated methicillin-resistant *Staphylococcus aureus* infection and their household contacts. Infect Control Hosp Epidemiol 2007;28:966–9.

FURTHER READING

Masterton RG, Coia JE, Notman AW, Kempton-Smith L, Cookson BD. Refractory methicillin-resistant *Staphylococcus aureus* carriage associated with contamination of the home environment. J Hosp Infect 1995;29:318–9.

Chapter 6

Food-Borne Transmission of Staphylococci

Antonello Paparella, Annalisa Serio, Chiara Rossi, Giovanni Mazzarrino, Clemencia Chaves López

Faculty of Bioscience and Technology for Food, Agriculture and Environment, University of Teramo, Teramo, Italy

6.1   INTRODUCTION

Staphylococcal food poisoning (SFP) is one of the most common food-borne diseases worldwide, and is due to the formation of staphylococcal enterotoxins (SEs) in foods.

For many years, identification of the microbial species involved in SFP has been based on coagulase detection, as *Staphylococcus aureus* was the only known species that produced coagulase.

Thermonuclease (TNase) was later added as a marker for identification of *S. aureus*. TNase, which can be extracted from foods, is considered an indicator of the presence of *S. aureus* above 1×10^6 cfu/g, even though Paparella and colleagues [1], in a challenge study with SEA-producing *S. aureus* ATCC 13565 in Italian Cacciatore salami, observed the production of TNase starting from 5×10^4 cfu/g.

Although most of the strains involved in SFP are coagulase and TNase positive, Park and colleagues [2] isolated a TNase negative enterotoxin-producing *S. aureus* from two samples of egg products. Most of the staphylococcal species do not produce coagulase, with the exception of *S. aureus*, *S. aureus* sp. *anaerobius,* the so-called *Staphylococcus intermedius* group (SIG; *S. intermedius*, *S. pseudintermedius*, and *S. delphini*), *Staphylococcus hyicus*, *Staphylococcus lutrae*, *Staphylococcus agnetis*, and *Staphylococcus schleiferi* subsp. *coagulans*. Moreover, TNase is produced by all the coagulase-positive staphylococci (CoPS), with the exception of *S. delphini* [3]. The classification of *S. aureus* is complicated by the fact that not all the strains show the typical characteristics of the species (e.g., ferment mannitol and are acetoin-positive).

Moreover, not all the strains of CoPS produce SEs, while these can be produced by coagulase-negative staphylococci (CoNS) [4], thus potentially leading to food poisoning. For these reasons, the European Commission Regulation on microbiological criteria for foodstuffs No. CE/2073/2005 of 15 November 2005, as amended by Regulations No. 1441/2007, 365/2010, 1086/2011, 209/2013, 1019/2013, and 217/2014, changed control strategy for SFP: the detection of SEs is considered a food safety criterion for dairy products, while the enumeration of CoPS is regarded as a process hygiene criterion for three categories of cheeses, described in the Regulation, milk powder and whey powder, cooked crustaceans, and shellfish. Only in cheeses, milk, and whey powder, if values $>10^5$ cfu/g are detected, the batch has to be tested for SEs.

SE production is linked to the activity of Agr, which is a *S. aureus* quorum sensing signal that causes a postexponential increase in the transcription on exotoxins [5]. This is the reason why the highest SE quantities are usually produced during the transition between the exponential and the stationary growth phase [6], and toxins production is often recognized with a *S. aureus* load higher than 10^5 cfu/g.

6.2   EFFECT OF FOOD HURDLES AND TECHNOLOGIES ON *S. AUREUS* GROWTH AND SURVIVAL

Staphylococci are ubiquitous in the environment. *S. aureus* is one of the most resistant nonspore-forming pathogens. As it can survive in a dry state, it is widely diffused in air, water, dust, sewage, surfaces, animals, and humans and has an important competitive advantage in Intermediate Moisture Foods (IMFs). Staphylococci have an efficient osmoprotectant system [7]; in particular, in low a_w conditions, glycine betaine overall, but also proline betaine, L-proline, choline, and taurine can accumulate in the cell, conferring resistance to external conditions. However, this process requires energy, and therefore toxins cannot be produced under extreme growth conditions. Furthermore, environmental stress conditions, as those that can be encountered in food products, activate the alternative sigma factor sigB, that is antagonist of Agr, thus impairing enterotoxins production [8]. For example, a_w values of 0.93 still permit extensive growth, while blocking the production of SEB. On the contrary, SEE production seems to be still active in media containing about 10% NaCl, corresponding to a_w values of 0.92 [9,10]. According to other authors, the accumulation of compatible solutes not only stimulates the growth, but also promotes toxins synthesis, as demonstrated by Qi and Miller [11], who observed the increase of SEB production at low a_w in presence of proline in the broth.

The principal factors limiting the growth and survival of *S. aureus* are reported in Table 6.1. Obviously, these limits can vary in relation to other parameters; for instance, increasing NaCl concentrations raise the minimum pH for growth. In addition, aerobic or anaerobic conditions also change the lower pH limit for growth and most of all for toxins production.

TABLE 6.1 Factors Affecting the Growth of *S. aureus* and SEs Production

Factors	Optimum for Growth	Range	Optimum for SE Production	Range	References
Temperature	35–41°C (37°C)	6–48°C[a]	34–40°C	10–46°C	[12,13]
pH	6.0–7.5[b]	4.2–9.8	7.0–8.0	5.0–9.6	[12,13]
Water activity	0.98	0.83 to >0.99 (a)	0.98 to >0.99	0.86 to >0.99 (a)	[14–16]
		0.90 to >0.99 (an)		0.92 to >0.99 (an)	[14,15]
NaCl (%)	0	0–20	0	0–12	[13]
Redox potential (E_h)	>+200mV	−200 to >+200mV	>+200m	+100 to >200mV	[15]
Atmosphere	Aerobic	Aerobic/anaerobic[c]	Aerobic (5%–20% O_2)	Anaerobic/aerobic[c]	[15]

(a) aerobiosis; (an) anaerobiosis.

[a] NaCl, monosodium glutamate and soy sauce extend the upper growth limit above 44°C.

[b] *S. aureus* growth can be hampered during lactic fermentation, by production of lactic acid, hydrogen peroxide and bacteriocins.

[c] High CO_2 concentrations reduce cell growth; high oxygen concentrations increase toxins production.

In fact, while detectable amounts of toxins are usually produced at pH as low as 5.1 in an aerobic atmosphere, on the contrary, in anaerobic conditions, their production is hardly any below pH 5.7 [17]. *S. aureus* is facultative anaerobic bacterium; however, its growth is accelerated in the presence of oxygen.

Regarding nutritional requirements, a minimal medium containing monosodium glutamate as C, N, and energy sources is sufficient for aerobic growth and enterotoxin production. This medium only contains three amino acids (arginine, cysteine, and phenylalanine) and group B vitamins [18]. Arginine is essential for enterotoxin B production [19].

The microorganism is a mesophile (Table 6.1), and the presence of NaCl, monosodium glutamate and soy sauce extends the upper growth limit above 44°C. As regards survival, the microorganism is usually killed at pasteurization t/T and during cooking, thus thermal treatments are the most common procedure applied to control microbial contamination in foods. However, *S. aureus* resistance increases in dry, high-fat, and high-salt foods, while toxins are extremely resistant. In fact, when a_w is comprised between 0.70 and 0.80, the bacterium increases its heat-resistance [10,20]. In addition, heat causes adverse effects in the sensory, nutritional, and functional properties of food. Furthermore, the increased consumer demand for fresh-like foods has promoted the development of nonthermal alternative methods for microbial control [21]. As regards enterotoxins, thermal inactivation is considered as the loss of their serological reactivity; nevertheless, biological activity could be lost before the serological activity [10].

Frozen storage does not influence *S. aureus* survival and does not inactivate SEs. Physical factors and preserving methods affecting *S. aureus* and SEs inactivation are reported in Table 6.2.

6.2.1 Organic Acids and Salts

Acetic and citric acids (≤0.5%), antimicrobial organic acids that are generally recognised as safe (GRAS), showed a maximum reduction at about 2 log cfu/mL of *S. aureus* at 10°C in tahini [34]. In foods, sorbate and benzoate also inhibit *S. aureus* with a MIC (Minimal Inhibitory Concentration) of 1000 mg/kg at pH 6.1 [35].

6.2.2 Other Chemicals

S. aureus has a high tolerance to compounds such as tellurite, mercuric chloride, neomycin, polymyxin and sodium azide, acriflavine, and borate, all used as selective agents in culture media. Also methyl and propyl parabens are effective.

Chlorin, halogens, quaternary ammonium compounds, and peracetic acid, routinely used in food industries, inactivate *S. aureus* on surfaces (except for resistant strains) [25,36].

TABLE 6.2 Physical Treatments Affecting the Inactivation of *S. aureus* and SEs

Factors	*S. aureus*	SEs	References
Drying, chilling, freezing	Resistant	Resistant	[15]
Heat[a]	Broth D_{60} 0.43–8.0 min	Broth (D_{121}) 3.0–8.0 min	[15]
		D_{121} 9.9–11.4 min	[22]
		D_{100} 70.0 min	[23]
		D_{149} 100 min at a_w 0.99	[24]
		D_{149} 225 min at a_w 0.90	[24]
		TSB[b] D_{53} 19.47–64.59	[23]
		TSB[b] D_{56} 5.17–8.78	[23]
		High a_w D_{60}1–6 min	[25]
		Phosphate buffer D_{60} 1–2.5 min	[25]
		Milk D_{60} 5.3 min	[25]
		Milk + 57% sucrose D_{60} 42.3 min	[25]
		Meat + 3%–4% NaCl D_{60} 6 min	[25]
		Meat + NaCl 8% D_{60} 25 min	[25]
	Buffer	100°C, 130 min (crude SEA)	[10]
		80°C, 3 min (purified SEA)	[10]
		100°C, 1 min (purified SEA)	[10]
	55°C for 5 min (1.25 ± 0.12 log reduction)		[26]
HPP	550 MPa 60 s×2 on beef jerky (1.32 log cfu/strip reduction)		[27]
	500 MPa 50°C 15 min in milk (8 log reduction)		[28]

Continued

TABLE 6.2 Physical Treatments Affecting the Inactivation of S. aureus and SEs—cont'd

Factors	S. aureus	SEs	References
	500 MPa 4°C 15 min in milk (5 log reduction)		[28]
Irradiation (D-kGy)	0.1–0.6	>30	[15]
	0.2–0.4 in meat and fish	Not affected	[14]
Pulsed electric field and heat	22–28 kV/cm, 50°C in milk (>6 log reduction)		[29]
TUVP	25 Mw/cm² for 30 s and 254 nm (>5 log reduction)		[30]
TUVP and HPP	8.45 J/cm² and 500 MPa in apple juice (6.7 log reduction)		[31]
Ultrasound	600 W for 15 min (0.31 ± 0.02 log reduction)		[26]
Ultraviolet radiation	Pulsed UV light (3 J/cm²) on dry-fermented salami (2.12 log cfu/g reduction)		[32]
	UVV-Led treatment: 0.2 mJ/cm² (>2 log reduction) 0.6 mJ/cm² (3–5 log reduction)		[33]

[a] D temperature is expressed in Celsius degrees (°C).
[b] Tryptone soy broth.

6.2.3   Natural Antimicrobials

Given the constant interest of the food industry for natural products with microbial growth inhibitory potential, different natural products such as spices, plant extracts, and derivatives from agricultural by-products have been studied. In this sense, García-Lomillo and colleagues [37] observed a bactericidal effect of red wine pomace seasoning (RWPS, 40 g/L) against S. aureus, and a bacteriostatic effect obtained using RWPS at the concentration of 20 g/L. Moreover, in another recent study [38] focused on the effect of oregano essential oil (OEO) and carvacrol (C) on biofilms of *S. aureus* from food-contact surfaces, the authors reported that minimum inhibitory concentrations of these antimicrobials were effective to inhibit planktonic cells and eradicated preformed biofilms on polystyrene. In particular, MIC were 5 and 2.5 μL/mL, respectively for OEO and C, against planktonic cells and onefold higher against sessile ones.

The bacteriocin nisin, an antimicrobial peptide known for its activity against several food-borne spoilage and pathogenic microorganisms, was recently tested against *S. aureus* in cashew, soursop, peach, and mango juices. After 24 h of incubation in the presence of nisin (5000 IU/mL), no viable cells were detected in the cashew, soursop, and peach juices, unlike mango juice, where a reduction of 4 log was observed [39].

The position of prenylation of isoflavonoids and stilbenoids can modulate the antimicrobial activity; in fact, some authors demonstrated that a legume extract rich in prenylated isoflavonoids and stilbenoids at a concentration between 0.05% and 0.1% w/v exerts antibacterial activity against MRSA [40].

Taking into account the food quality properties, novel food processing technologies (e.g., ohmic, microwave, pulsed electric field, ultrasound, irradiation, high pressure processing, plasma, ozonation) can also be used to control microbial growth in foods [41], including *S. aureus*.

6.2.4   Combined Hurdle Approach

The combination of different techniques can be applied in foods (hurdle effect) based on the assumption that they are able to get a maximum lethality against microorganisms without damaging sensory characteristics of foods [42]. In recent years, this hurdle approach and the synergistic preservation effect were investigated and applied in foods by different authors to control different microorganisms, including *S. aureus*.

Recently, Li et al. [26] conducted a study on the viability of *S. aureus* following ultrasound and mild heat treatments alone or in combination and showed a higher reduction in thermo-sonication compared with individual treatments.

The application on fresh pork of combined slightly acid electrolyzed water (30 mg/L), 0.5% fumaric acid and 40°C, reduced *S. aureus* by 2.38 log cfu/g [43], exploiting synergistic physical-chemical properties of sanitizers, and also extended the storage time of pork.

Ultrasound (400 W) and slightly acid electrolyzed water (2 mg/L for 10 min) exhibited a greater effect than the sum of individual treatments with an *S. aureus* count reduction of 3.68 log cfu/mL [44].

Another study revealed that the combined treatment of monolaurin with lauric and lactic acid exhibited synergistic activity against *S. aureus* in pork loin and when the agents were used in combination, the sensory quality was acceptable [45].

Ascorbic acid was previously used in combination with soybean bioactive compounds to extend shelf life of refrigerated burgers (from 5 to 8 days) and *S. aureus* was the most affected microorganism by soybean additives [46].

6.3 STAPHYLOCOCCAL BIOFILM AND FOOD SAFETY

The presence and persistence of *S. aureus* in food products is related to its resistance characteristics, but also to its ability to form biofilm on biotic and abiotic surfaces. In fact, it can colonize many surfaces, causing problems both in hospital environments (e.g., adhesion on prosthesis and medical equipment), and in food industries. Biofilm is a tridimensional structure principally made of esopolysaccharides and proteins, and it develops in consecutive stages, through a dynamic process. First, a reversible attachment of cells on a surface takes place, then this attachment becomes irreversible. After that, the biofilm structure starts to be developed by means of the formation of micro-colonies, hence the biofilm appears in its mature form and finally the attached cells could be dispersed again, and are free to attach to other surfaces. Cells are entrapped within the biofilm structure, thus becoming more resistant to heat, disinfectants, chemicals, and environmental stressing conditions, with respect to planktonic cells. The biofilm can be constituted by cells of a single species, or by microbial associations. It is influenced by several factors, such as strain, genomic features, kind of contact surface, and environmental factors (pH, available nutrients, temperature and presence of antimicrobial compounds) [47]. As well as toxins production, quorum-sensing also promotes biofilm formation. In detail, *S. aureus* is able to develop at least two different types of biofilm: ica-dependent or ica-independent. Intercellular adhesion (*ica*) locus (*icaR* regulatory, and *icaADBC*, biosynthetic genes) determines the production of polysaccharides of intercellular adhesion (PIA, mainly constituted of glycosaminylglycans), which is of primary importance in the biofilm accumulation [48]. It helps the bacteria to adhere to different surfaces, thus providing a stable tridimensional structure and protecting the cells embedded within the structure [49]. The *ica* operon is regulated by specific repressors, but also by genes involved in the response to environmental conditions often common in food products, such as anaerobic growth, general stress, presence of glucose or ethanol, osmolarity and temperature [50]. Several studies highlighted that glucose and salt induce *icaA* expression, apparently not increasing biofilm production [51]. On the contrary, biofilm formation was reduced by the addition of both glucose and NaCl in the culture medium. In fact, also an *ica*-independent biofilm formation mechanism exists, which seems to be strain-specific [48]. Besides, *S. aureus* expresses several surface proteins, including Bap, SasG, FnBPs and Spa [52]. The dispersion of mature biofilm is regulated by the *agr* gene, which is also involved in entetotoxin production, being responsible for Quorum-sensing.

For the preceding reasons, it is fundamental to recognize the factors able to stimulate or to reduce the biofilm production by *S. aureus* in food environments [51].

The implication of biofilm production in food industries and the environment is clear. As the biofilm can mature and then disperse, it acts as a reservoir of entoerotoxigenic staphylococci, thus contaminating food products, equipment, work surfaces and tools, and even workers. Moreover, these cells are protected from antimicrobials, chemicals, and cleaning and disinfection procedures. Furthermore, methicillin-resistant *S. aureus* (MRSA) can form biofilm, thus increasing their potential dissemination into the environment, and particularly through the food production chain [48]. Folsom and Frank [53] investigated the biofilm forming capability of common food-borne pathogens, revealing that *S. aureus* tends to develop more biofilm than *Salmonella typhimurium, Escherichia coli*, and *Bacillus cereus*. Some food industries are more exposed to biofilm formation; for example, in the dairy industry, sanitization is particularly difficult due to the high milk content of lipids and proteins that coagulate on surfaces, pipes, rubber seals, blind angles, and so on. Cleaning-in-place (CIP) procedures usually help in preventing biofilm formation, but often the accumulation of microorganisms in unfavorable points cannot be excluded, thus allowing biofilm formation [54].

In addition, the properties (roughness, porosity, hydrophobicity, etc.) of material strongly influence the biofilm-forming ability. Wood is porous and absorbent, and therefore is exposed to the deposit of organic material and microbial cells, resulting in a perfect basis for biofilm formations. Nevertheless, wood is uncommon in the food industry, whereas stainless steel, teflon, rubber, and polyurethane are most often employed. However, *S. aureus* has been reported to produce biofilm on stainless steel and polycarbonate [51]. According to Chia and colleagues [55], the adhesion strength on several contact surfaces was teflon, stainless steel, glass, nitrile, rubber, and polyurethane.

Once formed, biofilm has to be efficiently removed. Physical treatments such as ultrasounds, high pulsed electrical fields, or low electrical currents in combination with antibiotics have been successfully applied [56]. In the same way, also chemical methods, such as disinfectants, surfactants, detergents containing chelating agents, or biological methods, including bacteriocins application directly on food contact surfaces, or specific enzymes able to disrupt extracellular polymers have been used [56]. However a limitation in the application of these latter methods in the food industry exists, depending on their safety, absence of toxicity, capability to be completely rinsed or removed from equipment and surfaces, and toxicity of residues. The effect of deoxyribonuclease (DNaseI) on *S. aureus* biofilm was studied by Izano and colleagues [57], who highlighted a significant separation of the matrix from surfaces and a rapid dispersion of the biofilm in smaller cell aggregates, however without disrupting them in single cells. Interesting results have been reached by means of biopreservatives such as plant extracts. Nostro et al. [58] demonstrated the efficacy of *Punica granatum* L. and *Rhus coriaria* L. methanolic extracts in reducing *S. aureus* biofilm biomass of 90%–80% when applied in sublethal concentrations. Obviously, methanolic extracts are not suitable for food industries, but the results are promising and worthy of in-depth studies. In addition, *Ginkgo biloba* extract and ginkgolic acids inhibited biofilm formation of three different *S. aureus* strains [59].

Eight essential oils were demonstrated to significantly ($P<.01$) reduce the number of viable *S. aureus* cells in biofilm, but none of them could completely remove biofilms. Thyme and patchouli oils were the most effective, and in particular the application of sublethal quantities of thyme oil was shown to slow down biofilm formation, but none of the oils completely removed the formed biofilm [60]. Within biopreserving approaches, *Lactobacillus fermentum* TCUESC01, isolated during the fermentation of cocoa beans, significantly reduced the *S. aureus* biofilm thickness, and its supernatant modulated the expression of *icaA* and *icaR,* thus reducing biofilm by inhibiting PIA production [50].

Finally, chitosan is a natural compound derived from the deacetylation of chitin obtained from crustacean shells. It exerts antimicrobial activity when applied alone or in combination with other natural substances, but is particularly effective in the form of nanoparticles. Nevertheless, 1% chitosan was not effective in reducing *S. aureus* biofilm, in spite of a strong effectiveness on biofilms of other food-borne species, such as *Listeria monocytogenes* and *Pseudomonas fluorescens* [61].

On the contrary, chitosan-coated iron oxide nanoparticles inhibit biofilm formation and reduce the number of living *S. aureus* cells, thus combining the antimicrobial properties of chitosan with the generation of oxygen radicals and the disruption of cell membranes exerted by iron oxide nanoparticles [62].

6.4   PREVALENCE OF *S. AUREUS* IN FOODS

Staphylococci are ubiquitous; hence, they are commonly found in the environment, in mammals, and birds. Animals are considered a potential source of primary contamination; for example, ruminants could harbor staphylococci (also responsible for mastitis) that could be carried over from the udder into the milk [10]. In addition, kitchen personnel and unclean tools and instruments, because of poor hygiene practices, could be responsible for food contamination with enterotoxigenic staphylococci.

In fact, in all cases of SFP, the food or one of the ingredients was contaminated with an SE-producing *S. aureus* strain during processing, and was exposed to temperatures that allow its growth. In most cases, a failure in refrigeration or a temperature increase during processing was involved, permitting bacterial growth and inducing the enterotoxins production. Many foods can be a good growth medium for *S. aureus*, particularly those rich in proteins, and many of them have been implicated in SFP, including milk and cream, cream-filled pastries, butter, ham, cheeses, sausages, canned meat, salads, cooked meals, and sandwich fillings [63].

In a survey on 693 food samples associated with food-borne investigations from 2007 to 2010, Crago and coworkers [64] identified *S. aureus* in 10.5%

(73/693) of samples that included 29 meats, 20 prepared foods containing meat, 11 prepared foods not containing meat, 10 dairy samples, and 3 produce samples.

Several species of meat-producing animals have been implicated in *S. aureus* contamination, including pork [65,66], and poultry [67,68]. In retail poultry meat products, *S. aureus* can have either animal or human origin [69]. EFSA and ECDC [70], in a report on antimicrobial resistance in zoonotic and indicator bacteria in 2014, reported information on the high prevalence of MRSA from German turkeys. Also, Spanish fresh meat products were positive for *S. aureus* and in Switzerland 22/319 fresh broiler meat samples were positive for MRSA.

Gundogan and colleagues [71] analyzed 150 meat and chicken samples, and identified 80 *S. aureus* strains, 67.5% of which were methicillin-resistant (MRSA).

Other authors [72], in a study on the prevalence of *S. aureus* and its emetic enterotoxins in raw pork, salted meat, and ready-for-sale uncooked smoked ham, showed that about 51% of all samples were positive for *S. aureus* detection, 34.8% of which were enterotoxigenic. Enterotoxigenic *S. aureus* were also detected in coriander sauce, coconut slices, and ready-to-eat salads [73].

One of the common origins of *S. aureus* is from meat products, but it has been commonly detected in fresh produce. Due to the optimum moisture content for *S. aureus* growth (0.98) and the difficulty to reduce the moisture content of fresh produce unless it is processed [74], the bacterium can survive and grow on fresh produce. In fact 56.9%, 66.7%, and 54.2% of different vegetables, fruits, and sprouts respectively, purchased from street vendors in India, were contaminated with this pathogen [75]. Furthermore, *S. aureus* was found to be one of the most proliferating pathogens in vegetables purchased in Bangladesh shops [76] and 78% of 50 fresh vegetable samples analyzed were contaminated by this microorganism [77].

In a study carried out in China on 121 food samples collected in the market, 103 (85.1%) samples were found to be contaminated with *S. aureus*, 20 of which possessed *sea* gene, 2 *sed*, 8 *seg*, 4 *seh*, 24 *sei*, 5 *selj*, 24 *selk*, 44 *seln*, and 8 *ser* [78].

A recent study reported the detection of methicillin-resistant *Staphylococcus aureus* (MRSA) in 106/200 (53%) Egyptian raw milk and dairy products such as Damietta cheese, Kareish cheese, ice cream, and yogurt samples [79].

6.5 FOODS ASSOCIATED WITH STAPHYLOCOCCAL FOOD POISONING

SFP is caused by SEs that are produced during the massive growth of *S. aureus* or other enterotoxin-producing staphylococci in food. SEs are single-chain proteins, designated from SEA to SEE and SEG to SEIX [80]. In addition to classical major SEs (from SEA to SEE), also the more recently identified SEG, SEH,

SEI, SER, and SET are able to induce emesis. SEIs (SEIK, SEIL, SEIM, SEIN, SEIO, SEIP, SeIQ, SEIU, SEIV and SEIX) are considered as enterotoxins on the basis of their genetic structure; however, they have relatively weak emetic activity [81], do not cause emesis in primate models [82], or their activity potential has not been demonstrated yet [83].

SEA is the most reported classical enterotoxin implicated in SFP; nevertheless, outbreaks caused by *S. aureus* strains not producing classical enterotoxins have also been reported. Recently, strains harboring *seg* and *sei* genes [84], harboring *seg, sei, sem, sen, seo* and *selu* genes [82,85], and even deficient in all SE/SEIS [86] were recognized.

The first report of staphylococcal food poisoning was registered in Michigan (United States) in 1884 by Vaughan and Sternberg. It was caused by contaminated cheese [10]. Since then, many description of SFP cases have been reported, often related to foods of animal origin.

In the European Union, 0.07 cases of SFP per 100,000 people have been reported in 2012, ranging from <0.01 to 0.46 per 100,000 people between countries. This report confirmed the rates recorded in 2010 and 2011 [87,88]. The reporting of food-borne outbreaks has been mandatory in the European Union since 2005. Nevertheless, this illness is supposed to be significantly underreported, because of its usually mild nature, and as a consequence of the relatively short persistence of the symptoms (24–72 h from ingestion) that usually reduces the need to consult a doctor.

The foods that were most often involved in SFP differ widely from one country to another. For example, in the United Kingdom, 1%–6% of the total of food poisoning reports between 1969 and 1981, and 0.5%–1% between 1982 and 1990, were attributed to *S. aureus*. In a total of 2530 outbreaks from 1992 to 2009, *S. aureus* caused about 1.5%, being the sixth most common bacterial cause [74]. In the United Kingdom, 53% of the SFPs reported between 1969 and 1990 were due to meat products, meat-based dishes, and especially ham; 22% to poultry and poultry-based meals; 8% to milk products; 7% to fish and shellfish; and 3.5% to eggs [89]. In France, in a 2-year period (1999–2000), milk products, and especially cheeses, were responsible for 32% of the cases; meats for 22%; sausages and pies for 15%; fish and seafood for 11%; eggs and egg products for 11%; and poultry for 9.5% [90]. In the United States, in the period 1975–82, 36% were due to red meat, 12.3% to salads, 11.3% to poultry, 5.1% to pastries, and only 1.4% to milk products and seafood. In 17.1% of the cases, the food involved was unknown [91]. Differences among countries may be due to diversity in food habits. For example, the consumption of raw milk cheeses is much higher in France than in the Anglo-Saxon countries.

In any case, the main sources of contamination are humans, via either manual contact or the respiratory tract, by coughing and sneezing. However, in foods such as raw meat, sausages, raw milk, and raw milk cheese, contamination from animal origins is more frequent, due to either animal carriage or infections (e.g., mastitis).

SFP is commonly associated with the presence of CoPS, although some reports evidenced the possible role of CoNS. Moreover, *S. intermedius* is reported as the only non-*S. aureus* CoPS that has been clearly implicated in an SFP outbreak related to EA [92].

Nevertheless, *S. delphini* enteritis has been described in kits and minks (due to enterotoxin E), whereas enterotoxins have been found in *S. pseudintermedius* (SEA, SEB, SEC, SED, SEE); this emphasizes that not only *S. aureus* and *S. intermedius* may be potentially enterotoxigenic, among CoPS [93–96]. In addition, production of enterotoxin by CoNS strains belonging to *Staphylococcus simulans, Staphylococcus xylosus, Staphylococcus equorum, Staphylococcus lentus,* and *Staphylococcus capitis* isolated from goats' milk, whey, and cheese in quantities ranging from 10 to 90 ng/mL supernatant, has been reported [4]. On the other hand, some authors reported that strains of *Staphylococcus carnosus, Staphylococcus condimenti,* and *Staphylococcus equorum* were able to produce toxins, such as the most common SEs, the toxic shock syndrome toxin-1 (TSST-1), and the exfoliative toxin A (ETA) [97].

However, Even and colleagues [98] demonstrated the low toxigenic and enterotoxigenic capacities of CoNS, and consequently the low risk associated with CoNS in foodstuffs, especially compared with *S. aureus*. On the other hand, the emergence of members of the SIG as occasional opportunistic pathogens in humans [99] can stimulate further investigation into a possible role of SIG in SFP.

Various patterns of SFP have been described [17]. After a short incubation period of 1–7 h, symptoms may develop and include nausea, violent vomiting, abdominal cramps and diarrhea, malaise, pain, and sometimes prostration, sometimes requiring hospitalization [100].

Although SFP outbreaks by *S. aureus* tend to be sporadic, there can be person-to-person transmission, which generally occurs in the form of outbreaks in collective institutions [101]. The diagnosis of SFP is generally confirmed by at least one of the following: (i) the recovery of 10^5 of *S. aureus* or other enterotoxin-producing staphylococci/g from food remnants, (ii) the detection of SEs in food remnants, or (iii) the isolation of *S. aureus* or other enterotoxin-producing staphylococci of the same phage-type from both the patients and food remnants [102]. In some cases, confirmation of SFP is difficult because *S. aureus* is heat sensitive, whereas SEs are not. Thus, in heat-treated foods, *S. aureus* may be eliminated without inactivating SEs [10].

The most common factors contributing to outbreaks included inadequate temperature control, infected food handlers, contaminated raw ingredients, cross-contamination, and inadequate heat treatment. In one case, cheese was involved in an outbreak because it had been made from milk contaminated after pasteurization and before inoculation with a lactic starter culture that did not grow properly [17]. In 1985, chocolate milk was the source of an SFP in Kentucky, in the United States. The chocolate milk was contaminated and stored at a too high temperature for 4–5 h before pasteurization. Pasteurization killed staphylococci, but had no effect on the SEs. These, and many other examples, illustrate the importance of the elimination of any contamination sources during the processing and refrigeration of food and food ingredients whenever possible. Finally, in canned foods, bacteria and SEs are usually inactivated. Nevertheless, some cases of SFP involving canned mushrooms that were correctly processed were reported in the United States [103], which raises questions about the heat stability of SEs.

Numerous SFP outbreaks have been described since the end of the Second World War; Table 6.3 shows a selection of cases of SFP reported from 2000 to 2017.

TABLE 6.3 Food Poisoning by *Staphylococcus aureus* Reported in Literature

Country	Year	Incriminated Food	References
Japan	2000	Reconstituted milk	[10]
France	2001	Raw milk semihard cheese, raw ewe's milk cheese	[104]
France	2002	Potted meat	[104]
France	2003	Potato salad	[105]
Norway	2003	Mashed potato made with raw bovine milk	[106]
India	2005	Bhalla (fried potato balls in vegetable oil)	[107]
France	2006	Coconut pearls (Chinese dessert)	[10]
Austria	2006	Chicken and rice	[108]
Austria	2007	Pasteurized milk products, cacao milk, vanilla milk	[101]
Belgium	2007	Hamburger	[10]
Switzerland	2007	Robiola cheese	[85]
Paraguay	2007	UHT milk	[109]
Germany	2008	Pancakes filled with minced chicken	[110]
Argentina	2008	Vegetarian cannelloni	[111]
France	2008	Caribbean meals, pasta salad	[10]
France	2009	Soft cheese made from unpasteurized cow milk	[112]
Japan	2009	Crepes	[113]
Spain	2011	Macaroni and fresh tomatoes	[114]
Italy	2011	Seafood salad	[115]

TABLE 6.3 Food Poisoning by *Staphylococcus aureus* Reported in Literature—cont'd

Country	Year	Incriminated Food	References
Romania	2012	Milk	[116]
Italy	2012	Fried rice balls (arancini)	[117]
United States	2012	Perlo (chicken, sausage, and rice dish)	[118]
Australia	2012	Catered buffet	[119]
Germany	2013	Ice cream	[120]
Luxembourg	2014	Pasta salad with pesto	[121]
Switzerland	2014	Formagella d'alpe cheese	[85]
Switzerland	2014	Tomme cheese	[122]
United States	2014	Turkey	[123]
United States	2015	Chicken	[124]
Bulgaria	2015	Ready-to-eat roasted chicken legs	[125]
Japan	2016	Sushi, potato salad, fried shrimp and chicken, sandwiches, vegetables	[82]
United States	2017	Pulled pork	[126]

6.6 CONCLUDING REMARKS

Thermonuclease (TNase) is considered an indicator of the presence of *S. aureus* above 1×10^6 cfu/g, and therefore is at risk for the production of SEs; however, also TNase negative enterotoxin-producing *S. aureus* could be rarely isolated. On the contrary, TNAse is produced by almost all the coagulase-positive staphylococci that are not necessarily SEs producers. All these reasons, together with the phenotypic heterogeneity of SE-producing strains, strongly complicate detection and control strategies in foods.

Future research will have to be carried out to develop new rapid and cost-effective tools for routine investigation.

S. aureus is one of the most resistant nonspore-forming food-borne pathogens, and has a relevant competitive advantage in Intermediate Moisture Foods (IMFs). Nevertheless, quorum sensing, as well as environmental stress conditions often encountered in foods, could affect the pathogenesis, both stimulating or blocking enterotoxins production.

Continued

Enterotoxin-producing staphylococci are often inactivated during food processing, and different control methods including physical and/or chemical
treatments, application of natural antimicrobials, and also combined hurdle
approaches have been proposed and/or efficiently used against the pathogen.
Unfortunately, SEs have a strong resistance and heat tolerance; therefore, prevention of contamination and growth of these microorganisms is of paramount
importance in food manufacturing, distribution, and catering, as well as in
household environments.

Staphylococcal food poisoning (SFP) is one of the most common food-borne diseases worldwide. The persistence of S. aureus in food products and even in the
food processing environment is related to its ability to form biofilm that protects
the cells from heat, chemicals, and environmental stressing conditions and that
could promote the dissemination of free cells in the environment.

As well as for planktonic cells, it is necessary to develop control strategies
and effective, food-grade antimicrobial treatments to avoid or at least to reduce
S. aureus biofilm persistence, particularly in the food (but also in the hospital)
environment.

Animals are considered to be the potential source of primary contamination of
foods and also working personnel. For this reason, potentially all food products
could be contaminated by enterotoxigenic S. aureus, particularly those rich in
proteins, made of different ingredients mixed together, and exposed to temperatures that do not inactivate the microorganism or, in the worst case, that allow its
growth. Fresh meat, particularly of meat and pork, is the most often involved in
S. aureus contamination.

These data suggest the importance of supporting networks of cooperation among
veterinary microbiologists, food microbiologists, and clinical microbiologists.

In spite of SFP diffusion, the number of cases is likely to be severely underreported,
due to the mild symptoms that often do not require medical intervention. Varying
foods are more frequently associated with SFP, with important differences among
different countries, depending on food habits; nevertheless, foods of animal origin
are the most common. Recent findings on the high prevalence of S. aureus and,
particularly, MRSA in market food samples, together with SIG, need to be taken
into account, for their possible impact both on SFP and on the contamination of
food handlers.

REFERENCES

[1] Paparella A, Ianieri A, Tagino D, Poggetti V. Crescita di *Staphylococcus aureus* enterotossico e produzione di termonucleasi durante la stagionatura di salame italiano tipo cacciatore. In: Porretta S, editore. Ricerche e Innovazioni nell'Industria Alimentare, vol. IV. Pinerolo, Chiriotti, editori 1999; p. 66-76 [in Italian].

[2] Park CE, El Derea HB, Rayman MK. Evaluation of staphylococcal thermonuclease (TNase) assay as a means of screening foods for growth of staphylococci and possible enterotoxin production. Can J Microbiol 1978;24:1135–9.

[3] Savini V, Passeri C, Mancini G, Iuliani O, Marrollo R, Argentieri AV, et al. Coagulase-positive staphylococci: my pet's two faces. Res Microbiol 2013;164:371–4.

[4] Vernozy-Rozand C, Mazuy C, Prevost G, Lapeyre C, Bes M, Brun Y, et al. Enterotoxin production by coagulase-negative staphylococci isolated from goats' milk and cheese. Int J Food Microbiol 1996;30:271–80.

[5] Dunman PM, Murphy E, Haney S, Palacios D, Tucker-Kellogg G, Wu S, et al. Transcription profiling-based identification of *Staphylococcus aureus* genes regulated by the *agr* and/or *sarA* loci. J Bacteriol 2001;183:7341–53.

[6] Sihto HM, Stephan R, Engl C, Chen J, Johler S. Effect of food-related stress conditions and loss of *agr* and *sigB* on *seb* promoter activity in *S. aureus*. Food Microbiol 2017;65:205–12.

[7] Jablonsky L, Bohach GA. Staphylococcus aureus. In: Montville TJ, Matthews KR, editors. Food microbiology an introduction. Washington, DC: ASM Press; 2005. p. 175–86.

[8] Novick RP. Autoinduction and signal transduction in the regulation of staphylococcal virulence. Mol Microbiol 2003;48:1429–49.

[9] Thota FH, Tatini SR, Bennett RW. Effects of temperature, pH and NaCl on production of staphylococcal enterotoxins E and F. Abstr Ann Meet Am Soc Microbiol 1973;1:11.

[10] Hennekinne JA, De Buyser ML, Dragacci S. *Staphylococcus aureus* and its food poisoning toxins: characterization and outbreak investigation. FEMS Microbiol Rev 2012;36:815–36.

[11] Qi Y, Miller KJ. Effect of low water activity on staphylococcal enterotoxin A and B biosynthesis. J Food Prot 2000;63:473–8.

[12] Schelin J, Wallin-Carlquist N, Cohn MT, Lindqvist R, Barker GC, Rådström P. The formation of *Staphylococcus aureus* enterotoxin in food environments and advances in risk assessment. Virulence 2011;2:580–92.

[13] Le Loir Y, Baron F, Gautier M. *Staphylococcus aureus* and food poisoning. Genet Mol Res 2003;2:63–76.

[14] Baird-Parker TC. Staphylococcus aureus. In: Lund BM, Baird-Parker TC, Gould GW, editors. The microbiological safety and quality of food. Gaithersburg, MD: Aspen Publishers; 2000. p. 1317–35.

[15] Food Safety Authority of Ireland. Staphylococcus aureus. Microbial fact sheet series, 1, 2011. p. 1–5.

[16] Jay JM, Loessner MJ, Golden DA. Staphylococcal gastroenteritis. In: Jay JM, Loessner MJ, Golden DA, editors. Modern food microbiology. New York: Springer; 2005. p. 545–66.

[17] Bergdoll MS. Staphylococcus aureus. In: Doyle MP, editor. Food borne bacterial pathogens. New York: Marcel Dekker; 1989. p. 463–523.

[18] Miller RD, Fung DYC. Amino acid requirements for the production of enterotoxin B by *Staphylococcus aureus* S-6 in a chemically defined medium. Appl Microbiol 1973;25: 800–6.

[19] Wu CH, Bergdoll MS. Stimulation of enterotoxin B production. Infect Immun 1971;3: 784–92.

[20] Troller JA. Water relations of food borne bacterial pathogens-an updated review. J Food Prot 1986;49:656–70.

[21] Soni A, Oey I, Silcock P, Bremer P. *Bacillus* spores in the food industry: a review on resistance and response to novel inactivation technologies. Compr Rev Food Sci Food Saf 2016;15:1139–48.

[22] Sutherland JP, Bayliss AJ, Roberts TA. Predictive modelling of growth of *Staphylococcus aureus*: the effects of temperature, pH and sodium chloride. Int J Food Microbiol 1994;21:217–36.

[23] Notermans S, van Hoeij K. The food safety file: *Staphylococcus aureus*. Woerden: Food Doctors; 2008. 39. Available from http://www.fooddoctors.com/FSF/S_aureus.pdf.

[24] Dewanti-Hariyadi R, Hadiyanto J, Purnomo EH. Thermal resistance of local isolates of *Staphylococcus aureus*. Asian J Food Agro Ind 2011;4:213–21.

[25] Medveďová A, Valík L. *Staphylococcus aureus*: characterisation and quantitative growth description in milk and artisanal raw milk cheese production. In: Eissa AA, editor. Structure and function of food engineering. Rijeka: InTech; 2012. p. 71–102.

[26] Li J, Suo Y, Liao X, Ahn J, Liu D, Chen S, et al. Analysis of *Staphylococcus aureus* cell viability, sublethal injury and death induced by synergistic combination of ultrasound and mild heat. Ultrason Sonochem 2017;39:101–10.

[27] Scheinberg JA, Svoboda AL, Cutter CN. High-pressure processing and boiling water treatments for reducing *Listeria monocytogenes, Escherichia coli* O157:H7, *Salmonella* spp., and *Staphylococcus aureus* during beef jerky processing. Food Control 2014;39:105–10.

[28] Windyga B, Rutkowska M, Sokołowska B, Skąpska S, Wesołowska A, Wilińska M, et al. Inactivation of *Staphylococcus aureus* and native microflora in human milk by high pressure processing. High Pressure Res 2015;35:181–8.

[29] Sharma P, Bremer P, Oey I, Everett DW. Bacterial inactivation in whole milk using pulsed electric field processing. Int Dairy J 2014;35:49–56.

[30] Yoo S, Ghafoor K, Kim S, Sun YW, Kim JU, Yang K, et al. Inactivation of pathogenic bacteria inoculated onto a Bacto™ agar model surface using TiO_2-UVC photocatalysis, UVC and chlorine treatments. J Appl Microbiol 2015;119:688–96.

[31] Shahbaz HM, Yoo S, Seo B, Ghafoor K, Kim JU, Lee DU, et al. Combination of TiO_2-UV photocatalysis and high hydrostatic pressure to inactivate bacterial pathogens and yeast in commercial apple juice. Food Bioprocess Technol 2016;9:182–90.

[32] Rajkovic A, Tomasevic I, De Meulenaer B, Devlieghere F. The effect of pulsed UV light on *Escherichia coli* O157:H7, *Listeria monocytogenes Salmonella typhimurium Staphylococcus aureus* and staphylococcal enterotoxin A on sliced fermented salami and its chemical quality. Food Control 2017;73:829–37.

[33] Kim DK, Kim SJ, Kang DH. Bactericidal effect of 266 to 279nm wavelength UVC-LEDs for inactivation of Gram positive and Gram negative food borne pathogenic bacteria and yeasts. Food Res Int 2017;97:280–7.

[34] Olaimat AN, Al-Nabulsi AA, Osaili TM, Al-Holy M, Ayyash MM, Mehyar GF, et al. Survival and inhibition of *Staphylococcus aureus* in commercial and hydrated tahini using acetic and citric acids. Food Control 2017;77:179–86.

[35] El-Banna AA, Hurst A. Survival in foods of *Staphylococcus aureus* grown under optimal and stressed conditions and the effect of some food preservatives. Can J Microbiol 1983;29:297–302.

[36] Temiz M, Duran N, Duran GG, Eryilmaz N, Jenedi K. Relationship between the resistance genes to quaternary ammonium compounds and antibiotic resistance in staphylococci isolated from surgical site infections. Med Sci Monit 2014;20:544–50.

[37] García-Lomillo J, Gonzalez-SanJose ML, Del Pino-García R, RiveroPérez MD, Muñiz-Rodríguez P. A new seasoning with potential effect against food borne pathogens. LWT Food Sci Technol 2017;84:338–43.

[38] dos Santos Rodrigues JB, de Carvalho RJ, de Souza NT, de Sousa Oliveira K, Franco OL, Schaffner D, et al. Effects of oregano essential oil and carvacrol on biofilms of *Staphylococcus aureus* from food-contact surfaces. Food Control 2017;73:1237–46.

[39] de Oliveira Junior AA, de Araújo Couto HGS, Barbosa AAT, Carnelossi MAG, de Moura TR. Stability, antimicrobial activity, and effect of nisin on the physico-chemical properties of fruit juices. Int J Food Microbiol 2015;211:38–43.

[40] Araya-Cloutier C, den Besten HM, Aisyah S, Gruppen H, Vincken JP. The position of prenylation of isoflavonoids and stilbenoids from legumes (Fabaceae) modulates the antimicrobial activity against Gram positive pathogens. Food Chem 2017; 226:193–201.

[41] Khan I, Tango CN, Miskeen S, Lee BH, Oh DH. Hurdle technology: a novel approach for enhanced food quality and safety—a review. Food Control 2017;73:1426–44.

[42] Singh S, Shalini R. Effect of hurdle technology in food preservation: a review. Crit Rev Food Sci Nutr 2014;56:641–9.

[43] Mansur AR, Tango CN, Kim GH, Oh DH. Combined effects of slightly acidic electrolyzed water and fumaric acid on the reduction of food borne pathogens and shelf life extension of fresh pork. Food Control 2015;47:277–84.

[44] Li J, Ding T, Liao X, Chen S, Ye X, Liu D. Synergetic effects of ultrasound and slightly acidic electrolyzed water against *Staphylococcus aureus* evaluated by flow cytometry and electron microscopy. Ultrason Sonochem 2017;38:711–9.

[45] Tangwatcharin P, Khopaibool P. Inhibitory effects of the combined application of lauric acid and monolaurin with lactic acid against *Staphylococcus aureus* in pork. ScienceAsia 2012;38:54–63.

[46] Mahmoud MH, Abu-Salem FM. Synergistic effect of ascorbic acid and soybean bioactive compounds in extending shelf life of refrigerated burger. Res J Pharm, Biol Chem Sci 2015;6:17–26.

[47] Vitale M, Scatassa ML, Cardamone C, Oliveri C, Piraino G, Alduina R, et al. Staphylococcal food poisoning case and molecular analysis of toxin genes in *Staphylococcus aureus* strains isolated from food in Sicily, Italy. Foodborne Pathog Dis 2015;121:21–3.

[48] Doulgeraki AI, Di Ciccio P, Ianieri A, Nychas GE. Methicillin-resistant food-related *Staphylococcus aureus*: a review of current knowledge and biofilm formation for future studies and applications. Res Microbiol 2017;168:1–15.

[49] Singh R, Ray P, Das A, Sharma M. Penetration of antibiotics through *Staphylococcus aureus* and *Staphylococcus epidermidis* biofilms. J Antimicrob Chemother 2010;65:1955–8.

[50] Melo TA, dos Santos TF, de Almeida ME, Gusmão Fontes Junior LA, Ferraz Andrade E, Rezende RP, et al. Inhibition of *Staphylococcus aureus* biofilm by *Lactobacillus* isolated from fine cocoa. BMC Microbiol 2016;16:250.

[51] Miao J, Liang Y, Chen L, Wang W, Wang J, Li B, et al. Formation and development of *Staphylococcus* biofilm: with focus on food safety. J Food Saf 2017;37:e12358. https://doi.org/10.1111/jfs.12358.

[52] Archer NK, Mazaitis MJ, Costerton JW, Leid JG, Powers ME, Shirtliff ME. *Staphylococcus aureus* biofilms: properties, regulation, and roles in human disease. Virulence 2011;2: 445–59.

[53] Folsom JP, Frank JF. Chlorine resistance of *Listeria monocytogenes* biofilms and relationship to subtype, cell density, and planktonic cell chlorine resistance. J Food Prot 2006;69:1292–6.

[54] Mattila T, Manninen A, Kylasiurola AL. Effect of cleaning-in-place disinfectants on wild bacterial strains isolated from a milking line. J Dairy Sci 1990;57:33–9.

[55] Chia T, Goulter R, McMeekin T, Dykes G, Fegan N. Attachment of different *Salmonella* serovars to materials commonly used in a poultry processing plant. Food Microbiol 2009;26:853–9.

[56] Kumar CG, Anand SK. Significance of microbial biofilms in food industry: a review. Int J Food Microbiol 1998;42:9–27.

[57] Izano EA, Amarante MA, Kher WB, Kaplan JB. Differential roles of poly-N-acetylglucosamine surface polysaccharide and extracellular DNA in *Staphylococcus aureus* and *Staphylococcus epidermidis* biofilms. Appl Environ Microbiol 2008;74:470–6.

[58] Nostro A, Guerrini A, Marino A, Tacchini M, Di Giulio M, Grandini A, et al. In vitro activity of plant extracts against biofilm-producing food-related bacteria. Int J Food Microbiol 2016;238:33–9.

[59] Lee JH, Kim YG, Ryu SY, Cho MH, Lee J. Ginkgolic acids and *Ginkgo biloba* extract inhibit *Escherichia coli* O157:H7 and *Staphylococcus aureus* biofilm formation. Int J Food Microbiol 2014;174:47–55.

[60] Vázquez-Sánchez D, Cabo ML, Rodríguez-Herrera JJ. Antimicrobial activity of essential oils against *Staphylococcus aureus* biofilms. Food Sci Technol Int 2015;21:559–70.

[61] Orgaz B, Lobete MM, Puga CH, San Jose C. Effectiveness of chitosan against mature biofilms formed by food related bacteria. Int J Mol Sci 2011;12:817–28.

[62] Shi SF, Jia JF, Guo XK, Zhao YP, Chen DS, Guo YY, et al. Reduced *Staphylococcus aureus* biofilm formation in the presence of chitosan-coated iron oxide nanoparticles. Int J Nanomedicine 2016;11:6499–506.

[63] Dongyou L. Molecular detection of food borne pathogens. Boca Raton, FL: CRC Press; 2009. 246 p.

[64] Crago B, Ferrato C, Drews SJ, Svenson LW, Tyrrell G, Louie M. Prevalence of *Staphylococcus aureus* and methicillin-resistant *S. aureus* (MRSA) in food samples associated with food borne illness in Alberta, Canada from 2007 to 2010. Food Microbiol 2012;32:202–5.

[65] Hadjirin NF, Lay EM, Paterson GK, Harrison EM, Peacock SJ, Parkhill J, et al. Detection of livestock-associated meticillin-resistant *Staphylococcus aureus* CC398 in retail pork, United Kingdom, February 2015. Euro Surveill 2015;20:21156.

[66] Tang Y, Larsen J, Kjeldgaard J, Andersen PS, Skov R, Ingmer H. Methicillin-resistant and -susceptible *Staphylococcus aureus* from retail meat in Denmark. Int J Food Microbiol 2017;249:72–6.

[67] Teramoto H, Salaheen S, Biswas D. Contamination of post-harvest poultry products with multidrug resistant *Staphylococcus aureus* in Maryland-Washington DC metro area. Food Control 2016;65:132–5.

[68] Zaheer Z, Rahman SU, Zaheer I, Abbas G, Younas T. Methicillin-resistant *Staphylococcus aureus* in poultry—an emerging concern related to future epidemic. Matrix Sci Med 2017;1:15–8.

[69] Bortolaia V, Espinosa-Gongora C, Guardabassi L. Human health risks associated with antimicrobial-resistant enterococci and *Staphylococcus aureus* on poultry meat. Clin Microbiol Infect 2016;22:130–40.

[70] EFSA (European Food Safety Authority), ECDC (European Centre for Disease Prevention and Control). The European Union summary report on antimicrobial resistance in zoonotic and indicator bacteria from humans, animals and food in 2014. EFSA J 2016;14:1–207. https://doi.org/10.2903/j.efsa.2016.4380.

[71] Gundogan N, Citak S, Yucel N, Devren A. A note on the incidence and antibiotic resistance of *Staphylococcus aureus* isolated from meat and chicken samples. Meat Sci 2005;69:807–10.

[72] Atanassova V, Meindl A, Ring C. Prevalence of *Staphylococcus aureus* and staphylococcal enterotoxins in raw pork and uncooked smoked ham: a comparison of classical culturing detection and RFLP-PCR. Int J Food Microbiol 2001;68:105–13.

[73] Ghosh M, Wahi S, Kumar M, Ganguli A. Prevalence of enterotoxigenic *Staphylococcus aureus* and *Shigella* spp. in some raw street vended Indian foods. Int J Environ Health Res 2007;17:151–6.

[74] Wadamori Y, Gooneratne R, Hussain MA. Outbreaks and factors influencing microbiological contamination of fresh produce. J Sci Food Agric 2017;97:1396–403.

[75] Viswanathan P, Kaur R. Prevalence and growth of pathogens on salad vegetables, fruits and sprouts. Int J Hyg Environ Health 2001;203:205–13.

[76] Rahman F, Noor R. Prevalence of pathogenic bacteria in common salad vegetables of Dhaka metropolis. Bangladesh J Bot 2012;41:159–62.

[77] Hassan ZH, Purwani EY. Microbiological aspect of fresh produces as rettile and consumed in West Java, Indonesia. Int Food Res J 2016;23:350–9.

[78] Tang J, Zhang R, Chen J, Zhao Y, Tang C, Yue H, et al. Incidence and characterization of *Staphylococcus aureus* strains isolated from food markets. Ann Microbiol 2015;65:279–86.

[79] Al-Ashmawy MA, Sallam KI, Abd-Elghany SM, Elhadidy M, Tamura T. Prevalence, molecular characterization, and antimicrobial susceptibility of methicillin-resistant *Staphylococcus aureus* (MRSA) isolated from milk and dairy products. Foodborne Pathog Dis 2016;13:156–62.

[80] Martín MC, Fueyo JM, González-Hevia MA, Mendoza MC. Genetic procedures for identification of enterotoxigenic strains of *Staphylococcus aureus* from three food poisoning outbreaks. Int J Food Microbiol 2004;94:279–86.

[81] Sergelidis D, Angelidis AS. Methicillin-resistant *Staphylococcus aureus*: a controversial food-borne pathogen. Lett Appl Microbiol 2017;64:409–18.

[82] Umeda K, Nakamura H, Yamamoto K, Nishina N, Yasufuku K, Hirai Y, et al. Molecular and epidemiological characterization of staphylococcal food borne outbreak of *Staphylococcus aureus* herboring *seg, sei, sem, sen, seo* and *selu* genes without production of classical enterotoxins. Int J Food Microbiol 2017;256:30–5.

[83] Omoe K, Hu DL, Ono HK, Shimizu S, Takahashi-Omoe H, Nakane A, et al. Emetic potential of newly identified staphylococcal enterotoxin-like toxins. Infect Immun 2013;81:3627–31.

[84] Argudin MA, Mendoza MC, Rodicio MR. Food poisoning and *Staphylococcus aureus* enterotoxins. Toxins 2010;2:1751–73.

[85] Johler S, Giannini P, Jermini M, Hummerjohann J, Baumgartner A, Stephan R. Further evidence for staphylococcal food poisoning outbreaks caused by egc-encoded enterotoxins. Toxins 2015;7:997–1004.

[86] Kuramoto S, Kodama H, Yamada K, Inui I, Kitagawa E, Kawakami K. Food poisoning attributable to *Staphylococcus aureus* deficient in all of the staphylococcal enterotoxin gene so far reported. Jpn J Infect Dis 2006;59:347.

[87] EFSA. The European Union summary report on trends and sources of zoonoses, zoonotic agents and food borne outbreaks in 2011. EFSA J 2013;11:3129.

[88] EFSA. The European Union summary report on trends and sources of zoonoses, zoonotic agents and food borne outbreaks in 2012. EFSA J 2014;12:3547.

[89] Wieneke AA, Roberts D, Gilbert RJ. Staphylococcal food poisoning in the United Kingdom, 1969-1990. Epidemiol Infect 1993;110:519–31.

[90] Haeghebaert S, Le Querrec F, Gallay A, Bouvet P, Gomez M, Vaillant V. Les toxi-infections alimentaires collectives en France, en 1999 et 2000. Bull Epidemiol Hebd 2002;23:105–9. [in French].

[91] Genigeorgis CA. Present state of knowledge on staphylococcal intoxication. Int J Food Microbiol 1989;9:327–60.

[92] Khambaty FM, Bennett RW, Shah DB. Application of pulsed-field gel electrophoresis to the epidemiological characterization of *Staphylococcus intermedius* implicated in a food-related outbreak. Epidemiol Infect 1994;113:75–81.

[93] Gary JM, Langohr IM, Lim A, Bolin S, Bolin C, Moore I, et al. Enteric colonization by *Staphylococcus delphini* in four ferret kits with diarrhoea. J Comp Pathol 2014;151:314–7.

[94] Sledge DG, Danieu PK, Bolin CA, Bolin SR, Lim A, Anderson BC, et al. Outbreak of neonatal diarrhea in farmed mink kits (*Mustella vison*) associated with enterotoxigenic *Staphylococcus delphini*. Vet Pathol 2010;47:751–7.

[95] Tanabe T, Toyoguchi M, Hirano F, Chiba M, Onuma K, Sato H. Prevalence of staphylococcal enterotoxins in *Staphylococcus pseudintermedius* isolates from dogs with pyoderma and healthy dogs. Microbiol Immunol 2013;57:651–4.

[96] Yoon JW, Lee GJ, Lee SY, Park C, Yoo JH, Park HM. Prevalence of genes for enterotoxins, toxic shock syndrome toxin 1 and exfoliative toxin among clinical isolates of *Staphylococcus pseudintermedius* from canine origin. Vet Dermatol 2010;21:484–9.

[97] Zell C, Resch M, Rosenstein R, Albrecht T, Hertel C, Gotz F. Characterization of toxin production of coagulase-negative staphylococci isolated from food and starter cultures. Int J Food Microbiol 2008;127:246–51.

[98] Even S, Leroy S, Charlier C, Zakour NB, Chacornac JP, Lebert I, et al. Low occurrence of safety hazards in coagulase negative staphylococci isolated from fermented foodstuffs. Int J Food Microbiol 2010;139:87–95.

[99] Savini V, Barbarini D, Polakowska K, Gherardi G, Bialecka A, Kasprowicz A, et al. Methicillin-resistant *Staphylococcus pseudintermedius* infection in a bone marrow transplant recipient. J Clin Microbiol 2013;51:1636–8.

[100] Bone FJ, Bogie D, Morgan-Jones SC. Staphylococcal food poisoning from sheep milk cheese. Epidemiol Infect 1989;103:449–58.

[101] Schmid D, Fretz R, Winter P, Mann M, Höger G, Stöger A, et al. Outbreak of staphylococcal food intoxication after consumption of pasteurized milk products, June 2007, Austria. Wien Klin Wochenschr 2009;121:125–31.

[102] Bryan FL, Guzewich JJ, Todd ECD. Surveillance of food borne disease II. Summary and presentation of descriptive data and epidemiologic patterns; their value and limitations. J Food Prot 1997;60:567–78.

[103] Bennet RW. The biomolecular temperament of staphylococcal enterotoxin in thermally processed foods. J Assoc Off Anal Chem 1992;75:6–12.

[104] Kérouanton A, Hennekinne JA, Letertre C, Petit L, Chesneau O, Brisabois A, et al. Characterization of *Staphylococcus aureus* strains associated with food poisoning outbreaks in France. Int J Food Microbiol 2007;115:369–75.

[105] Bonnetain F, Carbonel S, Stoll J, Legros D. Toxi-infection alimentaire collective due à *Staphylococcus aureus*, Longevelle-sur-le-Doubs, juillet 2003. Bull Epidemiol Hebd 2003;47:231–2 [in French].

[106] Jørgensen HJ, Mathisen T, Løvseth A, Omoe K, Kristina QS, Loncarevic S. An outbreak of staphylococcal food poisoning caused by enterotoxin H in mashed potato made with raw milk. FEMS Microbiol Lett 2005;252:267–72.

[107] Nema V, Agrawal R, Kamboj DV, Goel K, Singh L. Isolation and characterization of heat resistant enterotoxigenic *Staphylococcus aureus* from a food poisoning outbreak in Indian subcontinent. Int J Food Microbiol 2007;117:29–35.

[108] Schmid D, Gschiel E, Mann M, Huhulescu S, Ruppitsch W, Bohm G, et al. Outbreak of acute gastroenteritis in an Austrian boarding school, September 2006. Euro Surveill 2007;12:224.

[109] Weiler N, Leotta GA, Zárate MN, Manfredi E, Alvarez ME, Rivas M. Food borne outbreak associated with consumption of ultrapasteurized milk in the Republic of Paraguay. Rev Argent Microbiol 2011;43:33–6.

[110] Johler S, Tichaczek-Dischinger PS, Rau J, Sihto HM, Lehner A, Adam M, et al. Outbreak of staphylococcal food poisoning due to SEA-producing *Staphylococcus aureus*. Foodborne Pathog Dis 2013;10:777–81.

[111] Brizzio AA, Tedeschi FA, Zalazar FE. Descripción de un brote de intoxicación alimentaria estafilocócica ocurrido en Las Rosas, Provincia de Santa Fe, Argentina. Rev Argent Microbiol 2011;43:28–32 [in Spanish].

[112] Ostyn A, De Buyser ML, Guillier F, Groult J, Félix B, Salah S, et al. First evidence of a food poisoning outbreak due to staphylococcal enterotoxin type E, France, 2009. Euro Surveill 2010;15. 19528.

[113] Kitamoto M, Kito K, Niimi Y, Shoda S, Takamura A, Hiramatsu T, et al. Food poisoning by *Staphylococcus aureus* at a university festival. Jpn J Infect Dis 2009;62:242–3.

[114] Solano R, Lafuente S, Sabate S, Tortajada C, Garcia de Olalla P, Hernando AV, et al. Enterotoxin production by *Staphylococcus aureus*: an outbreak at a Barcelona sports club in July 2011. Food Control 2012;33:114–8.

[115] Gallina S, Bianchi DM, Bellio A, Nogarol C, Macori G, Zaccaria T, et al. Staphylococcal poisoning food borne outbreak: epidemiological investigation and strain genotyping. J Food Prot 2013;76:2093–8.

[116] Coldea IL, Zota L, Dragomirescu CC, Lixandru BE, Dragulescu EC, Sorokin M, et al. *Staphylococcus aureus* harbouring egc cluster coding for non-classical enterotoxins, involved in a food poisoning outbreak, Romania, 2012/*Staphylococcus aureus* purtător de gene codante pentru enterotoxine non-clasice (cluster egc), implicat într-un focar de toxiinfecţie alimentară, România, 2012. Rev Romana Med Lab 2015;23:285–94.

[117] Bianchi DM, Gallina S, Macori G, Bassi P, Merlo P, Vencia W, et al. Enterotoxigenic strain of *Staphylococcus aureus* causing food-borne outbreak in a private context. Ital J Food Saf 2013;2(e32):113–6.

[118] Centers for Disease Control and Prevention (CDC). Outbreak of staphylococcal food poisoning from a military unit lunch party—United States, July 2012. Morb Mortal Wkly Rep 2013;62:1026–8.

[119] Pillsbury A, Chiew M, Bates J, Sheppeard V. An outbreak of staphylococcal food poisoning in a commercially catered buffet. Commun Dis Intell Q Rep 2013;37:E144–8.

[120] Fetsch A, Contzen M, Hartelt K, Kleiser A, Maassen S, Rau J, et al. *Staphylococcus aureus* food-poisoning outbreak associated with the consumption of ice-cream. Int J Food Microbiol 2014;187:1–6.

[121] Mossong J, Decruyenaere F, Moris G, Ragimbeau C, Olinger CM, Johler S, et al. Investigation of a staphylococcal food poisoning outbreak combining case-control, traditional typing and whole genome sequencing methods, Luxembourg, June 2014. Euro Surveill 2015;20. 30059.

[122] Johler S, Weder D, Bridy C, Huguenin MC, Robert L, Hummerjohann J, et al. Outbreak of staphylococcal food poisoning among children and staff at a Swiss boarding school due to soft cheese made from raw milk. J Dairy Sci 2015;98:2944–8.

[123] Centers for Disease Control and Prevention (CDC). Surveillance for food borne disease outbreaks. United States, 2014, Annual report, Atlanta, GA: US Department of Health and Human Services, CDC; 2016. Available from, https://www.cdc.gov/foodsafety/pdfs/foodborne-outbreaks-annual-report-2014-508.pdf.

[124] Centers for Disease Control and Prevention (CDC). Surveillance for food borne disease outbreaks. United States, 2015, Annual report, Atlanta, GA: US Department of Health and Human Services, CDC; 2017. Available from, https://www.cdc.gov/foodsafety/pdfs/2015FoodborneOutbreaks_508.pdf.

[125] Ivanova T, Gurova-Mehmedova E, Daskalov H. Case study of staphylococcal enterotoxin poisoning after consumption of ready-to-eat roasted chicken products Bulgarian J Agr Sci 3: 159-164.

[126] Food Poisoning Bulletin. *Staphylococcus aureus* sickened students at science Olympiad. Available from, https://foodpoisoningbulletin.com/2017/staphylococcus-aureus-sickened-students-at-science-olympiad/.

Chapter 7

The Staphylococcal Coagulases

Emilia Bonar*, Jacek Międzobrodzki[†], Benedykt Władyka*
*Department of Analytical Biochemistry, Faculty of Biochemistry, Biophysics and Biotechnology, Jagiellonian University, Krakow, Poland, [†]Department of Microbiology, Faculty of Biochemistry, Biophysics and Biotechnology, Jagiellonian University, Krakow, Poland

7.1 INTRODUCTION

Staphylococci are Gram-positive, commensal components of human and animal cutaneous and mucosal microflora. They are also leading causes of numerous skin, tissue, as well as systemic infections that are sometimes serious. The Janus-face nature of these bacteria implicates the need for a fast and precise species and strain recognition, as misidentification of clinical isolates might have serious consequences in terms of raised morbidity and mortality. Although currently considered to be artificial, classification of staphylococci based on their ability to clot rabbit plasma is the key step in the diagnostic algorithm both in clinical and veterinary microbiology laboratories [1–4]. Accordingly, staphylococci may be subdivided into two groups, namely coagulase-positive staphylococci or CoPS, comprising *Staphylococcus aureus* ssp. *aureus*, *Staphylococcus aureus* ssp. *anaerobius*, *Staphylococcus intermedius*, *Staphylococcus pseudintermedius*, *Staphylococcus delphini* (the last three species form the so called "*Staphylococcus intermedius* Group", or "SIG") *Staphylococcus schleiferi* ssp. *coagulans*, *Staphylococcus lutrae*, *Staphylococcus hyicus* (coagulase-variable), and *Staphylococcus agnetis* (coagulase-variable) [5–8] and coagulase-negative staphylococci (CoNS), among which are *Staphylococcus epidermidis*, *Staphylococcus saprophyticus*, and *Staphylococcus pasteuri* [9–11].

Historically, coagulase tests date back to 1903, when it was observed that inoculation of calcium-chelated plasma or blood with *S. aureus* resulted in rapid clotting [12]. Since then, this phenomenon has been studied with great interest, and the protein factors on which it relies, namely coagulase (Coa) and von Willebrand factor binding protein (vWbp), together with their respective genes, have been identified and characterized. Moreover, the molecular and structural basis behind the interaction with plasma proteins was elucidated. Finally, a series of in vivo studies showed that Coa and vWbp secretion is among the key virulence strategies that promote pathogenesis of staphylococcal abscesses, bacteraemia, and endocarditis, along with disease persistence [13].

Pet-to-Man Travelling Staphylococci: A World in Progress. https://doi.org/10.1016/B978-0-12-813547-1.00007-8
95

7.2 COAGULASE

Coagulation is a cascade of serine protease-driven reactions where thrombin is responsible for conversion of fibrinogen to fibrin, and plays a role as a defense mechanism against bacterial pathogens through immobilization of bacteria in a clot and their subsequent killing [14]. On the other hand, the phenomenon constitutes the target of bacterial immune evasive strategies.

The staphylococcal Coa, also referred to as staphylocoagulase, is an extracellular protein composed of a variable number of 600–700 amino acid residues, and encoded by the *coa* gene. Indeed, 10 serotypes of staphylocoagulase were identified, based on tests evaluating inhibition of the clotting activity by serotype-specific antibodies [15,16]. Later studies on the nucleotide sequence of *coa* explained the observed variations in the antigen composition. In fact, six regions can be distinguished in the Coa molecule; that is a signal sequence leading the coagulase to the extracellular milieu; D1 and D2 regions responsible for prothrombin binding; a central domain of unknown function; the region of fibrinogen binding repeats; and the very C-terminal region [17]. The first 33 amino acids in the N-terminal (including 26 amino acid of the signal peptide and the first 7 amino acids of the mature protein) and the last 5 amino acids in the C-terminal are exactly identical among the 10 serotypes. However D1 and D2 regions are quite divergent with amino acid average identity rates of 52.8% and 60.2%, respectively. The central region is instead more conserved (86.7% identity). The number (2–8) of tandem repeat units varies, but the 27-amino-acid sequences are relatively conserved, with average amino acid identities (among the 10 serotypes) accounting for 92.9%. Therefore, variation in the number of repeat units refers to the differences in molecular weight of Coa rather than its antigenicity [18]. The *coa* gene is located on the chromosome, between the lipase gene *geh* and the protein A gene *spa*. Interestingly, the flanking regions of *coa* are quite conserved, thus strongly suggesting that some *S. aureus* strains acquired *coa* by recombination, although the mechanism remains unknown.

The staphylocoagulase-associated clotting is quite different from the physiological coagulation pathway. Coa binds to prothrombin and its derivatives in a 1:1 stoichiometric ratio, to form a complex referred to as staphyloprothrombin [19,20]. During host infection, staphylocoagulase activates the central coagulation zymogen, that is, prothrombin. The Coa-prothrombin complex recognizes fibrinogen as a specific substrate, thereby converting it directly into fibrin [21]. Interestingly, the activation of protrombin is not caused by typical proteolitic cleavage; particularly, crystal structure of the active Coa-prothrombin complex revealed that Coa inserts its N-terminus into the cleft of the prethrombin 2 domain, thus inducing conformational changes that lead to formation of the functional active site [17,21].

The gene *coa* is selectively expressed during the exponential growth phase, and is then repressed by the regulatory system *agr*. Another regulatory locus, *sarA*, also influences the expression of exoproteins, including coagulase, and *sarA* mutants were shown to produce low levels of coagulase [22–26].

7.3 VON WILLEBRAND FACTOR-BINDING PROTEIN

In 2002, Bjerketorp and colleagues identified a staphylococcal protein able to bind the von Willebrand factor, a multifunctional glycoprotein mediating platelet adhesion [27]. Two years later, it became evident that the vWbp has yet to exhibit coagulase activity. vWbp is a secretory protein consisting of 482 amino acid residues in its mature form (508 aa with leader peptide). The open reading frame for vWbp has an alternative start codon (TGG, coding for leucine) and is located between the genes *clfA* and *emp*. *clfA* encodes clumping factor A, the main fibrinogen-binding adhesin in *S. aureus*, while *emp* encodes a cell surface, extracellular-matrix-protein-binding protein (Emp), showing multiple binding activities [28]. Similar to Coa, in vWbp, a removable signal sequence, and D1-D2 domains, sharing around 30% amino acid identities with Coa, could be identified. The remainder of vWbp is unique in sequence and includes a binding site for the von Willebrand factor. vWbp binds prothrombin and fibrinogen, the latter with a lower affinity than Coa [28–30]. Using recombinant variants of vWbp, it was elegantly shown that the first 250 amino acids were sufficient to induce coagulation, while the N-terminally truncated vWbp did not have any apparent coagulating activity [28]. Generally, vWbp activates prothrombin conformationally through a mechanism requiring the first two valine residues from N-terminus. However, in contrast to Coa, kinetic studies of prothrombin activation by vWbp demonstrated that the latter activates prothrombin via a substrate-dependent, hysteretic kinetic mechanism. vWbp binds weakly to prothrombin to form an inactive complex, which is activated through a slow conformational change by its specific physiological substrate, fibrinogen [30]. Thus, vWbp is recognized as a member of the bifunctional zymogen activator and adhesion protein (ZAAP) family for which Coa is the prototype [17,31]. vWbp is expressed early in the exponential phase, becomes more abundant during the exponential growth, and decreases in the stationary growth phase. Such a vWbp expression pattern supports its postulated role as a virulence factor, enabling staphylococci to escape the host defense system through formation of protective fibrinogen-platelet-bacteria, like in acute bacterial endocarditis vegetations [32].

7.4 DIFFERENCES IN COAGULATION TIME OF HUMAN AND ANIMAL PLASMAS

Soon after the discovery of coagulation phenomenon by Loeb in 1903 (the test is known today as the coagulase tube test), a slide test was developed [33]. Here, the clumps emerge on a slide immediately after mixing a bacterial suspension with plasma or serum. Later studies revealed that such clumps indicate fibrinogen precipitation around bacteria [34] (the biological context of this observation is discussed as follows). Nowadays, the slide and the tube tests are still in use [35]. However, the former is very quick to perform (with results within a minute) and is therefore the first step along the diagnostic algorithm of staphylococci, meaning that only slide test-negative isolates are further analyzed through the tube test. In this second case, clotting of plasma may occur in the first hour, but only isolates unable to clot plasma within 24 h are labeled as CoNS. Although coagulase tests are routinely performed on rabbit plasma, it is known that plasmas from different mammal species differ markedly in their test performance [36]. This observation may be understood after recognition that Coa and vWbp are responsible for the phenomenon, as well as after identification of their molecular target. Using recombinant variants of both proteins it was shown that coagulation time depends on their concentrations, leading a higher concentration to a shorter coagulation time. Again, even more importantly, strong tropism for plasmas from different species has been uncovered for each coagulating factor. Coa indiscriminately clots human and rabbit plasma within around 1 h, whereas vWbp efficiently coagulates both human and swine plasma within 1 min. Moreover, upon mixing with vWbp, goat and sheep plasmas coagulated in 10 min, equine plasma in 1 h, and bovine plasma in 6 h, while plasma from mice, rabbits, rats, and chickens did not coagulate within 18 h [28]. Interestingly, the addition of only 1% of human plasma to plasma from cows, horses, mice, and rabbits markedly reduced the vWbp-induced coagulation time to 10 min, while chicken plasma did not coagulate within 2 h even after human plasma addition. Indeed, further investigations later found that a different rate of prothrombin activation, both by vWbp and Coa, is responsible for differences in coagulation efficiency. Friedrich and colleagues, by using kinetics and structural studies, showed that a fully active Coa fragment, Coa-(1–325), binds to bovine and human prothrombin with similar affinity, but very poorly activates the bovine zymogen. The Coa-(1–325)-bovine prothrombin complex is around 5800-fold less active compared with its human counterpart. The crystal structure of the respective complexes confirmed that, in comparison with Coa-(1–325)-human prothormbin, the catalytic site in Coa-(1–325)-bovine prothorm-bin is incompletely formed. Thus, low catalytic activity of Coa-(1–325)-bovine prothrombin is mainly responsible for its low fibrinogen clotting activity [21].

In 1984, Devriese showed that most human *S. aureus* strains poorly clot ruminant plasmas, whereas the reverse is true for ruminant isolates, while ruminant strains do clot ruminant plasma; so this feature has been used to classify strains based on their adaptation to animal habitats [37]. Recently, it has been clarified that diverse coagulase activity is due to animal-specific alleles of the *vwb* gene coding for vWbp and that these *vwb* alleles are carried by highly mobile pathogenicity islands, named SaPIs. In contrast to chromosomal *vwb* as well as to *coa* genes, the SaPI-encoded vWbps possess a unique N-terminal region specific for activation of ruminant and equine prothrombin. Moreover, the SaPI-carried *vwb* genes (*vwbs*) are regulated differently from the chromosomal *vwb* genes (*vwbc*) of the same strains [38]. *vwbc* contain a conserved octanucleotide sequence within the promoter region, which is required for their expression [39]. This sequence is lacking in the promoter region of each of the *vwbs* genes, and its absence is required for *saeRS*-mediated regulation (the SaPI-carried *vwb* genes are substantially downregulated in the absence of the global regulator *saeRS*, whereas the chromosomal genes are unaffected) [38].

In conclusion, the diverse ability to clot plasma displayed by CoPS have their basis in molecular mechanisms of prothrombin activation by two chromosomally encoded factors, Coa and vWbp. Moreover, the SaPI-coded paralouges of the latter may play a relevant role in *S. aureus* host adaptation and pathogenicity.

7.5 ROLE OF COAGULASES IN STAPHYLOCOCCAL INFECTIONS

The ability to clot plasma provided the basis for classification of staphylococci into CoPS and CoNS species. Indeed, this subdivision also correlates with diverse virulence properties, and early studies emphasized contribution of coagulase(s) to pathogenesis of staphylococcal infections [40–42]. However, results of in vivo experiments are controversial. Deletion of *coa* does not affect *S. aureus* virulence in mouse models of subcutaneous infection or mastitis [43], or in a rat model of infective endocarditis [44,45]. In contrast, after intravenous *S. aureus* mice inoculation, a positive correlation was found between Coa titer and pulmonary bacterial load. Moreover, the number of viable bacteria recovered from a lung 7 days after infection was significantly higher for the wild-type strain than for the *coa* deletion mutant [46]. Nevertheless, the breakthrough in the field came with the discovery of vWbp as a second secreted factor that clotted blood [28]. Cheng and colleagues, using recombinant *S. aureus* strains, have shown that targeted deletion of either *coa* or *vwb* alone resulted in only a moderate decrease in virulence, whereas the *coa-vwb* double mutant displayed a huge reduction in the ability to cause abscesses or lethal sepsis in mice [29]. Thus, combined Coa and vWbp are required for abscess formation.

The initial step in staphylococcal infection pathogenicity is represented by bacterial invasion of tissues and/or the bloodstream via various routes, including surgical wounds and medical devices [47]. Then, bacteria disseminate rapidly, causing purulent lesions. Staphylococci grow as communities in the lesion center within pseudocapsules that protect the organism from immune system cells. Finally, abscesses grow in size and eventually rupture, enabling pathogens to enter blood vessels and disseminate to still uninfected sites [48]. Strikingly, histochemical staining has identified prothrombin and fibrin together with Coa within pseudocapsules and, at the periphery of abscess lesions, Coa, vWbp, prothrombin as well as fibrinogen/fibrin are co-localized. Therefore, Coa and vWbp are crucial contributors to the establishment of staphylococcal communities within pseudocapsules and subsequent abscess development. Moreover, as the establishment of abscess lesions can be inhibited through coagulase-specific antibodies, these virulence factors should be considered to design antistaphylo-coccal vaccines [29].

In summary, staphylococcal Coa and vWbp adopt the mammalian coagulation cascade as a pathogenic determinant. By nonenzymatic activation of prothrombin, these coagulases trigger the polymerization of fibrin. It is widely accepted that fibrin deposition and cross-linking of bacteria typically represent animal innate immune defenses that reduce the burden of infection. Paradoxically, *S. aureus* apply coagulases to induce plasma clotting, which in turn allows the bacteria to escape from the bactericidal activity of the host immune system. Thus, coagulases are among key virulence strategies promoting abscess pathogenesis, and infection persistence, as well as sepsis and endocarditis [13].

7.6 CONCLUSION

Staphylococcal coagulases (Coa, vWbp) are secretory proteins that cause blood clotting through the activation of prothrombin. Based on their ability to clot plasma staphylococci, they may be divided into coagulase-positive or coagulase-negative ones. The ability to coagulate the plasma is a first step in the diagnosis of these bacteria, as Coa and vWbp secretion is among key virulence strategies enabling staphylococci to overcome the host defense system. Nonenzymatic activation of prothrombin coagulases triggers the polymerization of fibrin, result-ing in cross-linking of bacteria in the clot, which in turn allows the bacteria to escape from the bactericidal activity of the host immune system. Development of additional vWbp increases coagulation kinetics of particular host plasma, indicating not only host-preference of staphylococcal strains, but also the im-portance of coagulases in virulence of these bacteria.

ACKNOWLEDGMENTS

This work was supported in part by the grants N303 813340 from the Polish Ministry of Science and Higher Education and UMO-2012/07/D/NZ2/04282 from National Science Centre (NCN, Poland) (to BW).

REFERENCES

[1] McAdow M, DeDent AC, Emolo C, et al. Coagulases as determinants of protective immune responses against *Staphylococcus aureus*. Infect Immun 2012;80:3389–98.

[2] Kasprowicz A, Heczko PB, Basta M. Identification of *Staphylococcus* strains obtained from various clinical specimens. Med Dosw Mikrobiol 1983;35:165–70.

[3] Bannerman T. *Staphylococcus, Micrococcus*, and other catalase–positive cocci that grow aero-bically. In: Murray PR, Baron EJ, Jorgensen JH, Pfaller MA, Yolken RH, editors. Manual of clinical microbiology. Washington, DC: ASM Press; 2003. p. 384–404.

[4] Luczak-Kadlubowska A, Krzyszton-Russjan J, Hryniewicz W. Characteristics of *Staphylo-coccus aureus* strains isolated in Poland in 1996 to 2004 that were deficient in species-specific proteins. J Clin Microbiol 2006;44:4018–24.

[5] Freney J, Kloos WE, Hajek V, et al. Recommended minimal standards for description of new staphylococcal species. Subcommittee on the taxonomy of staphylococci and streptococci of the International Committee on Systematic Bacteriology. Int J Syst Bacteriol 1999;49: 489–502.

[6] Devriese LA, Vancanneyt M, Baele M, et al. *Staphylococcus pseudintermedius* sp. nov., a coagulase-positive species from animals. Int J Syst Evol Microbiol 2005;55:1569–73.

[7] Blaiotta G, Fusco V, Ercolini D, et al. Diversity of *Staphylococcus* species strains based on partial kat (catalase) gene sequences and design of a PCR-restriction fragment length polymorphism assay for identification and differentiation of coagulase-positive species (*S. aureus, S. delphini, S. hyicus, S. intermedius, S. pseudintermedius,* and *S. schleiferi* subsp. *coagulans*). J Clin Microbiol 2010;48:192–201.

[8] Savini V, Passeri C, Mancini G, et al. Coagulase-positive staphylococci: my pet's two faces. Res Microbiol 2013;164:371–4.

[9] Gross H. Die Kulturfiltratstoffe der Staphylokokken und deren Beziehungen zueinander. Z Immunitats 1931;73:14–26.

[10] Fisher A. The plasma coagulating properties of staphylococci. Bull Johns Hopkins Hosp 1936;59:415–26.

[11] Cruikshank R. Staphylocoagulase. J Path Bact 1937;45:295–303.

[12] Loeb L. The influence of certain bacteria on the coagulation of the blood. J Med Res 1903;10:407–19.

[13] McAdow M, Missiakas DM, Schneewind O. *Staphylococcus aureus* secretes coagulase and von Willebrand factor binding protein to modify the coagulation cascade and establish host infections. J Innate Immun 2012;4:141–8.

[14] Loof TG, Morgelin M, Johansson L, et al. Coagulation, an ancestral serine protease cascade, exerts a novel function in early immune defense. Blood 2011;118:2589–98.

[15] Ushioda H, Terayama T, Sakai S, et al. In: Jeljaszewicz J, editor. Coagulase typing of *Staphylococcus aureus* and its application in routine work. Stuttgart: Gustav Fischer Verlag Staphyloco; 1981. p. 77–83.

[16] Kanemitsu K, Yamamoto H, Takemura H, et al. Relatedness between the coagulase gene 3'-end region and coagulase serotypes among *Staphylococcus aureus* strains. Microbiol Immunol 2001;45:23–7.

[17] Friedrich R, Panizzi P, Fuentes-Prior P, et al. Staphylocoagulase is a prototype for the mechanism of cofactor-induced zymogen activation. Nature 2003;425:535–9.

[18] Watanabe S, Ito T, Takeuchi F, et al. Structural comparison of ten serotypes of staphylocoagulases in *Staphylococcus aureus*. J Bacteriol 2005;187:3698–707.

[19] Morita T, Igarashi H, Iwanaga S. Staphylocoagulase. Methods Enzymol 1981;80:311–9.

[20] Hendrix H, Lindhout T, Mertens K, et al. Activation of human prothrombin by stoichiometric levels of staphylocoagulase. J Biol Chem 1983;258:3637–44.

[21] Panizzi P, Friedrich R, Fuentes-Prior P, et al. Fibrinogen substrate recognition by staphylocoagulase.(pro)thrombin complexes. J Biol Chem 2006;281:1179–87.

[22] Cheung AL, Koomey JM, Butler CA, et al. Regulation of exoprotein expression in *Staphylococcus aureus* by a locus (sar) distinct from agr. Proc Natl Acad Sci USA 1992;89:6462–6.

[23] Novick RP, Ross HF, Projan SJ, et al. Synthesis of staphylococcal virulence factors is controlled by a regulatory RNA molecule. EMBO J 1993;12:3967–75.

[24] Wojcik K. Regulation of virulence factors gene expression in *Staphylococcus aureus*. Post Mikrobiol 1997;24:77–85.

[25] Robinson DA, Monk AB, Cooper JE, et al. Evolutionary genetics of the accessory gene regulator (agr) locus in *Staphylococcus aureus*. J Bacteriol 2005;187:8312–21.

[26] Szymanek K, Mlynarczyk A, Mlynarczyk G. Regulatory systems of gene expression in *Staphylococcus aureus*. Post Mikrobiol 2009;48:7–22.

[27] Bjerketorp J, Nilsson M, Ljungh A, et al. A novel von Willebrand factor binding protein expressed by *Staphylococcus aureus*. Microbiology 2002;148:2037–44.

[28] Bjerketorp J, Jacobsson K, Frykberg L. The von Willebrand factor-binding protein (vWbp) of *Staphylococcus aureus* is a coagulase. FEMS Microbiol Lett 2004;234:309–14.

[29] Cheng AG, McAdow M, Kim HK, et al. Contribution of coagulases towards *Staphylococcus aureus* disease and protective immunity. PLoS Pathog 2010;6:e1001036.

[30] Kroh HK, Panizzi P, Bock PE. Von Willebrand factor-binding protein is a hysteretic conformational activator of prothrombin. Proc Natl Acad Sci USA 2009;106:7786–91.

[31] Panizzi P, Friedrich R, Fuentes-Prior P, et al. The staphylocoagulase family of zymogen activator and adhesion proteins. Cell Mol Life Sci 2004;61:2793–8.

[32] Mylonakis E, Calderwood SB. Infective endocarditis in adults. N Engl J Med 2001;345:1318–30.

[33] Much H. Über Versuche des Fibrinfermentes in Kulturen von *Staphylococcus aureus*. Biochem Z 1908;14:143.

[34] Birch-Hirschfeld D. Uber die Agglutination von Staphylokokken durch Bestandteile des Saugetierblutplasmas. Klin Wschr 1934;13:331–3.

[35] Goja A, Ahmed T, Saeed S, Dirar H. Isolation and identification of *Staphylococcus* sp. in fresh beef. Pakistan J Nutr 2013;12:114–20.

[36] Zajdel M, Wegrzynowicz Z, Jeljaszewicz J, Pulverer G. Mechanism of action of staphylocoagulase and clumping factor. Contrib Microbiol Immunol 1973;1:364–75.

[37] Devriese LA. A simplified system for biotyping *Staphylococcus aureus* strains isolated from animal species. J Appl Bacteriol 1984;56:215–20.

[38] Viana D, Blanco J, Tormo-Mas MA, et al. Adaptation of *Staphylococcus aureus* to ruminant and equine hosts involves SaPI-carried variants of von Willebrand factor-binding protein. Mol Microbiol 2010;77:1583–94.

[39] Harraghy N, Homerova D, Herrmann M, Kormanec J. Mapping the transcription start points of the *Staphylococcus aureus* eap, emp, and vwb promoters reveals a conserved octanucleotide sequence that is essential for expression of these genes. J Bacteriol 2008;190:447–51.

[40] Chapman GH, Berens C, Peters A, Curcio L. Coagulase and hemolysin tests as measures of the pathogenicity of Staphylococci. J Bacteriol 1934;28:343–63.

[41] Smith W, Hale JH, Smith MM. The role of coagulase in staphylococcal infections. Br J Exp Pathol 1947;28:57–67.

[42] Ekstedt RD, Yotis WW. Studies on staphylococci. II. Effect of coagulase on the virulence of coagulase negative strains. J Bacteriol 1960;80:496–500.

[43] Phonimdaeng P, O'Reilly M, Nowlan P, et al. The coagulase of *Staphylococcus aureus* 8325-4. Sequence analysis and virulence of site-specific coagulase-deficient mutants. Mol Microbiol 1990;4:393–404.

[44] Baddour LM. Virulence factors among gram-positive bacteria in experimental endocarditis. Infect Immun 1994;62:2143–8.

[45] Moreillon P, Entenza JM, Francioli P, et al. Role of *Staphylococcus aureus* coagulase and clumping factor in pathogenesis of experimental endocarditis. Infect Immun 1995;63:4738–43.

[46] Sawai T, Tomono K, Yanagihara K, et al. Role of coagulase in a murine model of hematogenous pulmonary infection induced by intravenous injection of *Staphylococcus aureus* enmeshed in agar beads. Infect Immun 1997;65:466–71.

[47] Lowy FD. *Staphylococcus aureus* infections. N Engl J Med 1998;339:520–32.

[48] Cheng AG, Kim HK, Burts ML, et al. Genetic requirements for *Staphylococcus aureus* abscess formation and persistence in host tissues. FASEB J 2009;23:3393–404.

Chapter 8

The Staphylococcal Hemolysins

Eugenio Pontieri

Department of Biotechnological and Applied Clinical Sciences, University of L'Aquila, L'Aquila, Italy

8.1 α-HEMOLYSIN (α-TOXIN, HLA)

Among the several cytolytic toxins secreted by *Staphylococcus aureus* and other coagulase-positive staphylococci (CoPS), the major compound is α-hemolysin (α-toxin, Hla), which has been largely studied and extensively reviewed over the past three decades [1–4] (Table 8.1). The toxin produces, on blood agar plates, a wide zone of complete hemolysis exerting toxic effects on a broad range of mammalian erythrocytes (rabbit, sheep, ox) [5]. In rabbits, it is observed that an LD_{50} (2 µg/kg body weight) is lethal, representing the lowest toxic dose tested in any animal species [6]. Moreover, the toxin has been recognized as the cause of skin necrosis, and lethal infection, as well as neurotoxic action both in humans and animals by exerting cytotoxic effects on several types of cells.

Alpha-hemolysin is encoded in *S. aureus* by the single copy chromosomal *hla* gene, conserved among *S. aureus* strains, and cloned in 1984 [7]. The *hla* locus encodes a 319 amino acid protein. The initial N-terminal 26 residues constitute a signal peptide. The mature protein is composed of 293 residues with a final molecular weight of approximately 33 kDa and pI of about 8.5. The mature protein is entirely constituted by β-sheets, and is considered the prototype for the class of β-barrel pore-forming cytotoxins [2,8,9].

Several global regulatory systems control the expression of the Hla [10]. The accessory gene regulator, the *agr* locus, is responsible for the control of Hla expression by a quorum-sensing system [11] together with a regulatory RNA molecule (the RNAIII) [12,13]. It increases the expression and secretion of Hla, while only 1% of total α-toxin remains cell-associated. Expression levels can also be modulated by both the regulatory systems *Sae* and *Sar* [12–17].

Alpha-toxin is hydrophilic; seven monomers aggregate, forming a cylindrical heptamer into the membrane of the target cell [18,19]. This oligomeric form is capable of lysing eukaryotic cells by creating 1–3 nm pores [20,21] and the ability to lyse erythrocytes is the defining characteristic. On the monomers, and consequently on the oligomers, three domains are well recognized: Cap, standing on

Pet-to-Man Travelling Staphylococci: A World in Progress. https://doi.org/10.1016/B978-0-12-813547-1.00008-X

TABLE 8.1 Different Properties Expressed by *Staphylococcus* Hemolysins

Hemolysin	Sensible Erythrocytes	Hot-Cold Phenomenon	Substrate of Action	Activity
Alpha-hemolysin	Rabbit, sheep, ox	–	ADAM10, lipids	Dermonecrotic, lethal, leukotoxic
Beta-hemolysin	Sheep, ox, man	+	Sphingomyelin	Slightly lethal
Gamma-hemolysin	Rabbit, human, sheep	–	?	Lethal
Delta-hemolysin	Human, rabbit, sheep, monkey, horse	–	Phosphatidylinositol	Slightly dermonecrotic, lethal, leukotoxic

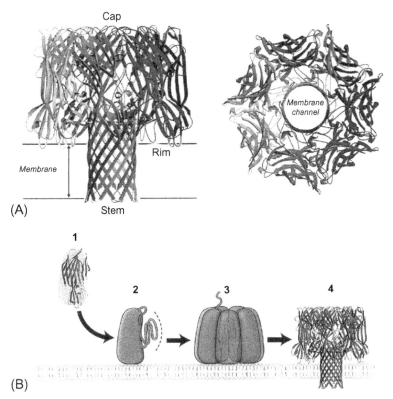

FIG. 8.1 Structure of the α-hemolysin, viewed perpendicular to membrane and along the seven-fold axis of symmetry (A); mechanism of alfa-hemolysin monomer assembly (B) [22].

the membrane; Rim, juxtaposed on the outer leaflet of host membrane; and Stem, that forms the membrane perforating β-barrel pore (Fig. 8.1) [22].

Two different ways have been described behind the binding and the incorporation of the monomers into the cell membranes [23]. At low concentration, a cell surface receptor binds the monomers, while at high concentration, the binding of the alpha-toxin to membrane lipids is nonspecific. Song et al. [24] by diffraction X-ray analysis found that the monomers penetrate in the cell membrane, diffusing in the bilayer, resulting in heptameric 1–2 nm pore formation, confirmed by Valeva et al. [19]. Conformational changes for pore activation have been suggested to be necessary in studies of N-terminal deletion mutants [25,26]. The toxin has several effects on the host, due principally to formation of the unregulated pores for ion transmission across the membranes of a variety of cell types [2]. The resulting pore enables a rapid efflux of K^+ and other small molecules and an influx of Na^+, Ca^+, and other small molecules with a molecular weight 1; the result is the lysis of the erythrocytes. Human erythrocytes are lysed to a lesser extent if compared with the rabbit erythrocytes. Nevertheless, the alpha-toxin exerts dermonecrotic, neurotoxic, and lethal activity in several animal-systems as well. Conversely, fibroblasts are able to repair the damage when exposed to low dose-alpha-toxin. Substantially, lethal disease, hemolysis, and dermonecrosis are the main alpha-toxin-related manifestations [27], although a complex interaction between toxins and a large range of host cells has been described (Table 8.2) [1,28,29]. The expression level and type of the effects have been induced to hypothesize the presence of a cellular membrane receptor for the α-toxin [23,30,31]. Several proposed models or experimental results, however, failed to explain the exquisite cell type and species specificity. Finally, Wilke and Bubek Wanderburg [32] reported the role of a disintegrin and metalloprotease 10 (ADAM10) as a candidate proteinaceous receptor for α-toxin. ADAM10 interacts with Hla and is required to initiate the sequence of events combining the toxin monomers into a cytolytic pore. The binding of Hla to eukaryotic cells needs the ADAM10 expression. ADAM10 is precipitated by Hla from the membrane of host cells, and ADAM10 results are necessary for toxin binding and oligomerization [30]. ADAM10 is needed in Hla-mediated cytotoxicity, mostly at low toxin concentrations [32], wherein the need for a high-affinity cellular receptor was predicted to be most relevant. A-toxin species specificity was demonstrated to correlate with ADAM10 expression on rabbit erythrocytes, while it is absent on human red cells surface. The observed interactions of α-toxin with both membrane lipids and a proteinaceous receptor probably indicate cooperation of the mentioned interactions in modulating toxin binding, assembly, and cytotoxicity [1].

8.2 β-HEMOLYSIN (SPHINGOMYELINASE C, HLB)

The β-hemolysin was identified by Glenny and Stevens [33]. Diversely from α-hemolysin, the β-toxin expresses highly hemolytic effects on sheep, ox, and human, but not rabbit, erythrocytes (Table 8.1), and neither dermonecrotic nor lethal effects have been observed in guinea pigs and mice, respectively. The toxin, on agar, produces a wide area of incomplete hemolysis [5]. The coding gene was sequenced by Projan et al. [34]. It is named *hlb*, is entirely chromosomal, and is located on a 4 kb *Cla*I digested DNA fragment. The protein consists of 330 amino acids with a 39,000 molecular weight, but, during secretion, a signal peptide represented by the first 34 amino acids is removed. The resulting peptide is the mature protein, with m.w. of 35,000 and isoelectric point above pH 9. The β-hemolysin is produced by a great number of *S. aureus* strains, in particular animal isolates, and by other CoPS such as *Staphylococcus pseudintermedius*, a veterinary agent of disease as well as an emerging zoonotic human pathogen. The protein shows sphingomyelinase activity, as lytic activity is limited to sphingomyelin and liso-phosphatidylcholine [35]. The different content of sphingomyelin found in different kinds of erythrocytes explains, therefore, the diverse hemolytic effects. Again, homology has been found in a

TABLE 8.2 Cell Types and Responses to Alpha-Toxin

Erythrocytes	Platelets	Monocytes	Neutrophils	T cells	Pneumocytes	Keratinocytes	Endothelial cells
Lysis (human)	Lysis	Lysis	Lysis	Lysis	Lysis	Lysis	Lysis
Lysis (mouse)	aggregation	apoptosis	chemokine secretion	apoptosis	apoptosis	↑ EGFR-dependent proliferation	apoptosis
	↓ Interaction with clotting factors/ECM	↓ Phagocytosis cytokine secretion	Inflammasome activation	Cytokine secretion	Focal adhesion disruption	Cytokine secretion	Neutrophil adhesion cytokine, NO release
Lysis (rabbit)	Activation	Inflammasome activation		Activation/ proliferation	Barrier disruption	p38 MAPK activation	Barrier disruption
						Barrier disruption	

α-Toxin concentration ▶

Modified from Berube BJ, Bubek Wandeburg J. *Staphylococcus aureus* α-toxin: nearly a century of intrigue. Toxins (Basel) 2013;5:1140–66. The effects depend on the cell type and the toxin concentration to which the cell is exposed.

200 residue-region of the sphingomyelinase of *Bacillus cereus* [36]. The toxin was defined "hot-cold" hemolysin after the observation that a second incubation below 10°C, after previous treatment at 37°C, enhances hemolysis, enabling the α-hemolytic band around colonies to get completely (β) hemolytic; as an explanation, it is supposed that low temperature permits phase separation and bilayer collapse in the perforated (by the toxin) membrane [37].

β-Hemolysin expression is supposed to be related to specific selective advantages in animal strains. *S. aureus* β-toxin-producing strains have enhanced growth ability on murine mammary glands if compared with isogenic organisms with the knockout *hlb* gene. Other CoPS that elaborate the lysin such as *Staphylococcus intermedius* and *S. pseudintermedius* are reported causes of infections in several animals (i.e., pyoderma in dogs and cats, canine otitis externa, and wound infections) [38–41]. Moreover, *S. intermedius* has been reported to cause human bacteremia, pneumonia, brain abscesses, wound infection (after canine bites), and food poisoning of dog origins. *S. pseudintermedius* accounts instead for only five disease episodes in man, including a skin ulcer and a cardioverter defibrillator infection, a bacteremia, an endocarditis, and a chronic rhinosinusitis [39,42–48]. In particular, the observation of *S. (pseud)intermedius* β-hemolysin-related "hot-cold," incomplete, α-hemolytic band (concomitantly to an absent or weak and delayed mannitol fermentation) has been suggested to be crucial for a presumptive recognition of a SIG species [49–52] (Fig. 8.2). As an exception, however, Awji et al. [53] reported a nonhemolytic *S. pseudintermedius* strain, although this paper provides unclear conclusions [51].

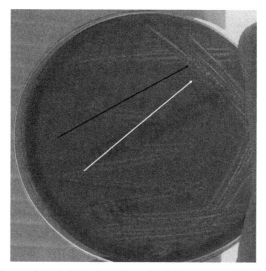

FIG. 8.2 Double zone hemolysis produced by a *Staphylococcus pseudintermedius* strain. The *black arrow* indicates the inner area of completely hemolytic band (β-hemolysis) while the *white arrow* indicates the external area of incompletely hemolytic band (α-hemolysis).

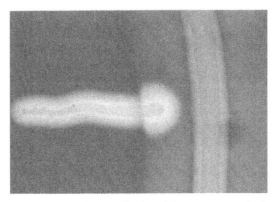

FIG. 8.3 CAMP test for presumptive identification of *Streptococcus agalactiae* (Group B streptococcus, GBS). A *Staphylococcus aureus* β-hemolysin-producing strain has been streaked onto a blood sheep agar plate. The presumptive GBS strain to be identified is inoculated perpendicularly, without touching. The arrow-shaped β-hemolytic zone shows test positivity.

Of interest, additionally, the CAMP test (after the names of Christie, Atkins, and Munch-Petersen who first observed and studied the CAMP reaction) relies on staphylococcal β-hemolysin release in the growth medium (Fig. 8.3) [54–56]. Traditionally, the assay is used to presumptively identify group B streptococci (*Streptococcus agalactiae*), *Listeria monocytogenes*, and *Rhodococcus equi*, and is performed by streaking the isolated to be tested perpendicularly to a β-hemolysin-producing *S. aureus* reference strain on sheep blood agar plate, without touching [57–59]. The test is positive if, after 24 h incubation at 37°C, a zone of candle flame- or arrow-shaped ("arrowhead") beta-hemolytic zone is observed. Alternatively, the test could be performed by dropping pure β-hemolysin or a toxin-containing supernatant on a layer of supposed group B streptococci and observing arrowhead on sheep blood agar after 30′ up to a few hours' incubation at 35°C [3]. The CAMP test could be performed inversely by using a sure CAMP positive bacteria as *S. agalactiae* for identifying a β-hemolysin producer bacteria. In this way, by inverted CAMP test, a *Staphylococcus delphini* was found to produce β-hemolysin [52], while *S. pseudintermedius*, constitutively an Hlb producer, could be useful for a CAMP-test [60].

8.3 γ-HEMOLYSIN (HLG)

γ-Hemolysin, first described by Smith and Price [61], is a member of a bi-component cytotoxin group comprehensive of PVL (Panton-Valentine Leukocydin), LukED and LukGH [62–65] isolated from various staphylococcal culture fluids. The toxin is composed by two nonassociated and water soluble peptides distinct by ion-exchange column elution in the S component (slow elution) and F component (fast elution) as described by Woodin [66]. γ-Hemolysin is produced by almost the totality of *S. aureus* strains. Gamma-hemolysin exerts cytolytic activity, especially on neutrophils (action at 10 ng/mL) and macrophages and on many types of mammalian erythrocytes [17,67,68] (Table 8.1). The hemolytic effects of gamma-hemolysin are not detected on blood agar plates because agar inhibits the toxic activity [67,69]. γ-Toxin is codified by the *hlg* locus, two S components, HlgA and HlgC, and one F component, HlgB. The three genes were found located in a cluster, in this order *hlg*A, *hlg*C, and *hlg*B [70] on a 4.5-kb *Sca*I-digested chromosome fragment. Hemolytic and leukotoxic activity have been showed by extracts from a clone containing this DNA fragment. *hlgA* is expressed separately, while *hlgC* and *hlgB* are transcribed on a single mRNA. The mature secreted proteins have molecular weights of 32,000 for HlgA, 34,000 for HlgB, and 32,500 for HlgC. Their pIs are estimated at 9.4, 9.1, and 9.0, respectively [3]. Previously, HlgB and HlgC have been termed respectively the γ_1 and the γ_2 components of gamma-hemolysin; together HlgB and HlgC were known to constitute the γ-hemolysin (F and S components respectively). Moreover, other two genes were identified in *S. aureus* encoding leukocidins, *luk*, isolated in *S. aureus* RIMD 310925 with *lukS* and *lukF* components [71] and *lukR*, in *S. aureus* P83, with *luk*F-R and *luk*S-R components [72]. These two leukocidins, diversely by PV-leukocidin disserted in Chapter 9 of this book, possess hemolytic activity; the first on rabbit, but not human, erythrocytes, the second on human erythrocytes [71,73]. In particular, the two products LukF-R and LukS-R are considered to correspond to HlgB and HlgC, respectively, because of the 97% of homology in their amino acid sequence [74]. In *S. aureus* P83, the sequence of LukS-R, designated as *hlg2*, was found to correspond to HlgA obtained in *S. aureus* 5R Smith. This Hlg2 (P83) showed a great minor hemolytic activity if combined with LukF, with the result depending on a single amino acid substitution ARG217LYS. The Hlg/Luk cluster has been found in almost the totality of *S. aureus* analyzed, but possesses a specific gene (S and F component), and depends on the specific staphylococcal strains analyzed. The LukF and HlgB (Hlg1) are identical. Differences in sequence have been found on the various S components, so that all these leukocidins could retain genetic variants of the gamma-hemolysin [17].

The three subunits of gamma-hemolysin assayed alone were found to lack hemolytic or leukotoxic activity; if added to erythrocytes as pairs, HlgA-HlgB had the highest hemolytic activity [75]. The lysis determined by gamma-hemolysin is mediated by the binding of each water-soluble component sequentially to the cell surface first as a monomer after oligomerizing in an hetero-octameric (four S components and four F components) form as transmembrane β-barrel pore across the cell membrane, determining increasing cation permeability and osmotic instability [68,76], leading host cell lysis. It is not clear if gamma-hemolysin is able to induce immunomodulatory activity, as the bi-component toxin PV-leukocidin, but [77] showed that the S subunit was more able in this respect, rather than the F subunit.

S. intermedius, a coagulase-positive veterinary pathogen, has been reported to produce Luk-I, a related bicomponent, and LukS-I and LukF-I toxin [78]. LukI exerts leukocytolytic activity on human PMNs. A slight hemolytic activity on rabbit erythrocytes has been also reported, and dermonecrotic activity, if tested on the rabbit skin model at a 100 ng dose. [79] showed a 4 residue sequence on the aminoacidic sequence of LukS-I at the root of the stem region, $K^{135}K^{136}I^{137}S^{138}$, identical to a phosphorylated segment of a protein phosphorylated by protein kinase A. A specific residue serine 138 was identified by mutational experiments to be crucial in the specific function of the LukS-I of the staphylococcal leukocidin, and the phosphorylation of this residue was an identified determinant for the leukocytolysis of the PMNs. Identical findings were previously showed for the staphylococcal leukocidin (Luk) at 4-residue $K^{243}R^{244}S^{245}T^{246}$, and identified as the minimum residue responsible for the LukS specific function of Luk [76,80,81] where the T^{246} was the phosphorylated residue.

8.4 δ-HEMOLYSIN

Delta-hemolysin (Delta-lysin, Hld), first reported by Williams and Harper [82], is an amphipathic small peptide composed of 26-amino-acids with α-helix structure able to damage the membrane of a variety of mammalian cells (Table 8.1). The toxin was detected forming a narrow zone of complete hemolysis on sheep blood agar plates around the producing colonies after adding of anti-α and anti-β hemolysins [5]. Toxin produced by human *S. aureus* strains has 9 amino acid differences compared with toxin produced by canine derived strains. However, both human derived and canine derived δ toxins have the same number of amino acid residues and molecular weights of approximately 3000 Da. Despite a few serological differences, charge and hemolytic properties are identical. It is assumed that 97% of *S. aureus* strains produce this protein, while *Staphylococcus hemolyticus* forms an immunologically related compound [3]. However, the toxin is produced by different *Staphylococcus* spp. as *S. intermedius*, among CoPs, and *S. epidermidis*, among CoNs. The *hld* gene, identified in 1989 [83] encodes for a 514-nucleotide transcript. The first 160 nucleotides contain an ORF of 45-codon, and the delta-toxin is located at the 3' end [11,84]. δ-Hemolysin is controlled by the accessory gene regulator *agr* and the highest expression of the protein is observed in the late-logarithmic phase (in broth cultures). The large transcript of *hld* gene seems to play an important role in regulation of the *agr* system, involved in the expression of cell-associated and virulence factors. No signal peptide is involved in the secretion of the peptide, and it has been suggested that it could be a good signal itself. Like *S. aureus*, 50%–70% of coagulase-negative staphylococci produce the toxin. δ-Hemolysin is able to lyse a broader range of mammalian erythrocytes more than other *S. aureus* hemolysins. Moreover, δ-hemolysin lyses are membranes of other mammalian cells and/or structures where cellular membranes are present, as described by Kreger et al. [85] and Freer and Birbeck [86]. Dermonecrotic activity and lethality in experimental animals (at high toxin concentrations) have been observed, but contamination with very small amounts of alpha-toxin, responsible for these activities, could be suspected in this case. Phospholipids inhibit the activity of delta-toxin [3]. As previously described [86–88], the action mechanism of delta-lysin consists of formation of α-helix expressing hydrophobic and hydrophilic domains on opposite sides. By acting as a surfactant, delta-toxin can disrupt the membrane structure, leading rapidly to cell lysis. Thelastam and Mollby [89] showed the toxin shares structure and mechanism similar to the bee venom mellitin and other similar proteins with additional antimicrobial activity, although the latter is absent in the delta-toxin. Monomers of the toxins co-aggregate on target cell membranes, thus forming ion channels on the lipid bilayer [85]. Also, it is supposed that six α-helicals co-aggregate to form a hexagonal cation-selective channel [90].

REFERENCES

[1] Berube BJ, Bubek Wandeburg J. *Staphylococcus aureus* α-toxin: nearly a century of intrigue. Toxins (Basel) 2013;5:1140–66.

[2] Bhakdi S, Tranum-Jensen J. Alpha-toxin of *Staphylococcus aureus*. Microbiol Rev 1991;55:733–51.

[3] Dinges MM, Orwin PM, Schlievert PM. Exotoxins of *Staphylococcus aureus*. Clin Microbiol Rev 2000;13:16–34.

[4] Freer JH, Arbuthnott JP. Toxins of *Staphylococcus aureus*. Pharmacol Ther 1983;19:55–106.

[5] Hébert GA, Hancock GA. Synergistic hemolysis exhibited by species of staphylococci. J Clin Microbiol 1985;22:409–15.

[6] Arbuthnott JP. Staphylococcal alpha-toxin. In: Montie TC, Kadis S, Ajl SI, editors. Microbial toxins, vol. III. New York, NY: Academic Press; 1970. p. 189–236.

[7] Gray GS, Kehoe M. Primary sequence of the a-toxin gene from *Staphylococcus aureus* Wood 46. Infect Immun 1984;46:615–8.

[8] Parker MW, Feil SC. Pore-forming protein toxins: from structure to function. Prog Biophys Mol Biol 2005;88:91–142.

[9] Prevost G, Mourey L, Colin D, et al. Alpha-helix and beta-barrel pore forming toxins (Leucocidins, alpha-, gamma-, and delta-cytolisins) of *Staphylococcus aureus*. In: Alouf JE, Freer JH, editors. Comprehensive sourcebook of bacterial toxins. London: Academic Press; 2005. p. 590–607.

[10] Recsei P, Kreiswirth B, O'Reilly M, Schlievert P, Gruss A, Novick RP. Regulation of exoprotein gene expression in *Staphylococcus aureus* by agar. Mol Gen Genet 1986;202:58–61.

[11] Peng HL, Novik RP, Kreiswirth B, Kornblum J, Schlievert P. Cloning, characterization and sequencing of an accessory gene regulator (*agr*) in *Staphylococcus aureus*. J Bacteriol 1988;170:98–103.

[12] Lina G, Jarraud S, Ji G, et al. Transmembrane topology and histidine protein kinase activity of *agr*C, the *agr* signal receptor in *Staphylococcus aureus*. Mol Microbiol 1998;2:655–62.

[13] Novick RP, Ross HF, Projan SJ, Kornblum J, Kreiswirth B, Moghazeh S. Synthesis of staphylococcal virulence factors is controlled by a regulatory RNA molecule. EMBO J 1993;12:3967–75.

[14] Koenig RL, Ray JL, Maleki SJ, Smeltzer MS, Hurlburt BK. *Staphylococcus aureus agr*A binding to the RNAIII-agr regulatory region. J Bacteriol 2004;186:7549–55.

[15] Lyon GJ, Wright JS, Muir TW, Novick RP. Key determinants of receptor activation in the *agr* autoinducing peptides of *Staphylococcus aureus*. Biochemistry 2002;41:10095–104.

[16] McNiven AC, Arbhunott JP. Cell-associated alpha-toxin from *Staphylococcus aureus*. J Med Microbiol 1972;5:123–7.

[17] Tomita T, Kamio Y. Molecular biology of the pore-forming cytolisins from *Staphylococcus aureus*, α- and γ-hemolysins and leukocidin. Biosci Biotechnol Biochem 1997;61:565–72.

[18] Gouau JE, Braha O, Hobaugh MR, et al. Subunit stoichiometry of staphylococcal alpha-hemolysin in crystals and on membranes: a heptameric transmembrane pore. Proc Natl Acad Sci U S A 1994;91:12828–31.

[19] Valeva A, Palmer M, Bhakdi S. Staphylococcal α-toxin: formation of the Heptameric pore is partially cooperative and proceeds through multiple intermediate stages. Biochemistry 1997;36:13298–304.

[20] Belmonte G, Cescatti L, Ferrari B, Nicolussi T, Ropele M, Menestrina G. Pore-formation by *Staphylococcus aureus* alpha-toxin in lipid bilayer: temperature and toxin concentration dependence. Eur Biophys J 1987;14:349–58.

[21] Freer JH, Arbuthnott JP, Bernheimer AW. Interaction of staphylococcal alpha-toxin with artificial and natural membranes. J Bacteriol 1968;95:1153–68.

[22] Montoya M, Gouaux E. β-Barrel membrane protein folding and structure viewed through the lens of α-hemolysin. Biochim Biophys Acta 2003;1609:19–27.

[23] Hildebrand A, Pohl M, Bhakdi S. *Staphylococcus aureus* alpha-toxin: dual mechanisms of binding to target cells. J Biol Chem 1991;266:17195–200.

[24] Song L, Hobaugh MR, Shustak C, et al. Structure of staphylococcal alpha-hemolysin, a heptameric transmembrane pore. Science 1996;274:1859–66.

[25] Gouaux E. α-Hemolysin from *Staphylococcus aureus*: an archetype of β-barrel, Channel-Forming Toxins. J Struct Biol 1998;121:110–22.

[26] Vandana S, Raje M, Krishnasastry V. The role of the amino terminus in the kinetics and assembly of α-hemolysin of *Staphylococcus aureus*. J Biol Chem 1997;272:24858–63.

[27] Burnet FM. The exotoxins of *Staphylococcus pyogens aureus*. J Pathol Bacteriol 1929;32:717–34.

[28] Jonas D, Walev I, Berger T, Liebetrau M, Palmer M, Bhakdi S. Novel path to apoptosis: small transmembrane pores created by staphylococcal alpha-toxin in nT lymphocytes evoke internucleosomal DNA degradations. Infect Immun 1994;62:1304–12.

[29] Suttorp N, Seeger W, Dewein E, Bhakdi S, Roka L. Staphylococcal α-toxin-induced-PGI2 in endothelial cells: role of calcium. Am J Phys 1985;248:C127–34.

[30] Cassidy P, Harshaman S. Studies on the binding of staphylococcal 125I-labeled alpha-toxin to rabbit erythrocytes. Biochemistry 1976;15:2348–55.

[31] Valeva A, Hellmann N, Walev I, et al. Evidence that clustered phosphocholine head groups serve ads sites for binding and assembly of an oligomeric protein pore. J Biol Chem 2006;281:26014–21.

[32] Wilke GA, Bubek Wanderburg J. Role of a disintegrin and metalloprotease 10 in *Staphylococcus aureus* alpha-hemolysin-mediated cellular injury. Proc Natl Acad Sci U S A 2010;107:13473–8.

[33] Glenny AT, Stevens NF. Staphylococcal toxins and antitoxins. J Pathol Bacteriol 1935;40:201–10.

[34] Projan SJ, Kornblum J, Kreiswirth B, Moghazeh SL, Eisner W, Novick RP. Nucletide sequence: the β-hemolysin gene of *Staphylococcus aureus*. Nucleic Acids Res 1989;17:3305.

[35] Doery HM, Magnuson BJ, Galasekharam J, Pearson JE. The properties of phospholipase enzymes in staphylococcal toxins. J Gen Microbiol 1965;40:283–96.

[36] Yamada AN, Tsukagoshi N, Udaka S, et al. Nucleotide sequence and expression in *Escherichia coli* of the gene coding for sphingomielinase of *Bacillus cereus*. Eur J Biochem 1988;175:213–20.

[37] Low DKR, Freer JH. The purification of β-lysin (sphingomyelinase C) from *Staphylococcus aureus*. FEMS Microbiol Lett 1977;2:139–43.

[38] Bond R, Loeffler A. What's happened to *Staphylococcus intermedius*? Taxonomic revision and emergence of multi-drug resistance. J Small Anim Pract 2012;53:147–54.

[39] Miedzobrodzki J, Kasptowicz A, Bialecka A, et al. The first case of a *Staphylococcus pseudintermedius* infection after joint prosthesis implantation in a dog. Pol J Microbiol 2010;59:133–5.

[40] Vanderhaeghen W, Vandendriessche S, Crombé S, et al. Species and staphylococcal cassette chromosome *mec* (SCC*mec*) diversity among methicillin-resistent non-*Staphylococcus aureus* staphylococci isolated from pigs. Vet Microbiol 2012;158:123–8.

[41] van Duijkeren E, Catry B, Greko C, et al. Review on methicillin-resistant *Staphylococcus pseudintermedius*. J Antimicrob Chemother 2011;66:2705–14.

[42] Casanova C, Iselin L, von Steiger N, Droz S, Sendi P. *Staphylococcus hyicus* bacteremia in a farmer. J Clin Microbiol 2011;49:4377–8.

[43] Chuang CY, Yang YL, Hsueh PR, Lee PI. Catheter-related bacteremia caused by *Staphylococcus pseudintermedius* refractory to antibiotic-lock therapy in a hemophilic child with dog exposure. J Clin Microbiol 2010;48:1497–8.

[44] O'Sullivan T, Friendship R, Blackwell T, et al. Microbiological identification and analysis of swine tonsils collected from carcasses at slaughter. Can J Vet Res 2011;75:106–11.

[45] Riegel P, Jesel-Morel L, Laventie B, Boisset S, Vandenesch F, Prevost G. Coagulase-positive *Staphylococcus pseudintermedius* from animals causing human endocarditis. Int J Med Microbiol 2011;301:237–9.

[46] Savini V, Barbarini D, Polakowska K, et al. Methicillin-resistant *Staphylococcus pseudintermedius* infection in a bone marrow transplant recipient. J Clin Microbiol 2013;51:1636–8.

[47] Stegman R, Burnens A, Maranta CA, Perreten V. Human infection associated with methicillin-resistant *Staphylococcus pseudintermedius* ST71. J Antimicrob Chemother 2010;65:2047–8.

[48] van Hoovels L, Vankeerberghen A, Boel A, Van Vaerenbergh K, De Beehouwer H. First case of *Staphylococcus pseudintermedius* infection in human. J Clin Microbiol 2006;44:4609–12.

[49] Devriese LA, Vancanneyt M, Baele M, et al. *Staphylococcus pseudintermedius* sp. Nov., a coagulase-positive species from animals. Int J Syst Evol Microbiol 2005;55:1569–73.

[50] Savini V, Passeri C, Mancini G, et al. Coagulse positive staphylococci: my pet's two faces. Res Microbiol 2013;164:371–4.

[51] Savini V, Carretto E, Polakowska K, et al. May *Staphylococcus pseudintermedius* be non-hemolytic? J Med Microbiol 2013;62(Pt8):1256–7. https://doi.org/10.1099/jmm.0.061952-0.

[52] Savini V, Koseka M, Marrollo R, et al. CAMP test detected *Staphylococcus delphini* ATCC 49172 β-haemolysin production. Pol J Microbiol 2013;62:465–6.

[53] Awji EG, Damte D, Lee SJ, Lee JS, Kim YH, Park SC. The in vitro activity of 15 antimicrobial agents against bacteria isolates from dogs. J Vet Med Sci 2012;74:1091–4.

[54] Christie R, Atkins NE, Munch-Petersen E. A note on a lytic phenomenon shown by group B streptococci. Aust J Exp Biol Med Sci 1944;22:197–200.

[55] Darling CL. Standardization and evaluation of the CAMP reaction for the prompt, presumptive identification of *Streptococcus agalactiae* (Lancefield group B) in clinical material. J Clin Microbiol 1975;1:171–4.

[56] Munch-Petersen E, Christie R, Simmons RT. Further notes on a lytic phenomenon shown by group B streptococci. Aust J Exp Biol Med Sci 1945;23:193–5.

[57] Ramsey KJ, Carter EC, McKee ML, Beek BJ. Reclassification of the Listeria-CAMP test strain ATCC 49444 *Staphylococcus aureus* as *Staphylococcus pseuintermedius*. J Food Prot 2010;73:1525–8.

[58] Savini V, Salutari P, Sborgia M, et al. Brief tale of a bacteremia by *Rhodococcus equi* with concomitant lung mass: what came first, the chicken or the egg? Mediterr J Hematol Infect Dis 2011;3:e2011006.

[59] Varaldo PE, Vancannyet M, Baele M, et al. *Staphylococcus delphini* sp. nov. a coagulase positive species isolated from dolphins. Int J Syst Bacteriol 1988;38:436–9.

[60] Savini V, Paparella A, Serio A, et al. *Staphylococcus pseudintermedius* for CAMP-test. Int J Clin Exp Pathol 2014;7(4):1733–4.

[61] Smith ML, Price SA. Staphylococcus γ-hemolysin. J Pathol Bacteriol 1938;47:379–93.

[62] Gravet A, Colin DA, Keller D, et al. Characterization of a novel structural member, LukE-LukD, of the bi-component staphylococcal leucotoxins family. FEBS Lett 1998;436:202–8.

[63] Morinaga N, Kaihou Y, Noda M. Purification, cloning and characterization of veriant LukE-LukD with strong leukocidal activity of staphylococcal bi-component leukotoxin family. Microbiol Immunol 2003;47:81–90.

[64] Panton PN, Valentine FCO. *Staphylococcal* toxin. Lancet 1932;219:506–8.

[65] Ventura CL, Malachowa N, Hammer CH, et al. Identification of a novel *Staphylococcus aureus* two-compèonent leukotoxin using cell surface proteomics. PLoS ONE 2010;5:e11634. https://doi.org/10.1371/journal.pone.0011634.

[66] Woodin AM. Fractionation of the two component of leukocidin from *Staphylococcus aureus*. Biochem J 1959;73:225–37.

[67] Prevost G, Couppie P, Prevost S, et al. Epidemiological data on *Staphylococcus aureus* strains producing synergohymenotropic toxins. J Med Microbiol 1995;42:237–45.

[68] Vandenesch F, Lina G, Henry T. *Staphylococcus aureus* hemolysins, bi-component leukocidins, and cytolytic peptides: a redundant arsenal of membrane-damaging virulence factors? Front Cell Infect Microbiol 2012;2:1–15.

[69] Prevost G, Cribier B, Couppie P, et al. Panton-Valentine leukocidin and gamma hemolysin from *Staphylococcus aureus* ATCC 49775 are encoded by distinct genetic loci and have different biological activities. Infect Immun 1995;63:4121–9.

[70] Cooney J, Kienle Z, Foster TJ, O'Toole PW. The gamma-hemolysin locus of *Staphylococcus aureus* comprises three linked genes, two of which are identical to the genes for the F and S components of leukocidin. Infect Immun 1993;61:768–71.

[71] Rahman A, Izaki K, Kato I, Kamio Y. Nucleotide sequence of leukocidin S-component gene (lukS) from methicillin resistant *Staphylococcus aureus*. Biochem Biophys Res Commun 1991;181:138–44.

[72] Supersac G, Prevost G, Piemont Y. Sequencing of leukocidin R from *Staphylococcus aureus* P83 suggests that staphylococcal leucocidins and gamma-hemolysin are members of a single, two-component family of toxins. Infect Immun 1993;61:580–7.

[73] Rahman A, Izaki K, Kato I, Kamio Y. Nucleotide sequence of leukocidin F-component gene (lukF) from methicillin resistant *Staphylococcus aureus*. Biochem Biophys Res Commun 1992;184:640–6.

[74] Rahman A, Izaki K, Kamio Y. Gamma-hemolysin genes in the same family with *lukF* and *lukS* genes in methicillin resistant*Staphylococcus aureus*. Biosci Biotechnol Biochem 1993;57:1234–6.

[75] Konig B, Prevost G, Konig W. Compositioin of staphylococcal bi-component toxins determines pathophysiological reactions. J Med Microbiol 1997;46:479–85.

[76] Kaneko J, Kamio Y. Bacterial two-component and hetero-heptameric pore-forming cytolytic toxins: structures, pore-forming mechanism, and organization of the genes. Biosci Biotechnol Biochem 2004;68:981–1003.

[77] Siqueira JA, Speeg-Schatz C, Frehas FIS, et al. Channel-forming leucotoxins from *Staphylococcus aureus* cause severe inflammatory reactions in a rabbit eye model. J Med Microbiol 1997;46:486–94.

[78] Prevost G, Boukhan T, Piemont Y, Monteil H. Characterization of a synergohymenotropic toxin produced by *Staphylococcus intermedius*. FEBS Lett 1995;376:135–40.

[79] Nishiyama A, Guerra MARV, Sugarawa N, Yokota K, Kaneko J, Kamio J. Identification of Serine138 residue segment K135K136I137S138 of LukS-I component of *Staphylococcus intermedius* leucocidin crucial for the LukS-I-specific function of *Staphylococcus leukocidin*. Biosci Biotechnol Biochem 2002;66:328–35.

[80] Nariya H, Nishiyama A, Kamio Y. Identification of the minimum segment in which the threonine246 residue is a potential phosphorylated site by protein kinase A foe the LukS-specific function of staphylococcal leukocidin. FEBS Lett 1997;415:96–100.

[81] Nishiyama A, Nariya H, Kamio Y. Phosphorylation of LukS by protein kinase A is crucial for the LukS-specific function of the staphylococcal leukocidin on human polymorphonuclear leukocytes. Biosci Biotechnol Biochem 1998;62:1834–8.

[82] Williams REO, Harper GJ. Staphylococcal hemolysins on sheep-blood agar with evidence for a fourth hemolysin. J Pathol Bacteriol 1947;59:69–78.

[83] Janzon L, Lofdahl S, Arvidson S. Identification of nucleotide sequence of the delta-lysin gene, *hld*, adjacent to the accessory gene regulator (*agr*) of *Staphylococcus aureus*. Mol Gen Genet 1989;219:480–5.

[84] Arvidson SL, Janzon L, Lofdahl S. The role of δ-hemolysin gene (*hld*) in the *agr*-dependent regulation of exoprotein synthesis in *Staphylococcus aureus*. In: Novick RP, editor. Molecular biology of the staphylococci. New York, NY: VCH Publisher Inc.; 1990. p. 419–31.

[85] Kreger AS, Kim KS, Zoboretsky F, Bernheim AW. Purification and properties of staphylococcal delta hemolysin. Infect Immun 1971;3:449–65.

[86] Freer JH, Birbeck TH. Possible conformation of delta-lysin, a membrane-damaging peptide of *Staphylococcus aureus*. J Theor Biol 1982;94:535–40.

[87] Lee KH, Fitton JE, Wuthrich K. Nuclear magnetic resonance investigation of the conformation of δ-hemolysin bound to dodecylphosphocholine mecelles. Biochim Biophys Acta 1987;911:144–53.

[88] Verdon J, Girardin N, Lacombe C, Berjeaud JM, Hechard Y. Delta-hemolysin, an update on a membrane-interacting peptide. Peptides 2009;30:817–23.

[89] Thelastam M, Mollby R. Determination of toxin induced leakage of different size nucleotides through the plasma membrane of human diploid fibroblast. Infect Immun 1975;11:640–8.

[90] Mellor IR, Thomas DH, Samson SP. Properties of ion channels formed by *Staphylococcus aureus* δ-toxin. Biochim Biophys Acta 1988;942:280–94.

Chapter 9

The Staphylococcal Panton-Valentine Leukocidin (PVL)

Paweł Nawrotek, Jolanta Karakulska, Karol Fijałkowski
Department of Immunology, Microbiology and Physiological Chemistry, Faculty of Biotechnology and Animal Husbandry, West Pomeranian University of Technology, Szczecin, Poland

9.1 INTRODUCTION

Staphylococcus aureus produces a number of molecules that allow bacteria to "deceive" the host's innate immune responses. These virulence factors include pore-forming cytotoxins, which are capable of causing lysis of different mammalian cells by forming pores in their cytoplasmic membrane. Cytotoxins that cause destruction of leukocytes, including phagocytes (neutrophils, monocytes and macrophages) and/or lymphocytes, are called leukotoxins. Leukocytotoxic activity enables infecting bacteria to survive inside the host organism during the first stages of the infection. One of these leukotoxins is the Panton-Valentine leukocidin (PVL) [1].

The PVL was first described by van de Velde in 1894 as a "substance leukocide" or leukocidin, a toxin produced by *S. aureus*, which causes lysis of leukocytes [2]. The name Panton-Valentine leukocidin comes from Sir Philip Noel Panton and Francis Valentine who, in 1932, distinguished PVL from hemolysins and demonstrated a correlation between the presence of PVL and severe infections of the skin and soft tissues, especially the boil gregarious [2,3].

The PVL is a bicomponent cytotoxin that belongs to staphylococcal synergohymenotropic (SHT) toxin family. It is encoded by two contiguous and cotranscribed genes located on the staphylococcus genome and carried on a prophage. Products of these genes are two secreted proteins—LukS-PV and LukF-PV—which form pores across the membrane of the mammalian leukocytes, including neutrophils, monocytes, and macrophages. The formed pores cause efflux of the cell content, leading to lysis. In addition, it was shown that at low (sublytic) concentrations, the PVL toxin can stimulate neutrophils to enhance production of pro-inflammatory factors [4,5].

PVL is produced by the majority of community-associated methicillin-resistant *S. aureus* (CA-MRSA) strains from humans [6–8], and there are also

Pet-to-Man Travelling Staphylococci: A World in Progress. https://doi.org/10.1016/B978-0-12-813547-1.00009-1
117

single reports of PVL-positive MRSA strains from infections in animals [9]. Although the role of PVL in *S. aureus* virulence has gained great interest and has undergone intensive research, its pathogenic potential in animal model infections has often appeared to be contradictory [10]. The reason for such controversial results has long been unclear. Finally, Löffler et al. [11] have provided a definitive explanation. They found, in fact, that PVL caused—in vitro—a rapid activation and apoptosis of human and rabbit neutrophils, although this effect was not observed in mouse and monkey neutrophils. Therefore, these authors questioned the value of murine and primate (nonhuman) infection models to explain PVL activity.

9.2 PVL STRUCTURE

PVL is a two-component and hetero-oligomeric pore-forming toxin. It consists of two subunits, proteins LukF-PV (~34 kDa) and LukS-PV (~32 kDa) [8,12], secreted as water-soluble monomers [1]. The crystal structure of each of the monomeric forms has been described and determined to be 2.0 Å. Subunits LukF-PV and LukS-PV have similar oblong elliptical shapes with molecular dimensions of ~70 Å × ~35 Å × ~25 Å. Both are rich in β-sheet structures, and are characteristically defined by a β-sandwich and a rim domain [13].

The upper part of each monomer consists of a β-sandwich or cap domain, which is formed by two antiparallel β-sheets. The same part of the monomer also includes a prestem region, which contains a β-hairpin that is incorporated across the membrane during the process of pore formation. In addition, the cap domain contains the triangle region that functions as a transition between the cap and the stem domains in the pore form. The lower part of these two structures includes a smaller rim domain, which is responsible for binding the membrane surface and is characterized by hydrophobic residues that anchor the toxin into the hydrophobic interior of the lipid bilayer [1]. Both LukF-PV and LukS-PV are greatly different in the conformation of the rim regions, and this probably determines their functional diversity as well [13]. However, so far, the three-dimensional structure of the oligomeric PVL has not been determined [1].

9.3 PVL-ENCODING GENES AND THEIR EXPRESSION

The ability to produce the PVL toxin relies on the presence of a *luk*-PV operon containing the *lukF*-PV and *lukS*-PV genes. These genes are located in the same genomic region of different PVL prophages of *S. aureus*, between the lysis module and the attachment site (*attP*) within the lysogeny genes. At least eight PVL-encoding phages (ΦPVL, ΦSLT, ΦSa2mw, ΦSa2USA300, ΦSLT-USA300_TCH1516, Φ108PVL, Φtp310-1, and Φ2985PVL) belonging to the *Siphoviridae* family have been characterized since PVL's discovery [5,14]. These prophages are classified into three groups, based on the replication/transcription regions and the morphogenesis module. PVL phages belonging to groups 1 and 3 have isometric capsids, while those from group 2 are elongated. The presence of the same PVL-encoding genes in morphologically different phages indicates that these genetic elements can be easily exchanged, for example, during co-infection [5]. Most of the virulence factors spread within the genus *Staphylococcus* through specific bacteriophage transduction and, since 1961, it is known that toxin (including PVL) production by *S. aureus* is associated with the phenomenon of lysogenic conversion [4]. However, phage/bacterial specificity factors can probably limit transmission of PVL phages between different strains of *S. aureus* [15].

Most phages harboring *pvl* genes are integrated into the *S. aureus* genome at a 29 bp *attB*-core sequence in the C-terminal region of an open reading frame (ORF) [14]. The genes encoding PVL may also be located within staphylococcal pathogenicity islands (SaPIs), which can be horizontally transferred through transducing bacteriophages. Mobile SaPIs containing *lukF* and *lukS* genes together with other staphylococcal virulence determinants, such as toxic shock syndrome toxin-1 (TSST-1), can significantly contribute to developing hypervirulent *S. aureus* strains [4]. This indicates that PVL can also indirectly affect the virulence by inducing expression of other virulence factors [16].

Among CA-MRSA mobile toxin genes, *lukF*-PV and *lukS*-PV are considered to be the most frequently transferred, and are labeled as the CA-MRSA markers around the world [17]. Moreover, the number of reports demonstrating the presence of *pvl* genes in methicillin- sensitive coagulase-negative staphylococci (MSCoNS) from cows and ewes with subclinical mastitis is greatly limited [18].

PVL synthesis and secretion, similarly to other exoproteins, occurs in a late exponential growth phase and during the stationary phase. At the same time, the production of proteins necessary for cell growth is reduced, and synthesis of adhesins is simultaneously inhibited [19]. This indicates the existence of several global regulatory loci where a single regulatory agent controls the expression of the other target genes [20]. In regulation of *lukF*-PV and *lukS*-PV expression, the two-component and sensitive to environmental signals *agr* regulatory system is involved and it has been proved that the occurrence of PVL-producing *S. aureus* strains is associated with the *agr*-I and/or *agr*-III phenotypes. In addition, in the regulation of PVL-encoding genes expression, the transcription factor MgrA may also be involved [19].

It should be noted that certain antibiotics can affect the level of PVL expression by inducing the *pvl* promoter. It has been shown that β-lactams (e.g., imipenem) increase the PVL production via transcriptional activation and promote the toxin expression in the infected tissues. Essential modulators of β-lactam-related PVL synthesis are the *S. aureus* regulatory systems such as the transcription factors SarA and Rot [21].

It was also shown that β-lactams can trigger the *S. aureus* SOS repair system and thereby promote replication and transfer of the temperate phages. Such findings may suggest that PVL production is dependent on the induction of PVL-encoding phages in response to various external factors, including certain antibiotics. Such effects result in clinical consequences following the use of such antimicrobials to treat particularly severe *S. aureus* infections [4,21].

9.4 MODE OF ACTION

PVL is a cytotoxin target of polymorphonuclear neutrophils (PMNs) and monocytes/macrophages. However, its mechanism of action during *S. aureus* infection and pathogenetic contribution is still unclear, and even controversial [22]. The PVL subunits, LukF-PV and LukS-PV, are secreted in the form of water-soluble monomers, and eventually turn into transmembrane pore-forming proteins [20]. Activity is synergistic and requires a specific sequence of events that take place on the target cell membrane surface [23].

The binding site or receptor for PVL has long been unknown. Nevertheless, recent studies led to the discovery of the human complement receptors C5aR and C5L2 as PVL host targets. It has been shown that C5aR and C5L2 mediate both toxin binding and cytotoxicity, and it was also proved that expression and interspecies differences in C5aR determine cell and species specificity of PVL. Particularly, the C5aR binding LukS-PV component is a potent inhibitor of C5a-induced immune cell activation [22].

Pore formation is a multistage process (Fig. 9.1). Initially, LukS-PV, then LukF-PV, bind to the cell surface, and form heterodimers [11,13,24]. Subsequently, the heterodimers undergo further oligomerization to heterotetramers, which are characterized by alternating LukF-PV and LukS-PV subunits. Then, heterotetramers form a disk-like, octameric prepore structure, which consists of LukS-PV and LukF-PV in a 1:1 stoichiometric ratio [13]. At this stage, PVL is the octamer in the prepore conformation and is not fully functional, nor can it transverse through the cell membrane. The prepore structure undergoes further conformational changes that eventually lead to the transmembrane pore formation, by extending the stem region of the two subunits across the plasma membrane [1,13].

From the molecular point of view, the pore is an octameric β-barrel molecular complex (with a diameter of 10–20 Å) that is located perpendicularly to the cell membrane and acts as an integral transmembrane protein, allowing the leakage of ions and small molecules (up to molecular weight around 1000 Da), leading to cell death [20].

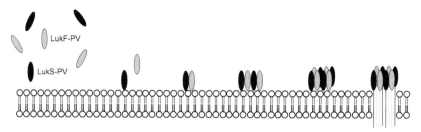

FIG. 9.1 Mechanism of PVL pore formation.

In vitro studies showed diverse and concentration-dependent activities of this toxin [25,26]. It was proved that sublytic concentrations of the PVL lead to the activation of the internal (mitochondrial) apoptosis pathway, while higher concentrations induce PMN necrosis. Concentration-dependent activity is associated with molecular properties of this toxin [27].

It has been shown that low concentrations of the PVL (~5 nM) trigger apoptosis of human neutrophils within 6 h [27]. At low concentrations, PVL binds to receptors on the cell membrane, thus forming small amounts of octameric pores [24]. At the mitochondrion level, instead, PVL might create pores in the outer membrane, and PVL-induced apoptosis has been shown to be associated with rapid disruption of mitochondrial homeostasis and activation of caspase-9 and caspase-3, suggesting that this process is preferentially mediated by the mitochondrial pathway [27].

In turn, higher PVL concentrations (~200 nM) cause lysis of neutrophils within 1 h [27]. At high concentrations, the toxin can nonspecifically adsorb to the lipid bilayer cell membrane, creating large, Ca^{2+}-permeable pores, or forming a great number of octameric pores, otherwise leading to the opening of Ca^{2+} channels [28].

After binding to the cell receptor associated with Ca^{2+} channels or directly to the Ca^{2+} channels in the cell membrane, the PVL subunits can also induce flow of divalent cations (Ca^{2+}, Mg^{2+}, Mn^{2+}, Zn^{2+}) by the Ca^{2+}-dependent channels; then, after forming pores, they enable the influx of monovalent ions (ethidium$^+$, Na^+, K^+). As a consequence, death of phagocytes occurs [27,29]. Furthermore, formation of large pores may lead to loss of ATP, which is essential for the initiation of most of the apoptotic processes [27,30].

It should be noted that PVL, with the exception of the ability to form pores in leukocytes membranes, is also a PMN priming agent [2,25,26]. Particularly, it was shown that PVL amounts below the concentration needed to create pores can stimulate neutrophils to enhanced production of reactive oxygen species and secrete pro-inflammatory molecules/chemotactic components such as interleukin-8 (IL-8), interleukin-6 (IL-6), and leukotriene B4 (LTB$_4$); also, they induce exocytosis of granules, then the release of myeloperoxidase, lysozyme and β-glucuronidase into the extracellular space, which can promote the host defense against infection. In addition, sublytic PVL concentrations induce the release of histamine from human basophils [20,26,31]. It is likely that PVL-mediated release of PMN proinflammatory molecules is caused by activation of signal transduction pathways, rather than cytolysis [26].

It should be emphasized that PVL-positive strains produce different amounts of PVL when comparing in vitro and in vivo growth [32]. PVL cytolytic properties are well-known from extensive work in vitro. However, there is insufficient evidence showing that the primary function of the PVL in vivo is leukocytes cytolysis [26]. It was proved, instead, that PVL in vivo is not sufficient to cause PMN lysis [32]. Hence, it was assumed that PVL sublytic concentrations can modify the organism's response to infections by strengthening the innate immune response and thereby intensifying the bacterial clearance [7]. Therefore, it can be summed up that sublytic concentrations act in vivo as a PMN-priming agent, modifying PMN gene expression and reinforcing phagocyte bactericidal capacity.

9.5 CLINICAL ASPECTS

Despite the relatively low percentage (2%–5%) of *S. aureus* isolates carrying the PVL-genes, there is strong epidemiological evidence for the association between PVL and the emergence of CA-MRSA infections in humans [11], although the pathogenic role of PVL in these cases remains unknown [1,2,33]. PVL-positive *S. aureus* strains are usually responsible for purulent skin diseases and subcutaneous tissue necrosis; also, they can cause severe infections such as septic arthritis, bacteremia, purpura fulminans, and community-acquired necrotizing pneumonia [1,8].

Clinical data showed a particular correlation between PVL and severe, even fatal, *S. aureus* necrotizing pneumonia cases, which are usually preceded by influenza or flu-like syndromes [33]. It is worth noting that the LukS-PV was detected in lung sections of patients with necrotizing pneumonia together with DNA fragmentation, suggesting that PVL induces apoptosis in vivo and is thereby directly involved in the disease pathophysiology [27].

The observation that PVL is associated with CA-MRSA major lineages suggests that this toxin may increase virulence of harboring strains and enhance their transmission [7,17,33]. However, it was demonstrated that the presence of PVL is not the major determinant *S. aureus* infection clinical course. Moreover, it was found that patients with PVL-positive infections were more susceptible to treatment, compared with those suffering from PVL-negative pathologies [34].

Of particular importance, especially from a clinical and epidemiological point of view, is the fact that not only human strains of *S. aureus* possess the *lukF*-PV and *lukS*-PV genes and are genetically determined to produce the PVL. There are several reports, in fact, concerning isolation of PVL-positive MRSA from severe infections in domestic animals including dogs, a cat, a rabbit and a parrot [9].

Furthermore, *pvl* genes were also detected in some MSCoNS, including *Staphylococcus haemolyticus*, *Staphylococcus simulans*, and *Staphylococcus warneri*, isolated from cows and ewes with subclinical mastitis [18]. Conversely, poor data are available about the isolation of PVL-positive *S. aureus* strains from mastitis in ruminants and, in this context, the importance of LukF-PV alone, among the toxin subunits, is emphasized. The latter, along with the leukocidin LukM subunit, are structural elements of the LukM/LukF'-PV toxin, which was shown to be the most active *S. aureus* leukotoxin against bovine neutrophils [35]. Therefore, it is considered that the LukM/LukF'-PV may play an important role in the pathogenesis of mastitis of cows, sheep, and goats [23,35,36].

To summarize, it should be highlighted that, despite numerous epidemiological studies, it is still impossible to define a direct role of PVL in the pathogenesis of *S. aureus* infections, or to determine whether the toxin directly contributes to the widespread distribution of MRSA clones [2,12]. Due to the possibility of a two-way transmission of MRSA between humans and animals, further studies on the clinical aspects, risk factors, and epidemiology of PVL-positive MRSA diseases represent an important issue both in human and veterinary medicine [9].

9.6 CONCLUSIONS

PVL shows activity against immune system cells, including neutrophils, monocytes, and macrophages, and affects human and rabbit, but not mouse and monkey phagocytes. The PVL toxin-producing *S. aureus* strains are often isolated from CA-MRSA infections in humans; whereas little is still known regarding isolation of PVL-producing strains from animals, and only a few published reports deal with the isolation of PVL-positive MRSA from veterinary sources. In vitro studies demonstrated a concentration-dependent toxic effect of PVL leading to either apoptosis or necrosis in neutrophils exposed to the different toxin amounts. It was also shown that the PVL effect is time-dependent as sublytic concentrations of the toxin trigger apoptosis within 6 h of treatment, whereas higher concentrations cause cell lysis within 1 h. However, it was also indicated that levels of PVL in vivo are insufficient to cause host cell lysis. On the other hand, it was proved that sublytic concentrations prime neutrophils to enhance production of reactive oxygen species (ROS), thus causing proinflammatory molecules secretion and granule exocytosis. Therefore, it is accepted, thus far, that PVL, in vivo, is primarily a factor that modulates the immune system rather than a toxin with cytolytic properties.

REFERENCES

[1] Aman MJ, Karauzum H, Bowden MG, Nguyen TL. Structural model of the pre-pore ring-like structure of Panton-Valentine leukocidin: providing dimensionality to biophysical and mutational data. J Biomol Struct Dyn 2010;28:1–12.

[2] Lo WT, Wang CC. Panton-Valentine leukocidin in the pathogenesis of community-associated methicillin-resistant *Staphylococcus aureus* infection. Pediatr Neonatol 2011;52:59–65.

[3] Panton PN, Valentine FCO. Staphylococcal toxin. Lancet 1932;219:506–8.

[4] Helbin WM, Polakowska K, Międzobrodzki J. Phage-related virulence factors of *Staphylococcus aureus*. Post Mikrobiol 2012;51:291–8.

[5] El Haddad L, Moineau S. Characterization of a novel Panton-Valentine leukocidin-encoding staphylococcal phage and its natural PVL-lacking variant. Appl Environ Microbiol 2013;79:2828–32.

[6] DeLeo FR, Otto M, Kreiswirth BN, Chambers HF. Community-associated methicillin-resistant *Staphylococcus aureus*. Lancet 2010;375:1557–68.

[7] Graves SF, Kobayashi SD, DeLeo FR. Community-associated methicillin-resistant *Staphylococcus aureus* immune evasion and virulence. J Mol Med 2010;88:109–14.

[8] Shallcross LJ, Fragaszy E, Johnson AM, Hayward AC. The role of the Panton-Valentine leucocidin toxin in staphylococcal disease: a systematic review and meta-analysis. Lancet Infect Dis 2013;13:43–54.

 [9] Rankin S, Roberts S, O'Shea K, et al. Panton-Valentine leukocidin (PVL) toxin positive MRSA strains isolated from companion animals. Vet Microbiol 2005;108:145–8.
[10] Villaruz AE, Bubeck Wardenburg J, Khan BA, et al. A point mutation in the *agr* locus rather than expression of the Panton-Valentine leukocidin caused previously reported phenotypes in *Staphylococcus aureus* pneumonia and gene regulation. J Infect Dis 2009;200:724–34.
[11] Löffler B, Hussain M, Grundmeier M, et al. *Staphylococcus aureus* Panton-Valentine leukocidin is a very potent cytotoxic factor for human neutrophils. PLoS Pathog 2010;6:e1000715.
[12] Chambers HF. Community-associated MRSA-resistance and virulence coverage. N Engl J Med 2005;352:1485–7.
[13] Guillet V, Roblin P, Werner S, et al. Crystal structure of leucotoxin S component: new insight into the Staphylococcal beta-barrel pore-forming toxins. J Biol Chem 2004;279:41028–37.
[14] Wirtz C, Witte W, Wolz C, Goerke C. Insertion of host DNA into PVL-encoding phages of the *Staphylococcus aureus* lineage ST80 by intra-chromosomal recombination. Virology 2010;406:322–7.
[15] Boakes E, Kearns AM, Ganner M, et al. Distinct bacteriophages encoding Panton-Valentine leukocidin (PVL) among international methicillin-resistant *Staphylococcus aureus* clones harboring PVL. J Clin Microbiol 2011;49:684–92.
[16] Shariati L, Validi M, Hasheminia AM, et al. *Staphylococcus aureus* isolates carrying Panton-Valentine leucocidin genes: their frequency, antimicrobial patterns, and association with infectious disease in Shahrekord city, Southwest Iran, Jundishapur. J Microbiol 2016;9:e28291.
[17] Vandenesch F, Naimi T, Enright MC, et al. Community-acquired methicillin-resistant *Staphylococcus aureus* carrying Panton-Valentine leukocidin genes: worldwide emergence. Emerg Infect Dis 2003;9:978–84.
[18] Ünal N, Çinar OD. Detection of stapylococcal enterotoxin, methicillin-resistant and Panton-Valentine leukocidin genes in coagulase-negative staphylococci isolated from cows and ewes with subclinical mastitis. Trop Anim Health Prod 2012;44:369–75.
[19] Szymanek K, Młynarczyk A, Młynarczyk G. Regulatory systems of gene expression in *Staphylococcus aureus*. Post Mikrobiol 2009;48:7–22.
[20] Bien J, Sokolova O, Bozko P. Characterization of virulence factors of *Staphylococcus aureus*: novel function of known virulence factors that are implicated in activation of airway epithelial proinflammatory response. J Pathog 2011; https://doi.org/10.4061/2011/601905.
[21] Dumitrescu O, Choudhury P, Boisset S, et al. β-Lactams interfering with PBP1 induce Panton-Valentine leukocidin expression by triggering *sarA* and *rot* global regulators of *Staphylococcus aureus*. Antimicrob Agents Chemother 2011;55:3261–71.
[22] Spaan AN, Henry T, van Rooijen WJM, et al. The staphylococcal toxin Panton-Valentine leukocidin targets human C5a receptors. Cell Host Microbe 2013;13:584–94.
[23] Kaneko J, Kamio Y. Bacterial two-component and heteroheptameric pore-forming cytolytic toxins: Structures, pore-forming mechanism, and organization of the genes. Biosci Biotechnol Biochem 2004;68:981–1003.
[24] Gauduchon V, Werner S, Prevost G, et al. Flow cytometric determination of Panton-Valentine leucocidin S component binding. Infect Immun 2001;69:2390–5.
[25] Kobayashi SD, DeLeo FR. An update on community-associated MRSA virulence. Curr Opin Pharmacol 2009;9:545–51.
[26] Graves SF, Kobayashi SD, Braughton KR, et al. Sublytic aconcentrations of *Staphylococcus aureus* Panton-Valentine leukocidin alter human PMN gene expression and enhance bactericidal capacity. J Leukoc Biol 2012;92:361–74.

[27] Genestier AL, Michallet MC, Prevost G, et al. *Staphylococcus aureus* Panton-Valentine leukocidin directly targets mitochondria and induces Bax-independent apoptosis of human neutrophils. J Clin Invest 2005;115:3117–27.

[28] Baba Moussa L, Werner S, Colin DA, et al. Discoupling the Ca(2+)-activation from the pore-forming function of the bi-component Panton-Valentine leucocidin in human PMNs. FEBS Lett 1999;461:280–6.

[29] Staali L, Monteil H, Colin DA. The staphylococcal pore-forming leukotoxins open Ca^{2+} channels in the membrane of human polymorphonuclear neutrophils. J Membr Biol 1998;162:209–16.

[30] Leist M, Single B, Castoldi AF, et al. Intracellular adenosine triphosphate (ATP) concentration: a switch in the decision between apoptosis and necrosis. J Exp Med 1997;185:1481–6.

[31] Konig B, Prevost G, Piemont Y, Konig W. Effects of *Staphylococcus aureus* leukocidins on inflammatory mediator release from human granulocytes. J Infect Dis 1995;171:607–13.

[32] Badiou C, Dumitrescu O, George N, et al. Rapid detection of *Staphylococcus aureus* Panton-Valentine leukocidin in clinical specimens by enzyme-linked immunosorbent assay and immunochromatographic tests. J Clin Microbiol 2010;48:1384–90.

[33] Gillet Y, Issartel B, Vanhems P, et al. Association between *Staphylococcus aureus* strains carrying gene for Panton-Valentine leukocidin and highly lethal necrotising pneumonia in young immunocompetent patients. Lancet 2002;359:753–9.

[34] Bae IG, Tonthat GT, Stryjewski ME, et al. Presence of genes encoding the Panton-Valentine leukocidin exotoxin is not the primary determinant of outcome in patients with complicated skin and skin structure infections due to methicillin-resistant *Staphylococcus aureus*: results of a multinational trial. J Clin Microbiol 2009;47:3952–7.

[35] Barrio MB, Rainard P, Prévost G. LukM/LukF'-PV is the most active *Staphylococcus aureus* leukotoxin on bovine neutrophils. Microbes Infect 2006;8:2068–74.

[36] Rainard P, Corrales JC, Barrio MB, et al. Leucotoxic activities of *Staphylococcus aureus* strains isolated from cows, ewes, and goats with mastitis: importance of LukM/LukF'-PV leukotoxin. Clin Diagn Lab Immunol 2003;10:272–7.

Chapter 10

The Staphylococcal Exfoliative Toxins

Michał Bukowski*, Benedykt Władyka*, Adam Dubin*,
Grzegorz Dubin†,‡
*Department of Analytical Biochemistry, Faculty of Biochemistry, Biophysics and Biotechnology,
Jagiellonian University, Krakow, Poland †Malopolska Centre of Biotechnology, Jagiellonian
University, Krakow, Poland ‡Department of Microbiology, Faculty of Biochemistry, Biophysics and
Biotechnology, Jagiellonian University, Krakow, Poland

10.1 INTRODUCTION

Among the numerous factors that are directly involved in *Staphylococcus aureus* virulence (and are described in detail in other chapters of this book), exfoliative toxins (ETs) represent one of the best known, and the mechanism on which their action relies is well understood nowadays. ETs are small (~30 kDa) chymotrypsin-like serine proteases produced and secreted by the bacteria into the external environment in the postexponential phase of growth, leading to systemic effects in the human host that may even be fatal. These molecules are typical toxins, as their effect can be produced, in vitro, by the purified toxin, in the absence of bacterial cells. Although acting at a systemic level, ETs only target a single host protein, that is the cell adhesion molecule desmoglein-1 (Dsg-1), located in the epidermal layer of the skin. Precise hydrolysis of only a single peptide bond within Dsg-1 is responsible, alone, for the observed toxic effect, represented by skin blistering ("exfoliation" or "acantholysis"). The ETs-associated toxicoses, known as staphylococcal scalded skin syndrome (SSSS), primarily affects infants, and results in severe dehydration and secondary infections which, although infrequent, are of significant concern in neonatal care.

10.2 MOLECULAR MECHANISM OF ACTION

ETs are serine proteases sharing significant structural homology with chymotrypsin [1–3]. Skin blistering observed in SSSS is perfectly explained by ET-catalyzed hydrolysis of Dsg-1, and the specificity of this process, which does not affect Dsg-3 or other desmogleins. Dsg-1 and its isoform Dsg-3 are calcium ions stabilized [4] desmosomal cadherins present in the most external

Pet-to-Man Travelling Staphylococci: A World in Progress. https://doi.org/10.1016/B978-0-12-813547-1.00010-8
127

layer of skin, the epidermis [4,5]. Dsg-1 is responsible for the integrity of the desmosomes (intracellular junctions) in the *stratum granulosum* of the epidermal layer. Dsg-3 is predominantly present, instead, in deeper epidermis strata (*stratum spinosum* and *stratum basale*) but absent in the *stratum granulosum*. Moreover, Dsg-1 is not present in mucosal epithelium. This perfectly explains the observed pattern of the SSSS-related epidermal damage. Skin blistering occurs in *stratum granulosum*, where Dsg-1 is the toxin's target (it is hydrolyzed only at a single peptide bond, that is between EC3 and EC4 extracellular domains) [6–9]. In other strata, Dsg-1 is affected as well, but the effect is balanced by the presence of unaffected desmogleins other than Dsg-1. Mucosal membranes are not involved in ETs spectrum of activity as they contain Dsg-3 [10]. High proteolysis specificity is clearly structure-guided, because hydrolysis is not observed when Dsg-1 is disrupted by calcium ion depletion [11].

The fact that toxins are secreted far from their site of action and spread systemically, presumably through the bloodstream [12] and, seemingly, without affecting any proteins other than the target desmoglein (even closely related desmogleins are not hydrolyzed), indicates a tight control of their proteolytic activity. This is thought to be related to unique characteristics of ETs uncovered by structural studies as described herein. Like all other active serine proteases of the chymotrypsin family, ETs are characterized by a canonical catalytic triad, a key-feature for catalyzing peptide bond hydrolysis. However, the oxyanion hole, a further crucial structural element is not preformed in these proteases. It is speculated that only target substrate binding leads to oxyanion rearrangement, which releases proteolytic activity (this has, however, not been experimentally demonstrated as yet) [3,13,14]. Further, ETs contain an N-terminal extension found only in proteolytically inactive precursors of other homologous proteases [15]. Homologous zymogens are activated by proteolytic removal of the N-terminal extension, but the toxins remain active without processing. Although relying on a currently unknown mechanism, this observation may add to the specificity of these proteases.

It must be noted here that ETs have also been reported to exhibit superantigen activity. Superantigens bridge T-cell receptors and MHC-II molecules of antigen-presenting cells in an antigen-independent manner. This leads to antigen-unrelated T-cells activation, polyclonal proliferation, and production of a high amount of cytokines [16]. Such a process allows the pathogen to blind the immune system. The findings concerning superantigen properties of ETs are, however, highly contradictory [17–21], and they were strongly influenced, presumably, by studies on other, well defined staphylococcal superantigens, such as the toxic shock syndrome toxin (TSST) [22–24]. Because superantigens are very efficient, even minute impurities in ET preparations could influence results. The issue is, however, much more complicated, and interested readers may find further elucidations in the original, published works on the topic [17–21,25,26].

10.3 DISEASE

SSSS was first described by Baron Gottfried Ritter von Rittershain, in 1878 [27], and is therefore also known as 'Ritter's diseases' Affected patients suffer from erythematous rush, along with the formation of fluid-filled blisters that gradually increase in size. Blisters are fragile rupture easily, which results in epidermal lesions [28,29]. In general, signs and symptoms resemble those of a skin burn. The localized form of SSSS, limited to the infection site, is known as *bullous impetigo*.

The association between S. *aureus* infection and SSSS was not immediately evident. Only in the 1960s Lyell and colleagues demonstrated that strains of this species isolated from SSSS patients could induce blistering [30]. In 1972, Melish showed that sterile blister fluid could reproduce the same effects when administered intraperitoneally, orally, or subcutaneously [31]. After such a finding, isolation and identification of ETs was only a matter of time [32–36] and the crowning achievement was the observation that purified recombinant *Escherichia coli*-produced ETs could reproduce the syndrome [37–39]. However, full demonstration of the molecular mechanism of action took almost three further decades [6,7] and involved studies on a disease showing a completely unrelated etiology.

The link between SSSS and Dsg-1 arose, in fact, from studies dealing with *pemphigus foliaceus* (PF), and autoimmune disease characterized by a very similar clinical presentation [40–42]. It was demonstrated that, in PF, skin blistering and exfoliation is caused by auto-antibodies against Dsg-1. In a closely related disease, *pemphigus vulgaris*, auto-antibodies against Dsg-3 are produced instead, and damage is limited to mucosal membranes [42]. It was not long after when ET-related hydrolysis was found to only involve Dsg-1, but not other desmogleins or cell adhesion molecules [6,7].

The high susceptibility to ETs, which is observed in neonates (rather than in older children or adults), is thought to be related to a still immature immune system, along with an age-related reduced renal clearance [43]. This explanation is corroborated by the findings that compromised renal function, immune system impairment (i.e., HIV patients), pharmacological immunosuppression, and chemotherapy for lymphoma that predispose elderly and immunocompromised people to SSSS [44–49]. Although mortality among newborns is relatively low and does not exceed 5%, it reaches more than 50% in old patients, although this is mostly associated with severe underlying conditions other than SSSS [47,50]. Again, healthy adults are not susceptible to the syndrome, probably due to an efficient neutralization of ETs by immunoglobulins [51]. However, reasons behind a greater susceptibility of neonates than adult subjects still have to be more deeply understood.

10.4 GLOBAL DISTRIBUTION AND SPECIES-SPECIFICITY

The most common serotypes of S. *aureus* ETs are ETA and ETB, which are produced by 1.5% and 0.5% human strains, respectively [52]. ETC was described in a strain of S. *aureus* from a horse with phlegmon [53]. Finally, the relation between the rare ETD serotype and SSSS is not yet clearly defined [52,54,55]. Toxins sharing significant homology with *S. aureus* ETs were also isolated from other staphylococcal species. These include SCET and ExhB, from *Staphylococcus chromogenes* [56,57]; SHETA and SHETB, as well as ExhA, ExhB, ExhC, and ExhD, from *Staphylococcus hyicus* [58–61]; the EXI toxin, from canine *Staphylococcus pseudintermedius* strains [62,63]; ExhC, from *Staphylococcus sciuri* [64,65]. Interestingly, toxins produced by different species of staphylococci have divergent host specificity; in particular (and not unexpectedly), animal hosts that represent the main niches for certain species are also the primary target for toxins produced by those species. This is related to differences in the amino acid sequence of the Dsg-1 target region. For example, ETA and ETB produced by human strains cause exfoliation in mice, but not in rats. In the case of SHETA and SHETB, instead, both toxins induce exfoliation in piglets and chicks, but not in mice [66]. EXI, finally, was demonstrated to cleave canine and chicken Dsg-1 [62].

10.5 CONCLUSION

Virulence of several bacteria relies on production of particular toxins, and strains deprived of the toxin usually become harmless. In general, in staphylococci, virulence relies on production and concert action of multiple factors that usually do not confer significant effect when singled out. Nonetheless, staphylococci may also produce certain toxins, including ETs, that are directly responsible for particular diseases. Hence, ETs represent ideal targets in prevention/treatment of SSSS. However, both the very low syndrome prevalence and significant costs associated with development of novel treatments make it unlikely that ETs will ever be explored in that direction.

ACKNOWLEDGMENTS

The present work was partially supported by grants from the Ministry of Science and Higher Education (MNiSW, Poland): Iuventus Plus No. 0108/IP1/2011/71 (to GD) and No. N303 813340 (to BW); and by a grant from the National Science Centre (NCN, Poland) No. UMO-2012/07/D/NZ2/04282 (to BW).

REFERENCES

[1] Dancer SJ, Garratt R, Saldanha J, Jhoti H, Evans R. The epidermolytic toxins are serine proteases. FEBS Lett 1990;268(1):129–32.

[2] Barbosa JA, Saldanha JW, Garratt RC. Novel features of serine protease active sites and specificity pockets: sequence analysis and modelling studies of glutamate-specific endopeptidases and epidermolytic toxins. Protein Eng 1996;9(7):591–601.

[3] Cavarelli J, Prévost G, Bourguet W, et al. The structure of *Staphylococcus aureus* epidermolytic toxin A, an atypic serine protease, at 1.7 A resolution. Structure 1997;5(6):813–24.

[4] Getsios S, Huen AC, Green KJ. Working out the strength and flexibility of desmosomes. Nat Rev Mol Cell Biol 2004;5(4):271–81.

[5] Angst BD, Marcozzi C, Magee AI. The cadherin superfamily: diversity in form and function. J Cell Sci 2001;114(Pt 4):629–41.

[6] Amagai M, Matsuyoshi N, Wang ZH, Andl C, Stanley JR. Toxin in bullous impetigo and staphylococcal scalded-skin syndrome targets desmoglein 1. Nat Med 2000;6(11):1275–7.

[7] Amagai M, Yamaguchi T, Hanakawa Y, Nishifuji K, Sugai M, Stanley JR. Staphylococcal exfoliative toxin B specifically cleaves desmoglein 1. J Investig Dermatol 2002;118(5):845–50.

[8] Hanakawa Y, Schechter NM, Lin C, et al. Molecular mechanisms of blister formation in bullous impetigo and staphylococcal scalded skin syndrome. J Clin Invest 2002;110(1):53–60.

[9] Hanakawa Y, Schechter NM, Lin C, Nishifuji K, Amagai M, Stanley JR. Enzymatic and molecular characteristics of the efficiency and specificity of exfoliative toxin cleavage of desmoglein 1. J Biol Chem 2004;279(7):5268–77.

[10] Payne AS, Hanakawa Y, Amagai M, Stanley JR. Desmosomes and disease: pemphigus and bullous impetigo. Curr Opin Cell Biol 2004;16(5):536–43.

[11] Hanakawa Y, Selwood T, Woo D, Lin C, Schechter NM, Stanley JR. Calcium-dependent conformation of desmoglein 1 is required for its cleavage by exfoliative toxin. J Investig Dermatol 2003;121(2):383–9.

[12] Melish ME, Glasgow LA, Turner MD, Lillibridge CB. The staphylococcal epidermolytic toxin: its isolation, characterization, and site of action. Ann N Y Acad Sci 1974;236(0):317–42.

[13] Vath GM, Earhart CA, Rago JV, et al. The structure of the superantigen exfoliative toxin A suggests a novel regulation as a serine protease. Biochemistry 1997;36(7):1559–66.

[14] Vath GM, Earhart CA, Monie DD, Iandolo JJ, Schlievert PM, Ohlendorf DH. The crystal structure of exfoliative toxin B: a superantigen with enzymatic activity. Biochemistry 1999;38(32):10239–46.

[15] Khan AR, James MN. Molecular mechanisms for the conversion of zymogens to active proteolytic enzymes. Protein Sci 1998;7(4):815–36.

[16] Proft T, Fraser JD. Bacterial superantigens. Clin Exp Immunol 2003;133(3):299–306.

[17] Morlock BA, Spero L, Johnson AD. Mitogenic activity of staphylococcal exfoliative toxin. Infect Immun 1980;30(2):381–4.

[18] Fleischer B, Bailey CJ. Recombinant epidermolytic (exfoliative) toxin A of *Staphylococcus aureus* is not a superantigen. Med Microbiol Immunol 1992;180(6):273–8.

[19] Monday SR, Vath GM, Ferens WA, et al. Unique superantigen activity of staphylococcal exfoliative toxins. J Immunol 1999;162(8):4550–9.

[20] Rago JV, Vath GM, Bohach GA, Ohlendorf DH, Schlievert PM. Mutational analysis of the superantigen staphylococcal exfoliative toxin A (ETA). J Immunol 2000;164(4):2207–13.

[21] Nishifuji K, Sugai M, Amagai M. Staphylococcal exfoliative toxins: "molecular scissors" of bacteria that attack the cutaneous defense barrier in mammals. J Dermatol Sci 2008;49(1):21–31.

[22] Miethke T, Duschek K, Wahl C, Heeg K, Wagner H. Pathogenesis of the toxic shock syndrome: T cell mediated lethal shock caused by the superantigen TSST-1. Eur J Immunol 1993;23(7):1494–500.

[23] Dinges MM, Orwin PM, Schlievert PM. Exotoxins of *Staphylococcus aureus*. Clin Microbiol Rev 2000;13(1):16–34. table of contents.

[24] Vaishnani J. Superantigen. Indian J Dermatol Venereol Leprol 2009;75(5):540–4.

[25] Schrezenmeier H, Fleischer B. Mitogenic activity of staphylococcal protein A is due to contaminating staphylococcal enterotoxins. J Immunol Methods 1987;105(1):133–7.

[26] Choi YW, Kotzin B, Herron L, Callahan J, Marrack P, Kappler J. Interaction of *Staphylococcus aureus* toxin "superantigens" with human T cells. Proc Natl Acad Sci U S A 1989;86(22):8941–5.

[27] Von Rittershain GR. Die exfoliative dermatitis jungener senglinge. Z Kinderheilkd 1878;2:3–23.

[28] Lyell A. Toxic epidermal necrolysis. Nurs Mirror Midwives J 1973;136(2):42–5.

[29] Ladhani S, Joannou CL, Lochrie DP, Evans RW, Poston SM. Clinical, microbial, and biochemical aspects of the exfoliative toxins causing staphylococcal scalded-skin syndrome. Clin Microbiol Rev 1999;12(2):224–42.

[30] Lyell A. A review of toxic epidermal necrolysis in Britain. Br J Dermatol 1967;79(12): 662–71.

[31] Melish ME, Glasgow LA, Turner MD. The staphylococcal scalded-skin syndrome: isolation and partial characterization of the exfoliative toxin. J Infect Dis 1972;125(2):129–40.

[32] Arbuthnott JP, Billcliffe B, Thompson WD. Isoelectric focusing studies of staphylococcal epidermolytic toxin. FEBS Lett 1974;46(1):92–5.

[33] Melish ME, Glasgow LA. The staphylococcal scalded-skin syndrome. N Engl J Med 1970;282(20):1114–9.

[34] Kondo I, Sakurai S, Sarai Y. Purification of exfoliatin produced by *Staphylococcus aureus* of bacteriophage group 2 and its physicochemical properties. Infect Immun 1973;8(2):156–64.

[35] Johnson AD, Metzger JF, Spero L. Production, purification, and chemical characterization of *Staphylococcus aureus* exfoliative toxin. Infect Immun 1975;12(5):1206–10.

[36] Dimond RL, Wuepper KD. Purification and characterization of a staphylococcal epidermolytic toxin. Infect Immun 1976;13(2):627–33.

[37] O'Toole PW, Foster TJ. Molecular cloning and expression of the epidermolytic toxin A gene of *Staphylococcus aureus*. Microb Pathog 1986;1(6):583–94.

[38] Lee CY, Schmidt JJ, Johnson-Winegar AD, Spero L, Iandolo JJ. Sequence determination and comparison of the exfoliative toxin A and toxin B genes from *Staphylococcus aureus*. J Bacteriol 1987;169(9):3904–9.

[39] O'Toole PW, Foster TJ. Nucleotide sequence of the epidermolytic toxin A gene of *Staphylococcus aureus*. J Bacteriol 1987;169(9):3910–5.

[40] Jordon RE, Sams Jr. WM, Diaz G, Beutner EH. Negative complement immunofluorescence in pemphigus. J Investig Dermatol 1971;57(6):407–10.

[41] Stanley JR, Koulu L, Thivolet C. Distinction between epidermal antigens binding pemphigus vulgaris and pemphigus foliaceus autoantibodies. J Clin Invest 1984;74(2):313–20.

[42] Eyre RW, Stanley JR. Human autoantibodies against a desmosomal protein complex with a calcium-sensitive epitope are characteristic of pemphigus foliaceus patients. J Exp Med 1987;165(6):1719–24.

[43] Plano LR, Adkins B, Woischnik M, Ewing R, Collins CM. Toxin levels in serum correlate with the development of staphylococcal scalded skin syndrome in a murine model. Infect Immun 2001;69(8):5193–7.

[44] Richard M, Mathieu-Serra A. Staphylococcal scalded skin syndrome in a homosexual adult. J Am Acad Dermatol 1986;15(2 Pt 2):385–9.

[45] O'Keefe R, Dagg JH, MacKie RM. The staphylococcal scalded skin syndrome in two elderly immunocompromised patients. Br Med J (Clin Res Ed) 1987;295(6591):179–80.

[46] Petzelbauer P, Konrad K, Wolff K. Staphylococcal scalded skin syndrome in 2 adults with acute kidney failure. Hautarzt 1989;40(2):90–3.

[47] Gemmell CG. Staphylococcal scalded skin syndrome. J Med Microbiol 1995;43(5):318–27.

[48] Hardwick N, Parry CM, Sharpe GR. Staphylococcal scalded skin syndrome in an adult. Influence of immune and renal factors. Br J Dermatol 1995;132(3):468–71.

[49] Scheinpflug K, Schalk E, Mohren M. Staphylococcal scalded skin syndrome in an adult patient with T-lymphoblastic non-Hodgkin's lymphoma. Onkologie 2008;31(11):616–9.

[50] Cribier B, Piemont Y, Grosshans E. Staphylococcal scalded skin syndrome in adults. A clinical review illustrated with a new case. J Am Acad Dermatol 1994;30(2 Pt 2):319–24.

[51] Melish ME, Chen FS, Sprouse S, Stuckey M, Murata MS. Epidermolytic toxin in staphylococcal infection: toxin levels and host response. Zentralbl Bakteriol, Suppl 1981;10:287–98.

[52] Becker K, Friedrich AW, Lubritz G, Weilert M, Peters G, Von Eiff C. Prevalence of genes encoding pyrogenic toxin superantigens and exfoliative toxins among strains of *Staphylococcus aureus* isolated from blood and nasal specimens. J Clin Microbiol 2003;41(4):1434–9.

[53] Sato H, Matsumori Y, Tanabe T, Saito H, Shimizu A, Kawano J. A new type of staphylococcal exfoliative toxin from a *Staphylococcus aureus* strain isolated from a horse with phlegmon. Infect Immun 1994;62(9):3780–5.

[54] Yamaguchi T, Nishifuji K, Sasaki M, et al. Identification of the *Staphylococcus aureus* etd pathogenicity island which encodes a novel exfoliative toxin, ETD, and EDIN-B. Infect Immun 2002;70(10):5835–45.

[55] Yamasaki O, Tristan A, Yamaguchi T, et al. Distribution of the exfoliative toxin D gene in clinical *Staphylococcus aureus* isolates in France. Clin Microbiol Infect 2006;12(6):585–8.

[56] Sato H, Hirose K, Terauchi R, et al. Purification and characterization of a novel *Staphylococcus chromogenes* exfoliative toxin. J Vet Med B Infect Dis Vet Public Health 2004;51(3):116–22.

[57] Andresen LO, Ahrens P, Daugaard L, Bille-Hansen V. Exudative epidermitis in pigs caused by toxigenic *Staphylococcus chromogenes*. Vet Microbiol 2005;105(3–4):291–300.

[58] Sato H, Tanabe T, Kuramoto M, Tanaka K, Hashimoto T, Saito H. Isolation of exfoliative toxin from *Staphylococcus hyicus* subsp. *hyicus* and its exfoliative activity in the piglet. Vet Microbiol 1991;27(3–4):263–75.

[59] Andresen LO. Differentiation and distribution of three types of exfoliative toxin produced by *Staphylococcus hyicus* from pigs with exudative epidermitis. FEMS Immunol Med Microbiol 1998;20(4):301–10.

[60] Sato H, Watanabe T, Murata Y, et al. New exfoliative toxin produced by a plasmid-carrying strain of *Staphylococcus hyicus*. Infect Immun 1999;67(8):4014–8.

[61] Ahrens P, Andresen LO. Cloning and sequence analysis of genes encoding *Staphylococcus hyicus* exfoliative toxin types A, B, C, and D. J Bacteriol 2004;186(6):1833–7.

[62] Futagawa-Saito K, Makino S, Sunaga F, et al. Identification of first exfoliative toxin in *Staphylococcus pseudintermedius*. FEMS Microbiol Lett 2009;301(2):176–80.

[63] Iyori K, Futagawa-Saito K, Hisatsune J, Yamamoto M, Sekiguchi M, Ide K, et al. *Staphylococcus pseudintermedius* exfoliative toxin EXI selectively digests canine desmoglein 1 and causes subcorneal clefts in canine epidermis. Vet Dermatol 2013;22(4):319–26.

[64] Li H, Li X, Lu Y, Wang X, Zheng SJ. *Staphylococcus sciuri* exfoliative toxin C is a dimer that modulates macrophage functions. Can J Microbiol 2011;57(9):722–9.

[65] Li H, Wang Y, Ding L, Zheng SJ. *Staphylococcus sciuri* exfoliative toxin C (ExhC) is a necrosis-inducer for mammalian cells. PLoS ONE 2013;6(7):e23145.

[66] Sato H, Kuramoto M, Tanabe T, Saito H. Susceptibility of various animals and cultured cells to exfoliative toxin produced by *Staphylococcus hyicus* subsp. *hyicus*. Vet Microbiol 1991;28(2):157–69.

Chapter 11

Extracellular Proteases of *Staphylococcus* spp.

Natalia Stach*, Paweł Kaszycki[†], Benedykt Władyka[§], Grzegorz Dubin*[,‡]
*Department of Microbiology, Faculty of Biochemistry, Biophysics and Biotechnology, Jagiellonian University, Krakow, Poland [†]Biochemistry Unit, Institute of Plant Biology and Biotechnology, Faculty of Biotechnology and Horticulture, University of Agriculture, Krakow, Poland [‡]Malopolska Centre of Biotechnology, Jagiellonian University, Krakow, Poland [§]Department of Analytical Biochemistry, Faculty of Biochemistry, Biophysics and Biotechnology, Jagiellonian University, Krakow, Poland

11.1 INTRODUCTION

Depending on the growth phase, staphylococci produce various surface-associated and secreted proteins that interact with the host tissues. In an early stage, expression of surface proteins, mainly acting as adhesive factors, predominates. With the beginning of the exponential phase, instead, the surface protein synthesis is suppressed and replaced by secretion of soluble extracellular factors. The transition is controlled by global regulatory systems of gene expression, primarily *agr* (accessory gene regulator) and *sar* (staphylococcal accessory regulator) [1], and is believed to reflect the transition from the early colonization state of disease to the late invasive state.

The *agr* is a general system that regulates the production of surface adhesins (downregulation) and extracellular proteases (upregulation) and a variety of other genes [2], examples of which include phenol-soluble modulins (PSM) or toxins [3]. The second global regulator, *sar,* acts somewhat in opposition to *agr*. While *agr* promotes the migration of bacteria at the late stages of infection, *sar* promotes the adhesion to host tissues (downregulation of proteases), controlling the early stage of infection [4].

Staphylococci produce three catalytic groups of proteases, which include serine proteases, cysteine proteases, and metalloproteases [1]. There is a lasting and still mostly unsettled debate as to whether staphylococcal proteases function to provide nutrients and facilitate the spread of bacteria by nonspecifically degrading the host tissue and bacterial surface proteins [5] or if the proteases rather have specific, targeted functions [6]. However, recent

Pet-to-Man Travelling Staphylococci: A World in Progress. https://doi.org/10.1016/B978-0-12-813547-1.00011-X
135

findings suggest that the latter case is more likely. These and other findings are described in more detail herein.

11.2 THE SERINE PROTEASES

11.2.1 V8 Protease

Glutamyl endopeptidase, also known as V8 protease, was the first staphylococcal protease to be purified and characterized. Studies have shown that it is produced by 67% of strains [1]. The individual strains are characterized by varying molecular weights of the protease, depending on the number of tandem repeats of a tripeptide $(Pro-Asn/Asp-Asn)_n$ located at the C-terminus, but the function of this variability is unknown. The additional heterogeneity results from autoproteolysis at the C-terminal part of the molecule. The gene coding for V8 protease is located in a single operon (*sspABC*) with the genes encoding the extracellular cysteine protease—staphopain B and its specific inhibitor—staphostatin B. Protease V8 is produced as a zymogen and its full activation requires processing by aureolysin [7–10]. V8 exhibits restricted substrate specificity, which is manifested by its activity to hydrolyze peptide bonds almost exclusively after glutamic acid residues [11]. Because of the exclusive specificity, the enzyme has found a routine use for protein fragmentation in mass spectrometry. Numerous human proteins are thought to be natural substrates of V8 protease. It has been shown that human antimicrobial peptide LL-37 is inactivated by hydrolysis [12]. Moreover, the enzyme inactivates α1-proteinase inhibitor (α1-PI), which under normal conditions, controls the activity of host proteases such as plasminogen, trypsin, and neutrophil elastase, thus deregulating the proteolytic balance at the site of infection. V8 can also degrade human immunoglobulins: IgD, IgE, IgG, and IgM. It has been proposed that this activity may facilitate breaching of the immune defenses [13–15]. It must be mentioned, however, that the majority of these findings were obtained under in vitro conditions, and their real physiological significance remains unknown.

11.2.2 Serine Protease-Like Proteins (Spl)

Enzymes of the *spl* operon are currently the least characterized group of staphylococcal proteases. The earliest information was contributed by Rieneck and collaborators who examined the reactivity of human serum obtained from *Staphylococcus aureus* endocarditis patients with staphylococcal proteins to identify SplC as a single protein eliciting a strong humoral response. Such a response was not observed in infections by the closely related species *Staphylococcus epidermidis*. On this basis, the authors postulated that the SplC protein participates in the etiology of endocarditis. They also noted the sequence homology with V8 protease, thus suggesting SplC was a proteolytic enzyme [16]. The full *spl* operon was described a few years later and it was demonstrated that in different strains, it encodes for just a single protease, up to six homologous serine proteases (SplA-SplF). The *spl* operon is located on a staphylococcal pathogenicity island [17] and, like the other proteases, its expression is under the control of *agr* [18].

Spl proteases (~22 kDa) are produced with a signal sequence guiding extracellular secretion, but, unlike other staphylococcal proteases, along with the majority of other serine proteases of the family S1, without any kind of propeptides. Spl proteases share an amino acid sequence homology from 44% between SplA and SplD to 95% between SplD and SplF [18].

Despite the location of the *spl* operon on the staphylococcal pathogenicity island among the genes encoding enterotoxins and leucocidins, which would suggest a role of Spl proteases in virulence, their actual role in this process remains unknown. In recent years, we have provided a biochemical and structural characterization of several encoded enzymes. Interestingly, the enzymes characterized to date exhibit high substrate specificity. SplA recognizes the following consensus: (W/Y)-L-Y*(T/S) (asterisk indicates the cleavage site) [19]. SplB recognizes the consensus of W-E-L-Q* [20], while SplD of R-(Y/W)-(P/L)-(T/L/I/V)*S [21]. So far, the substrate specificities of SplE and SplF have not been determined. SplC shows no activity in in vitro assays but it has been proposed that a specific substrate may activate the enzyme; however, such a substrate has not been demonstrated as yet [22]. The structural studies have revealed certain interesting mechanistic details regarding the process of activation and the catalytic mechanism; however, the discussion of these topics is beyond the scope of this review and the reader is directed to the original works [19–22]. As mentioned herein, the role of Spl proteases in staphylococcal virulence remains unknown; however, the strict specificity suggests a toxin-like function directed at particular proteins rather than a general digestive role. It is also of note that the homolog of the *spl* operon was found in the genome of the canine pathogen *Staphylococcus pseudintermedius*.

11.2.3   Serine Protease (Esp)—*S. epidermidis*

Esp is an extracellular serine protease produced by *S. epidermidis*. The mature enzyme has a 59% sequence identity with V8 protease [23,24]. By analogy to V8 protease, it is predicted that the N-terminal propeptide of Esp is processed by SepP1, the Aur homolog in *S. epidermidis* [24,25].

Interestingly, Iwase and collaborators demonstrated that Esp inhibits *S. aureus* biofilm formation during nasal colonization and destroys the preformed *S. aureus* biofilms [26]. Using proteomic and immunological assays, Sugimoto and colleagues [27] demonstrated that Esp degraded at least 75 proteins of *S. aureus*, including 11 biofilm formation- and colonization-associated proteins, such as the extracellular adherence protein, the extracellular matrix protein-binding protein, and fibronectin-binding protein A. Moreover, Esp degraded several human receptor proteins that participate in colonization and infection of *S. aureus* (e.g., fibronectin, fibrinogen, and vitronectin). It has long been known that colonization with

S. epidermidis is important for health. It is possible that the preceding processes may constitute one of the responsible mechanisms [27].

Esp is more stable than V8 and has a strong preference for cleavage after glutamic acid, but not after aspartic acid. Therefore, Esp and its engineered derivatives with a hyperproteolytic activity may, in the future, replace V8 protease in proteomics [23,28].

11.2.4 Epidermolytic Toxins

Staphylococci produce yet another group of serine proteases, the epidermolytic toxins, which have a clearly defined role in virulence. The epidermolytic toxins are described in greater detail in another chapter of this book.

11.3 THE CYSTEINE PROTEASES

11.3.1 Staphopains

Staphopains are papain-like, cysteine proteases, and, in *S. aureus*, these enzymes include staphopains A, B, and C. All staphopains are encoded in operons with their specific cytoplasmic inhibitors, the role of which is to protect the cell against the deleterious activity of misdirected staphopains. Staphopain A (*scpA*) is encoded in *scp* operon together with staphostatin A; staphopain B (*sspB*) is encoded in a single operon (*ssp*) together with V8 protease (*sspA*) and staphostatin B (*sspC*). The organization of staphopain C operon is similar to that of staphopain A, but the operon is located on a large plasmid (rather than in the chromosome) associated with staphylococcal strains from poultry [29,30]. Staphopains are secreted as proenzymes [4]. Prostaphopain B is activated by V8 protease [4] while prostaphopain A and C maturation involves autocatalytic processing [31–33]. Despite structural similarity, staphopains A and B have different substrate specificities [4].

Staphopains B and C seem to play an important role in human and avian diseases, respectively, whereas staphopain A seems not to be related to virulence, with a possible housekeeping function [34], as suggested by animal studies and indirect analyses. Staphopain C has been postulated as a poultry-specific virulence factor [30]. Indeed, the gene encoding the protease was found in poultry-borne strains exhibiting high virulence in a chicken embryo model [35]. Moreover, biochemical studies revealed that staphopain C is efficiently inhibited in human plasma by alpha-1-antichymotrypsin, whereas it remains fully active in the chicken blood plasma [32,33]. The substrate preferences of staphopains B and C are comparable, while the preference of staphopain A differs significantly. It is tempting to speculate that these properties reflect the distinct physiological roles of the enzymes [34], but this needs to be further confirmed, especially that the role of staphopain A in virulence is far from being clarified. For example, Laarman and collaborators identified staphopain A as a chemokine receptor blocker. Cleavage of CXCR2 made neutrophils unresponsive to the relevant chemokines. Based on these findings, the authors argued that staphopain A is an important immunomodulatory protein that inhibits neutrophil recruitment [36].

Ohbayashi and collaborators showed that staphopains A and B were able to degrade collagen and fibrinogen, thereby delaying the clotting of blood, and for that reason, they may be responsible for bleeding tendency. Staphopain B cleaved the fibrinogen A-chain at the C-terminal region very efficiently, while staphopain A was rather inefficient [37]. Both staphopains also degraded host extracellular cystatins (cysteine protease inhibitors) C, D, and E/M, thus deregulating the proteolytic balance at the site of their action. As a result the activity of cathepsins increases, which in turn leads to degradation of antimicrobial peptides [38]. It was also demonstrated that staphopains A and B potentially deregulated the kallikrein/kinin pathway by degrading the kinin precursor - kininogen. Moreover, staphopains were found to degrade surface proteins of phagocytic cells leading, them to apoptotic death and removal by macrophages [10].

11.3.2   *S. epidermidis* Cysteine Protease (Ecp)

Ecp was purified by Sloot and collaborators, and by our group, from *S. epidermidis* culture supernatants [1,39]. It seems that two forms of the enzyme coexist: the cell-associated and the soluble ones. The identity of the cell-associated form is not clear. Ecp does not have a membrane-anchoring signal (LPXTG) and is not a sortase substrate [40]. Rather, the enzyme is only loosely attached to the cell wall because it can be released with mild detergents [40,41].

Ecp is secreted with a pro-fragment, but little is known of its maturation [23,40]. The mature form shares 75% similarity with S. *aureus* staphopain A [9,40,42,43].

It was demonstrated that Ecp can degrade human IgA, IgM, serum albumin and all three subunits of fibrinogen [1,39], but the physiological significance of these observations is unknown. Because Ecp degrades elastin, it was implicated in perifollicular macular atrophy, a skin disease characterized by selective loss of elastic fibers around hair follicles [44]. The specific cysteine proteases inhibitors (cystatins) of the host remain inactive against Ecp [1,23].

11.4   THE METALLOPROTEASES

11.4.1   Aureolysin

Aureolysin, a zinc metalloprotease, belongs to the thermolysin family [45,46]. In S. *aureus* the enzyme is present in two allelic forms that exert different specific activities [47–49]. Moreover, its homologs were identified in S. *epidemidis* and S. *pseudintermedius* [50]. Aureolysin has low substrate specificity [1,47,48]. Regarding its potential role in pathogenesis, aureolysin was demonstrated to degrade the antimicrobial peptides cathelicidin LL-37 and dermicidin [51].

Aureolysin cleaves complement protein C3, creating active C3a and C3b. However, in contrast with convertase, which acts at the bacterial surface, the conversion by aureolysin is carried out at sites distant from the bacteria, and does not promote opsonization. Thus, the activated C3b is degraded by the host regulators FH and FI. Concurrently, the process deregulates complement-dependent responses such as neutrophils activation [52]. A recent investigation by Cassat and colleagues suggests that aureolysin modulates the pathogenesis of osteomyelitis [53]. Aureolysin cleaves staphylococcal surface-associated proteins (fibronectin-binding protein, protein A and clumping factor B), which is thought to support the transition from an adherent to an invasive phenotype [43].

11.4.2 Hyicolysins

Hyicolysins are extracellular metalloproteases produced by *Staphylococcus hyicus*. Two zinc metalloproteases (ShpI and ShpII) were isolated from *S. hyicus* by Ayora and collaborators [54]. The enzymes are produced as zymogens. The amino acid sequence of the mature proteases shows low but significant similarity with the metalloproteases from *S. epidermidis* and *S. aureus*. Production of hyicolysins correlates with the growth phase—the enzymes are produced in the late logarithmic phase. So far, only the role of ShpII in prolipase conversion was documented [1,54,55], but it is expected that hyicolysins can also play additional roles. The role of these proteases, if any, in *S. hyicus* pathogenesis remains unknown [56].

11.5 LESSONS LEARNED FROM STUDIES IN ANIMAL MODELS

The role of proteases in staphylococcal infection was studied in a number of animal models. However, only exfoliative toxins were convincingly demonstrated to play a crucial role in staphylococcal scalded skin syndrome (SSSS), a disease involving skin blistering in infants [57,58]. Despite the fact that multiple indirect studies briefly reviewed herein suggest the involvement of *S. aureus* proteases in virulence [1], the evidence gained from animal models is largely contradictory. Calander and colleagues studied the virulence of strains lacking *aur* or *sspB* genes as well as both *sspA* and *sspB* genes in a mouse model of septic arthritis. The authors have demonstrated that deletion of none of the studied genes attenuated *S. aureus* virulence [59]. Similar results were obtained for *sspA*-knockout *S. aureus* in a rat model [18]. However, an independent study demonstrated an impaired virulence of the *sspA*- and *sspB*-mutated strain in three different murine models [60]. In addition, it has been shown that both cysteine proteases induce vascular leakage and shock in a guinea pig model of infection [61]. Finally, a number of studies have reported generation of antibodies against staphylococcal extracellular proteases during infection, strongly supporting their expression in the event of disease [62–64]. The recent investigation of Kolar and colleagues casts new light on the preceding contradictory evidence. The authors defined the collective impact of all staphylococcal proteases on pathogenesis. By using a strain hyperproducing the secreted proteases and a variant lacking all 10 of these enzymes it was shown that these are collectively required for growth in peptide-rich environments, serum, in the presence of antimicrobial peptides, and in human blood. Moreover, the proteases were shown to be of importance in resisting phagocytosis by human leukocytes. Using different in vivo murine infection models, model specific roles were documented: the exo-protease deletion decreases abscess formation and impairs organ invasion; however, when mortality is evaluated, the protease-null strain demonstrates hypervirulence compared with the wild type. The different roles of staphylococcal proteases in different settings were clarified by proteomics that revealed that exoproteases are key mediators of secreted and cell-wall associated virulence determinants' stability [65].

11.6 CONCLUSION

Unlike many bacterial species whose virulence relies on a single or a limited number of toxins, staphylococcal pathogenesis depends on the production of a large number of factors. The majority of those factors acting alone do not significantly affect the host but, when acting in concert, they provide *S. aureus* with a relevant pathogenic potential. This is true in particular for staphylococcal proteases, which were implicated in multiple infectious processes in the human host, but none of those enzymes, alone, seems to be essential for virulence (apart from exfoliative toxins). Nevertheless, based on the reports presented herein, it may be concluded that *S. aureus* extracellular proteases play a variety of deleterious roles that infection relies on. Specifically, they aid in protection against the innate immune system, at both cell-dependent and independent levels, which in turn has a strong impact on the progression of localized and systemic staphylococcal infections. They also play an important role in surface proteins' turnover and processing of secreted factors. Currently, it is most challenging to demonstrate how the combination of these different mechanisms and activities contributes to the overall *S. aureus* pathogenic potential and which is the role of homologous enzymes in nonpathogenic staphylococci.

ACKNOWLEDGMENT

This work was supported by the Grant UMO-2011/01/D/NZ1/01169 (to G. D.) from the National Science Center.

REFERENCES

[1] Dubin G. Extracellular proteases of *Staphylococcus* spp. Biol Chem 2002;383(7–8):1075–86.
[2] Kong KF, Vuong C, Otto M. *Staphylococcus* quorum sensing in biofilm formation and infection. Int J Med Microbiol 2006;296(2–3):133–9.

[3] Queck SY, Jameson-Lee M, Villaruz AE, Bach THL, Khan BA, Sturdevant DE, et al. RNAIII-independent target gene control by the agr quorum-sensing system: insight into the evolution of virulence regulation in *Staphylococcus aureus*. Mol Cell 2008;32(1):150–8.

[4] Kantyka T, Shaw LN, Potempa J. Papain-like proteases of *Staphylococcus aureus*. Adv Exp Med Biol 2011;712:1–14.

[5] Travis J, Potempa J, Maeda H. Are bacterial proteinases pathogenic factors? Trends Microbiol 1995;3(10):405–7.

[6] Ricklin D, Tzekou A, Garcia BL, Hammel M, McWhorter WJ, Sfyroera G, et al. A molecular insight into complement evasion by the staphylococcal complement inhibitor protein family. J Immunol 2009;183(4):2565–74.

[7] Carmona C, Gray GL. Nucleotide sequence of the serine protease gene of *Staphylococcus aureus*, strain V8. Nucleic Acids Res 1987;15(16):6757.

[8] Drapeau GR. Unusual COOH-terminal structure of staphylococcal protease. J Biol Chem 1978;253(17):5899–901.

[9] Rice K, Peralta R, Bast D, De Azavedo J, McGavin MJ. Description of staphylococcus serine protease (ssp) operon in *Staphylococcus aureus* and nonpolar inactivation of sspA-encoded serine protease. Infect Immun 2001;69(1):159–69.

[10] Smagur J, Guzik K, Magiera L, Bzowska M, Gruca M, Thøgersen IB, et al. A new pathway of staphylococcal pathogenesis: apoptosis-like death induced by staphopain B in human neutrophils and monocytes. J Innate Immun 2009;1(2):98–108.

[11] Sørensen SB, Sørensen TL, Breddam K. Fragmentation of proteins by *S. aureus* strain V8 protease. Ammonium bicarbonate strongly inhibits the enzyme but does not improve the selectivity for glutamic acid. FEBS Lett 1991;294(3):195–7.

[12] Sieprawska-Lupa M, Mydel P, Krawczyk K, Wójcik K, Puklo M, Lupa B, et al. Degradation of human antimicrobial peptide LL-37 by *Staphylococcus aureus*-derived proteinases. Antimicrob Agents Chemother 2004;48(12):4673–9.

[13] Burchacka E, Skoreński M, Sieńczyk M, Oleksyszyn J. Phosphonic analogues of glutamic acid as irreversible inhibitors of *Staphylococcus aureus* endoproteinase GluC: an efficient synthesis and inhibition of the human IgG degradation. Bioorg Med Chem Lett 2013;23(5):1412–5.

[14] Miedzobrodzki J, Tadeusiewicz R, Porwit-Bóbr Z. Evaluation of the effect of staphylococcal serine proteinase on phagocytosis. Arch Immunol Ther Exp 1987;35(6):877–85.

[15] Takai T, Ikeda S. Barrier dysfunction caused by environmental proteases in the pathogenesis of allergic diseases. Allergol Int 2011;60(1):25–35.

[16] Rieneck K, Renneberg J, Diamant M, Gutschik E, Bendtzen K. Molecular cloning and expression of a novel *Staphylococcus aureus* antigen. Biochim Biophys Acta 1997;1350(2):128–32.

[17] Kuroda M, Ohta T, Uchiyama I, Baba T, Yuzawa H, Kobayashi I, et al. Whole genome sequencing of meticillin-resistant *Staphylococcus aureus*. Lancet 2001;357(9264):1225–40.

[18] Reed SB, Wesson CA, Liou LE, Trumble WR, Schlievert PM, Bohach GA, et al. Molecular characterization of a novel *Staphylococcus aureus* serine protease operon. Infect Immun 2001;69(0019–9567):1521–7.

[19] Stec-Niemczyk J, Pustelny K, Kisielewska M, Bista M, Boulware KT, Stennicke HR, et al. Structural and functional characterization of SplA, an exclusively specific protease of *Staphylococcus aureus*. Biochem J 2009;419:555–64.

[20] Dubin G, Stec-Niemczyk J, Kisielewska M, Pustelny K, Popowicz GM, Bista M, et al. Enzymatic activity of the *Staphylococcus aureus* SplB serine protease is induced by substrates containing the sequence Trp-Glu-Leu-Gln. J Mol Biol 2008;379(2):343–56.

[21] Zdzalik M, Kalinska M, Wysocka M, Stec-Niemczyk J, Cichon P, Stach N, et al. Biochemical and Structural Characterization of SplD Protease from *Staphylococcus aureus*. PLoS ONE 2013;8(10):e76812.

[22] Popowicz GM, Dubin G, Stec-Niemczyk J, Czarny A, Dubin A, Potempa J, et al. Functional and structural characterization of Spl proteases from *Staphylococcus aureus*. J Mol Biol 2006;358(1):270–9.

[23] Dubin G, Chmiel D, Mak P, Rakwalska M, Rzychon M, Dubin A. Molecular cloning and biochemical characterisation of proteases from *Staphylococcus epidermidis*. Biol Chem 2001;382(11):1575–82.

[24] Ohara-Nemoto Y, Ikeda Y, Kobayashi M, Sasaki M, Tajika S, Kimura S. Characterization and molecular cloning of a glutamyl endopeptidase from *Staphylococcus epidermidis*. Microb Pathog 2002;33(1):33–41.

[25] Teufel P, Gotz F. Characterization of an extracellular metalloprotease with elastase activity from *Staphylococcus epidermidis*. J Bacteriol 1993;175(13):4218–24.

[26] Iwase T, Uehara Y, Shinji H, Tajima A, Seo H, Takada K, et al. *Staphylococcus epidermidis* Esp inhibits *Staphylococcus aureus* biofilm formation and nasal colonization. Nature 2010;465(7296):346–9.

[27] Sugimoto S, Iwamoto T, Takada K, Okuda K-I, Tajima A, Iwase T, et al. *Staphylococcus epidermidis* Esp degrades specific proteins associated with *Staphylococcus aureus* biofilm formation and host-pathogen interaction. J Bacteriol 2013;195(8):1645–55.

[28] Sugimoto S, Iwase T, Sato F, Tajima A, Shinji H, Mizunoe Y. Cloning, expression and purification of extracellular serine protease Esp, a biofilm-degrading enzyme, from *Staphylococcus epidermidis*. J Appl Microbiol 2011;111(6):1406–15.

[29] Bukowski M, Lyzen R, Helbin WM, Bonar E, Szalewska-Palasz A, Wegrzyn G, et al. A regulatory role for *Staphylococcus aureus* toxin–antitoxin system PemIKSa. Nat Commun 2013;4:2012.

[30] Lowder BV, Guinane CM, Ben Zakour NL, Weinert LA, Conway-Morris A, Cartwright RA, et al. Recent human-to-poultry host jump, adaptation, and pandemic spread of Staphylococcus aureus. Proc Natl Acad Sci 2009;106(46):19545–50.

[31] Nickerson N, Ip J, Passos DT, McGavin MJ. Comparison of Staphopain A (ScpA) and B (SspB) precursor activation mechanisms reveals unique secretion kinetics of proSspB (Staphopain B), and a different interaction with its cognate Staphostatin, SspC. Mol Microbiol 2010;75(1):161–77.

[32] Wladyka B, Dubin G, Dubin A. Activation mechanism of thiol protease precursor from broiler chicken specific *Staphylococcus aureus* strain CH-91. Vet Microbiol 2011;147(1–2):195–9.

[33] Wladyka B, Kozik AJ, Bukowski M, Rojowska A, Kantyka T, Dubin G, et al. α1-Antichymotrypsin inactivates staphylococcal cysteine protease in cross-class inhibition. Biochimie 2011;93(5):948–53.

[34] Kalińska M, Kantyka T, Greenbaum DC, Larsen KS, Władyka B, Jabaiah A, et al. Substrate specificity of *Staphylococcus aureus* cysteine proteases—Staphopains A, B and C. Biochimie 2012;94(2):318–27.

[35] Polakowska K, Lis MW, Helbin WM, Dubin G, Dubin A, Niedziolka JW, et al. The virulence of *Staphylococcus aureus* correlates with strain genotype in a chicken embryo model but not a nematode model. Microbes Infect 2012;14(14):1352–62.

[36] Laarman AJ, Mijnheer G, Mootz JM, van Rooijen WJM, Ruyken M, Malone CL, et al. *Staphylococcus aureus* Staphopain A inhibits CXCR2-dependent neutrophil activation and chemotaxis. EMBO J 2012;31(17):3607–19.

[37] Ohbayashi T, Irie A, Murakami Y, Nowak M, Potempa J, Nishimura Y, et al. Degradation of fibrinogen and collagen by staphopains, cysteine proteases released from *Staphylococcus aureus*. Microbiology 2011;157(3):786–92.

[38] Vincents B, Önnerfjord P, Gruca M, Potempa J, Abrahamson M. Down-regulation of human extracellular cysteine protease inhibitors by the secreted staphylococcal cysteine proteases, staphopain A and B. Biol Chem 2007;388(4):437–46.

[39] Sloot N, Thomas M, Marre R, Gatermann S. Purification and characterisation of elastase from *Staphylococcus epidermidis*. J Med Microbiol 1992;37(3):201–5.

[40] Oleksy A, Golonka E, Bańbuła A, Szmyd G, Moon J, Kubica M, et al. Growth phase-dependent production of a cell wall-associated elastinolytic cysteine proteinase by *Staphylococcus epidermidis*. Biol Chem 2004;385(6):525–35.

[41] Cossart P, Jonquières R. Sortase, a universal target for therapeutic agents against gram-positive bacteria? Proc Natl Acad Sci U S A 2000;97(10):5013–5.

[42] Rzychon M, Sabat A, Kosowska K, Potempa J, Dubin A. Staphostatins: an expanding new group of proteinase inhibitors with a unique specificity for the regulation of staphopains, *Staphylococcus* spp. cysteine proteinases. Mol Microbiol 2003;49(4):1051–66.

[43] Shaw L, Golonka E, Potempa J, Foster SJ. The role and regulation of the extracellular proteases of *Staphylococcus aureus*. Microbiology 2004;150(1350–0872):217–28.

[44] Varadi David P, Saqueton Angelito C. Elastase from *Staphylococcus epidermidis*. Nature 1968;218:468–70.

[45] Arvidson S. Studies on extracellular proteolytic enzymes from *Staphylococcus aureus*. II. Isolation and characterization of an EDTA-sensitive protease. Biochim Biophys Acta 1973;302(1):149–57.

[46] Banbula A, Potempa J, Travis J, Fernandez-Catalén C, Mann K, Huber R, et al. Amino-acid sequence and three-dimensional structure of the *Staphylococcus aureus* metalloproteinase at 1.72 å resolution. Structure 1998;6(9):1185–93.

[47] Sabat AJ, Wladyka B, Kosowska-Shick K, Grundmann H, van Dijl JM, Kowal J, et al. Polymorphism, genetic exchange and intragenic recombination of the aureolysin gene among *Staphylococcus aureus* strains. BMC Microbiol 2008;8:129.

[48] Sabat A, Kosowska K, Poulsen K, Kasprowicz A, Sekowska A, Van Den Burg B, et al. Two allelic forms of the aureolysin gene (aur) within *Staphylococcus aureus*. Infect Immun 2000;68(2):973–6.

[49] Takeuchi, S., Saito, M., Imaizumi, K., Kaidoh, T., Higuchi, H. and Inubushi, S. (2002) Genetic and enzymatic analyses of metalloprotease (aureolysin) from *Staphylococcus aureus* isolated from domestic animals. Vet Microbiol, 84(1–2), pp. 135–142.

[50] Wladyka B, Bista M, Sabat AJ, Bonar E, Grzeszczuk S, Hryniewicz W, et al. A novel member of the thermolysin family, cloning and biochemical characterization of metalloprotease from *Staphylococcus pseudintermedius*. Acta Biochim Pol 2008;55(3):525–36.

[51] Beaufort N, Wojciechowski P, Sommerhoff CP, Szmyd G, Dubin G, Eick S, et al. The human fibrinolytic system is a target for the staphylococcal metalloprotease aureolysin. Biochem J 2008;410(1):157–65.

[52] Potempa M, Potempa J. Protease-dependent mechanisms of complement evasion by bacterial pathogens. Biol Chem 2012;873–88.

[53] Cassat JE, Hammer ND, Campbell JP, Benson MA, Perrien DS, Mrak LN, et al. A secreted bacterial protease tailors the *Staphylococcus aureus* virulence repertoire to modulate bone remodeling during osteomyelitis. Cell Host Microbe 2013;13(6):759–72.

[54] Ayora S, Lindgren PE, Gotz F. Biochemical properties of a novel metalloprotease from *Staphylococcus hyicus* subsp. hyicus involved in extracellular lipase processing. J Bacteriol 1994;176(11):3218–23.

[55] Ayora S, Götz F. Genetic and biochemical properties of an extracellular neutral metalloprotease from *Staphylococcus hyicus* subsp. hyicus. Mol Gen Genet 1994;242(4):421–30.

[56] Takeuchi S, Murase K, Kaidoh T, Maeda T. A metalloprotease is common to swine, avian and bovine isolates of *Staphylococcus hyicus*. Vet Microbiol 2000;71(1–2):169–74.

[57] Bukowski M, Wladyka B, Dubin G. Exfoliative toxins of *Staphylococcus aureus*. Toxins 2010;2(5):1148–65.

[58] Plano LRW, Adkins B, Woischnik M, Ewing R, Collins CM. Toxin levels in serum correlate with the development of staphylococcal scalded skin syndrome in a murine model. Infect Immun 2001;69(8):5193–7.

[59] Calander AM, Jonsson IM, Kanth A, Arvidsson S, Shaw L, Foster SJ, et al. Impact of staphylococcal protease expression on the outcome of infectious arthritis. Microbes Infect 2004;6(2):202–6.

[60] Coulter SN, Schwan WR, Ng EYW, Langhorne MH, Ritchie HD, Westbrock-Wadman S, et al. *Staphylococcus aureus* genetic loci impacting growth and survival in multiple infection environments. Mol Microbiol 1998;30(2):393–404.

[61] Imamura T, Tanase S, Szmyd G, Kozik A, Travis J, Potempa J. Induction of vascular leakage through release of bradykinin and a novel kinin by cysteine proteinases from *Staphylococcus aureus*. J Exp Med 2005;201(10):1669–76.

[62] Calander AM, Dubin G, Potempa J, Tarkowski A. *Staphylococcus aureus* infection triggers production of neutralizing, V8 protease-specific antibodies. FEMS Immunol Med Microbiol 2008;52(2):267–72.

[63] Holtfreter S, Nguyen TTH, Wertheim H, Steil L, Kusch H, Truong QP, et al. Human immune proteome in experimental colonization with *Staphylococcus aureus*. Clin Vaccine Immunol 2009;16(11):1607–14.

[64] Zdzalik M, Karim AY, Wolski K, Buda P, Wojcik K, Brueggemann S, et al. Prevalence of genes encoding extracellular proteases in *Staphylococcus aureus*—important targets triggering immune response in vivo. FEMS Immunol Med Microbiol 2012;66(2):220–9.

[65] Kolar SL, Antonio Ibarra J, Rivera FE, Mootz JM, Davenport JE, Stevens SM, et al. Extracellular proteases are key mediators of *Staphylococcus aureus* virulence via the global modulation of virulence-determinant stability. Microbiologyopen 2013;2(1):18–34.

Chapter 12

Staphylococcal Lipases

Aldo Lepidi
Department of Life, Health and Environmental Sciences (MeSVA), L'Aquila University, L'Aquila, Italy

12.1 LIPASES AND THEIR MOLECULAR RELATIVES

Lipases are water soluble enzymes that catalyze hydrolysis, synthesis, and modification of triglycerides, displaying maximal activity toward water-insoluble long-chain triglycerides. Hydrolysis of estereous bonds is the usual catalytic activity of lipases within and outside the cell where such enzymes exert a universal and basic role in the cellular systems. The reverse reactions of esterification and transesterification are catalyzed as well, provided that proper conditions for reaction are realized. This property constitutes a powerful tool in several successful modern biotechnologies. Other substrates for lipases are phospholipids and cholesterol esters. The systematic name for these enzymes according to The International Union of Biochemistry and Molecular Biology is triacylglycerol acylhydrolase, ascribed to the subclass EC.1.1.1.3. The reaction they reversibly catalyze is shown in Fig. 12.1.

Lipase, or "true" lipase, is a common name describing lipolytic enzymes altogether, sometimes including carboxyl-ester hydrolases, or carboxylesterases, (EC 3.1.1.1) which hydrolyze small ester-containing molecules that are at least partly soluble in water, and phospholipase (EC 3.1.1.4) releasing fatty acids from the second carbon group of glycerol-based phospholipids, basic components of biological membranes.

Lipases are members of a large group of enzymes possessing the α/β hydrolase fold, one of the most versatile and widespread folds found in cell proteins [1]. In their canonical form, lipase proteins hold the α/β domain largely conserved in the hydrolase enzymes, composed of a central core, done by eight parallel (on occasion antiparallel) β-sheets, intermingled with α helices. This intriguing structure is outlined in Fig. 12.2, which points out the preserved arrangement of the catalytic residue (His-Ser-Asp or Cys) located in a tight loop after the $\beta5$ strand. This active loop is accompanied by a widely conserved histidine residue after the last strand, by an acidic residue after the strand $\beta7$, and by a consensus motif frequently identified in the sequence Gly-Xaa-Ser-Xaa-Gly [2].

Pet-to-Man Travelling Staphylococci: A World in Progress. https://doi.org/10.1016/B978-0-12-813547-1.00012-1

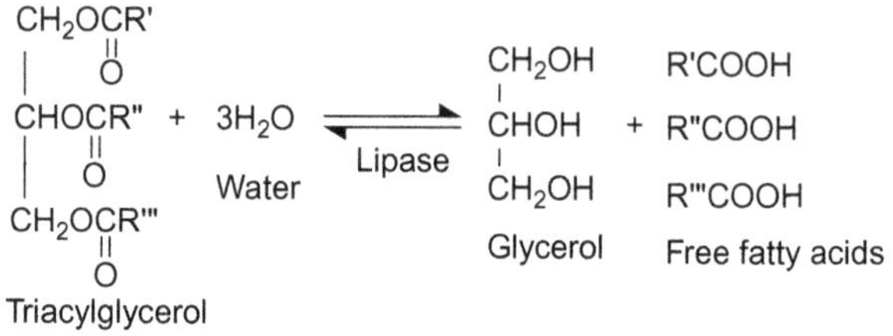

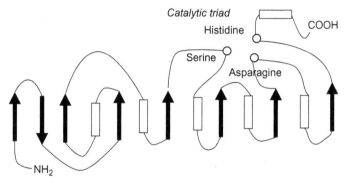

Triacylglycerol

FIG. 12.1 The leading reversible reaction catalyzed by lipases.

FIG. 12.2 Diagram of the secondary structure of staphylococcal lipases. α helices and β strands are represented respectively by white rectangles and black arrows (progressive numbers of strands and helices from left to right; notice the antiparallelism of the second β sheet). Amino acids of the catalytic triad are evidenced. *(Reshaped from Nardini M, Dijstra BW. α/β hydrolase fold enzymes: the family keeps growing. Curr Opin Struct Biol 1999;9:732–7.)*

Since 1992, when the α/β hydrolytic domain was identified and described [1], catalytic folds of sheets and helices supporting active sites formed by an aminoacidic triad or dyad similar to the one present in lipases have been found in a very large number of enzymes, including proteases, esterases, dehalogenases, peroxidases, and epoxide hydrolases. The α/β hydrolase protein superfamily is therefore a very significant example of biological evolution through mutation, gene duplication, and horizontal gene transfer.

This led some authors to postulate that these enzymes constitute a multigenic superfamily sharing structural similarity, and preserving a similar arrangement of the catalytic sites and deriving from common evolutionary ancestors [3]. During such a long and complex evolution, several kinds of mutations, insertions, and deletions have been successfully introduced into the protein sequences, and some parts of the gene have been duplicated and/or acquired by horizontal gene transfer [4].

Such a versatile protein superfamily arose through evolution in a biological context where hydrolases are ubiquitous tools useful in diverse metabolic steps with different substrates (i.e., degrading biopolimers, opening estereous-ethereous bonds, hydrating molecules). Different hydrolases, showing diverse specificities, originated by evolution while the general protein shape and the catalytic efficacy of the active site were thoroughly kept [5].

In the catalytic triad of lipases, aspartic acid forms a hydrogen bond with histidine, increasing the pKa of the histidine imidazole nitrogen. This gives the histidine a powerful basic reaction that removes a proton from serine. The deprotonated serine confers a negative charge to the ester carbonyl of triglycerides so that a water molecule donates a proton to histidine, thus strengthening the attack on the carbonyl carbon of the lipid, which unties the ester bond, diffusing the fatty acid monomer away [6].

12.2 LIPASE GENE FAMILY

The lipase protein family is identified by precise amino acid sequences revealing conspicuous similarities among lipases from very different origins such as bacteria, fungi, animals, and plants. Such sequence homologies are in accordance with other structural and functional similarities linking together the lipase proteins and supporting by and large the current acceptance of a phylogenetic relationship among a conspicuous part of the enzymes dealing with the metabolism of fatty compounds (triglycerides, etc.).

Due to the widespread biological role of lipases and to their applied interest in health care, industry, household facilities, and the environment, a database (called LIPABASE) has been devoted to general, taxonomic, physicochemical, structural, and molecular data on lipases. The site is free and available at http://www.lipabase-pfba-tun.org.

Within the α/β hydrolase enzymes superfamily, lipases are linked together in one (or a few) protein families, deeply rooting up to the beginning of life when membranes based on triglycerides became part of the cell structures. Further on, lipases differentiated to cope with the numerous and various requirements and roles of lipids and related compounds such as membrane bound activities (i.e., beginning of life, photosynthesis, respiration) [7], eukaryotization [8] and, in the blooming success of differentiation, in cell-to-cell dialog, hormone regulation, defense of cells, tissues, and organisms of plants and animals [9]. The phylogenetic tree of bacterial lipases, constructed using several dozens of registered sequences present in the genomes of strains pertaining to various systematic branches, shows a rich differentiation among diverse bacterial strains [3,4]. The level of genetic diversification in bacterial lipases is confirmed all over in the living world with a blooming of lipase diversification in differentiated organisms such as vertebrates [9].

12.2.1 Human Lipases

Human lipases are of key concern for human health and are therefore submitted to extensive investigation. By now, more than a dozen diverse proteins (and genes) have been described in mammals encoding for lipases of the groups EC 3.1.1.3 (true lipases), EC 3.1.1.34 (lipoprotein lipases) and EC 3.1.1.43 (lecithin-cholesterol acyltransferases). The diverse isozymes are largely expressed in lipocytes and at the pancreatic, hepatic, and gastro-lingual level [10]. Phylogenetic analysis based on comparative structures and sequencing indicates that, in vertebrate evolution, those of lipases are very ancient genes deriving, through duplications and recombination, from ancestral genes that, in their turn, primarily encode for adipose triglyceride lipase having a pivotal role in lipocytes [9]. All the members of this protein group, present also in several bacteria including Staphylococci, conserve noticeable homologies in amino acid sequences and the active site formed by the usual triads (Ser-His-Asp) or, occasionally, a catalytic dyad (Ser-Asp) [11]. Number and distribution of exons is also largely conserved in vertebrate lipases which, apart from a few fish, predominantly contain nine coding exons [9].

In humans, the various isoenzymes of the lipase complex have multifunctional properties, broad substrate specificity, and regiospecificity. They do hydrolyze triglycerides, mostly in adipocytes, perform hydrolysis and transacylation of several acylglycerols [12], regulate various hormone-sensitive metabolisms [13], intervene during fasting [14] and training of skeletal muscles [15]. An intriguing feature of true human lipases is the high degree of homology and structural similarities they share with patatin, a glycoprotein originally found as the main component of the storage protein in potato tubers [16]. In potatoes, the genes for patatin are duplicated 10–15 times for the haploid genome and are associated with plant defense against parasites and cell-to-cell signal transduction [17]. In 2001, Hirschberg et al. observed that extensive domains are conserved between patatin and human phospholypases, the most conserved ones being the regions centered around the Ser-His-Asp triad. Further research demonstrated that significant sequence homologies and structural/functional similarities are largely widespread among genes and proteins of the entire group of acyl hydrolyzing enzymes spanning the whole field of hydrolases, regardless of which group of living beings is involved, from bacteria to man [18].

12.3 STAPHYLOCOCCAL LIPASES IN HUMAN AND ANIMAL HEALTH

Lipolytic enzymes are used by several pathogens (bacteria, molds/yeasts and occasionally viruses) to begin and consolidate the infection process during localized and systemic diseases [19–21]. Extracellular lipases and phospholipases are secreted by some pathogenic bacteria to damage host cells with extensive disruptions starting from the cell membrane. Lipases of specific pathogenic concern are released by *Candida albicans*, whose genome holds several diverse lipase genes that are selectively expressed during infections in order to liberate free fatty acids enabling the yeast cells to adhere to host cells and tissues [22].

Staphylococcal lipases have been implicated as possible virulence factors in diverse pathologies, from localized superficial skin infections (boils, pimples, abscesses), to invalidating diseases (arthritis), to life threatening evils (endocarditis, septicemia) [23,32]. Similarly, studies by Lowe et al. [24] showed that streptococcal lipases are expressed in vitro during abscess development in a murine model.

Staphylococci encode and express diverse lipases differentiated by molecular and cytological characters. Specific lipases are found in cell cytoplasm, cell membranes, the cell surface, and extra-cellular space [25]. Trans-membrane lipases have been recently found in bacteria, and such a location joins lipases to other trans-membrane proteins sharing the module of multiple β-sheets [26].

Extra-cellular lipases in Staphylococci are excreted via the general secretory pathway or via ABC transporters; sometimes folding and disulphide-bonds are needed for excretion. Staphylococcal lipases are produced as pre-pro-enzymes with a signal peptide in the preregion and are secreted as pro-enzymes needing a specific cleavage for mature configuration [27] (Fig. 12.3).

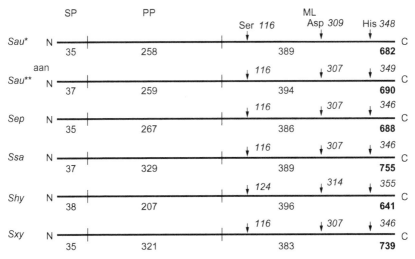

FIG. 12.3 Phylogenetic relationship among lipases, phospholipases, and surface associated lipases of the genus *Staphylococcus*. SP = signaling peptide; PP = pro-peptide; ML = mature lipase; Sau * = *Staphylococcus aureus* NCTC 8530; Sau** = *S. aureus* PS54; Sep = *Staphylococcus epidermidis* RP62A; Ssa = *Staphylococcus saprophyticus* RN 7108; Shy = *Staphylococcus hyicus* DSM 20459; Sxy = *Staphylococcus xylosus* DSM 20266. *(Data from Gotz F, Verheij HM, Rosenstein R. Staphylococcal lipases: molecular characterization, secretion and processing. Chem Phys Lipids 1998;93:15–25; Rosenstein R, Gotz F. Staphylococcal lipases: biochemical and molecular characterization. Biochemie 2000;82:1005–14; Sakinc T, Woznowski M, Ebsen M, Gatermann SG. The surface associated protein of Staphylococcus saprophyticus is a lipase. Infect Immun 2005;73:6419–28.)*

Pathogenicity of Staphylococci is associated with several genes located both on chromosomes and on mobile elements encoding for a variety of pathogenic factors that are thoroughly treated in other chapters of this volume. Staphylococcal lipases have been found to contribute to the success of pathogenic bacterial-host interrelationships as true pathogenic factors, but the mechanisms by which they contribute to a successful pathogenic event are fully addressed only in some cases [23].

Lipases of Staphylococci (true lipases and phospholipases) are reported to contribute to a successful infection by:

- disturbing the immune response [28];
- interfering with granulocyte function, aggregation, and recruitment [29,30];
- disrupting the membranes of the host cells [31];
- interfering with host cell signaling [32];
- disturbing phagocytosis [30].

12.3.1 Staphylococcal Lipases and Fatty Acid Modifying Enzymes

Large amounts of monoglycerides and unsaturated long chain free fatty acids are released in focal lesions infected by several staphylococcal species [33]. These molecules exert bactericidal effects against bacterial cells and are therefore part of the first defense shield of the host against invading pathogens [34]. At the same time, *Staphylococcus aureus* strains able to successfully infect the skin, internal lesions, the bladder, and lactating cows' udders release a proteic factor in culture media counteracting both kinds of bactericidal lipids [35]. This factor catalyzes esterification of free fatty acids to short chain, primary alcohols and cholesterol, with the latter as the preferred substrate, as well as transfer of the acyl group of the monoglycerides to an appropriate alcoholic radical; therefore it is known as "fatty acid modifying enzyme" (FAME). Only FAME-producing strains are found in infected tissues, the FAME defective ones being inactivated [34]. Staphylococcal FAME, however, is strongly inhibited by triglycerides released by the host cells in the infection site with the effect of saturating and inactivating the enzyme [33]. It has been observed that only *Staphylococcus* strains producing both lipase and FAME do establish a successful infection [34]. It is thought therefore that *Staphylococcus* lipases liberate free fatty acids from the glyceride backbone to restore esterification of the free fatty acids by FAME [35].

Similar mechanisms, likely exerting the same function, are found in several Staphylococci, including coagulase-negative species, such as *Staphylococcus schleiferi* subsp. *schleiferi*, *Staphylococcus saprophyticus*, and *Staphylococcus simulans* [2,35]. Expression of the FAME gene is not found in isolates of the coagulase-positive *S. schleiferi* subsp. *coagulans* [36].

12.3.2   Staphylococcal Lipases in Biofilm Formation

Biofilm formation is a key issue in several infectious events, mainly those involving respiratory and urinary systems as well as catheters and other medical and surgical implants. Pathogenic Staphylococci develop biofilms on cultures of collagen cells, mimicking the host extracellular protein framework where bacteria are used to adhere when starting and developing infections [37]. Mutants of biofilm-producing strains defective for lipase biosynthesis lose pathogenicity, including biofilm formation ability [38]. Antilipase serum inhibits biofilm formation by Staphylococci both in vitro and in vivo [39]. Lipases are likely to contribute to bacterial cell nutrition and proper gene expression so that prevention of biofilm settlement through lipase inhibition is considered a promising tool for control of pathogenic multidrug-resistant strains [23]. Staphylococcal lipase in biofilms affecting the urinary tract interacts with ureases, thus increasing alkalinization that further enhances, in its turn, lipase activity [34].

12.3.3   Lipase in Surface Associated Proteins

A particular mechanism associating staphylococcal diseases to lipases was described by Sakinc et al. [40] for *S. saprophyticus*. Strains of this species are important causes of urinary tract infections. Some molecules produced by bacterial cells to bind fibronectin and laminarin and to agglutinate erythrocytes are included among the pathogenic factors on which these diseases rely. Gatermann et al. [41] reported that most clinical *S. saprophyticus* isolates produce large quantities of a surface-associated protein (Ssp) that is absent, and vice versa, in harmless strains. This protein is different from adhesins and agglutinins and it forms fuzzy surface appendages interpreted primarily as adherence tools by which bacterial cells recognize and bind host cells. However, this protein does not bind fibronectin, fibrinogen, collagen, and laminarin, so clearly, a different explanation is needed.

Sakinc et al. [40] cloned the *ssp* gene of *S. saprophyticus* through an Ssp-specific antiserum. They characterized the gene product, including the amino acid sequence, and found extensive and significant homology to staphylococcal lipase genes such as the catalytic triad of lipases (serine, aspartic acid, and histidine) and the P-loop consensus sequence. These Authors confirmed the lipase nature of the Ssp protein by demonstrating that only Ssp-producing strains possess lipolytic activity and that lipase-negative Staphylococci do produce the enzyme after transformation by a suitable vector of the *ssp* gene.

12.4   BIOTECHNOLOGICAL APPLICATIONS OF STAPHYLOCCCOCAL LIPASES

Lipases are relevant in several food processes, many of which take part in very long traditions. Therefore, lipase-producing microbes have been studied since the beginning of microbiology. In the context of modern technologies, bacterial lipases have become important biocatalysts, finding momentous applications in food, dairy, detergent, pharmaceutical, and energy industries, and supporting the expansion of commercial ventures [42]. Extracellular bacterial lipases are among the most widely used enzymes in biotechnological industries. The most relevant sources of these enzymes include *Achromobacter, Alcaligenes, Arthrobacter, Burkholderia, Chrombacterium, Pseudomonas, Bacillus,* and *Staphylococcus* [43–46]. New lipases are frequently uncovered, and technological properties of these compounds are continuously enhanced by the novel potentialities of scientific research, such as molecular imprinting, directed evolution, and solvent and protein engineering [25,47].

The use of enzymes represents an important mean in the biotechnology industry with a rapid increase in worldwide usage approaching 4 billion dollars, according to the 2011 Report of Global Industry Analysts.

Hydrolytic enzymes, such as proteases, amylases, and lipases, hold the major share of the industry enzyme market (Table 12.1). Lipases emerged as key factors in fast growing industries (see Table 12.2) owing to their all-around properties, their capability to carry out hydrolysis and the reverse reaction of esterification, and their ability to stay active in aqueous and nonaqueous media and within a wide temperature range. Lipases are active toward very diverse substrates (such as esters and ethers) and, in the meantime, they are highly specific as to regio- and enantioselective catalysis [25,57]. New lipases for industry are created by site directed mutagenesis [58,59] and by directed evolution [60,61]; they have been recently found by means of in silico characterization [62] and of metagenomic analysis [63].

TABLE 12.1 *Staphylococcus* Species Acting as Sources of Lipases for Industry (Both Clinical and Environment Isolates)

Species	Reference(s)
Staphylococcus arlettae	[64]
Staphylococcus aureus	[27,65]
Staphylococcus epidermidis	[27,58,59]
Staphylococcus haemolyticus	[51]
Staphylococcus hyicus	[27,54]
Staphylococcus lipolyticus	[66]
Staphylococcus simulans	[67]
Staphylococcus warneri	[68]
Staphylococcus xylosus	[54]

TABLE 12.2 Relevant Areas of Industrial Application of Staphylococcal and Other Bacterial Lipases

Industrial field	Action	Products and results	Reference(s)
Bakery	Emulsification, softening	Bread, cakes, quality, and shelf life improvement	[46,48]
Biodiesel production	Transesterification	Biodiesel fuel from vegetable and waste oils	[49,50]
Biosensors	Analysis of acylglycerols, ancillary reactant in other analyses	Analysis of blood and other clinical samples, foods, drugs, and pesticides	[49,51,52]
Chemical/ oleochemical industry	Selective transesterification-interesterification, region selective modification	Structured lipids for synthesis, pharmaceutical PUFA, biodegradable soaps and plastics	[49,53]
Cooking oils	Degumming, hydrolysis, transesterification	Koji, tempeh, soybean sauces, cocoa specialties	[45,51]
Detergent industry	Hydrolysis of fat and fat containing aggregates	Household/industrial laundry and waste/sewage treatment	[49,54]
Diary industries	Hydrolysis, transesterification	Butter/cheese preproducing and ripening, human milk substitutes	[46,51]
Dietetic foods	Incorporation/ removal of specific fatty acids	Low caloric fats, resistance to high temperatures	[51,55]
Egg yolk treatment	Lecithin hydrolysis	Mayonnaise, sauces, cafes	[56]
Fats	Glyceride replacing by trans-interesterification	Margarine, cocoa butter equivalents, nutritionally tailored fatty products	[46,51]
Food dressing	Synthesis of special esters, regio-selective modification	Enhancement of nutritional value and flavor	[45,46]

Continued

TABLE 12.2 Relevant Areas of Industrial Application of Staphylococcal and Other Bacterial Lipases—cont'd

Industrial field	Action	Products and results	Reference(s)
Meat/fish and derivatives	Removal of fats and/ or of specific fatty components	Defatted foods	[46]
Pharmaceuticals and fine chemicals	Resolution of racemic mixtures, synthesizing chiral blocks, enantioselective esterification in polar and nonpolar solvents	Vitamin antibiotic and other drugs synthesis and refinement, removal of allergens, liposome functionalizing	[45,46,51,54]
Textile industry	Soft degreasing, excess reagent removal	Strengthened fabrics, better washability and staining (natural and synthetic fibers)	[54]

REFERENCES

[1] Ollis DL, Cheah E, Cygler M, Dijkstra B, et al. The α/β hydrolase fold. Protein Eng 1992;5:197–211.

[2] Rosenstein R, Gotz F. Staphylococcal lipases: biochemical and molecular characterization. Biochemie 2000;82:1005–14.

[3] Yadav SK, Dubey AK, Bisht D, Darmwal NS, Yadav D. Aminoacid sequences based phylogenetic and motif assessment of lipases from different organisms. J Bioinform 2012;13:400–17.

[4] Messaudi A, Baguith H, Ghram I, Hamida JB. LIPABASE: a database for true lipasi family enzymes. Int J Bioinforma Res Appl 2011;7:390–401.

[5] Arpigny JL, Jaeger KE. Bacterial lipolytic enzymes: lassification and properties. Biochem J 1999;343:177–83.

[6] Gabor I, Szabadka Z, Grolmusz V, Náray-Szabó G. Four spatial points that define enzyme families. Biochem Biophys Res Commun 2009;383:417–20.

[7] Fedonkin MA. Eukeryotization of the early biosphere: a geobiochemical aspect. Geochem Int 2009;47:1265–333.

[8] Hengeveld R, Fedonkin MA. Causes and consequences of eukaryotization through mutualistic endosymbiosis ans compartimemtalization. Acta Biotheor 2004;52:105–54.

[9] Holmes RS, VandeBerg JL, Cox LA. Vertebrate endothelial lipase: comparative studies of an ancient gene and protein in vertebrate evolution. Genetica 2011;139:291–304.

[10] Lord CC, Thomas G, Brown JM. Mammalian α/β hydrolase domain proteins: lipid metabolizing enzymes at the interface of cell signalling and energy metabolism. Biochim Biophys Acta 2013;1831:792–802.

[11] Rydel TJ, Williams JM, Krieger E, Moshiri F, et al. The crystal structure, mutagenesis, and activity studies reveal that patatin is a lipid acyl hydrolase with a Ser-Asp catalytic dyad. Biochemistry 2003;42:6696–708.

[12] Zimmermann R, Lass A, Haemmerle G, Zechner R. Fate of fat: the role of adipose triglyceride lipase in lipolysis. Biochim Biophys Acta 2009;1791:494–500.

[13] Be Zaire V, Mairal A, Ribet C, Lefort C, et al. Contribution of adipose triglyceride lipase and hormone-sensitive lipase to lipolysis in hMADS adipocytes. J Biol Chem 2009;284:18282–91.

[14] Lake AC, Sun Y, Li JL, Kim JE, Johnson JW, et al. Expression, regulation, and triglyceride hydrolase activity of adiponutrin family members. J Lipid Res 2005;46:2477–87.

[15] Alsted TJ, Nybo L, Schweiger M, Fledelius C, et al. Adipose triglyceride lipase in human skeletal muscle is upregulated by exercise training. Am J Physiol Endocrinol Metab 2009;296:445–53.

[16] Andrews DL, Beames B, Summers MD, Park WD. Characterization of the lipid acyl hydrolase activity of the major potato (*Solanum tuberosum*) tuber protein, patatin, by cloning and abundant expression in baculovirus vector. Biochem J 1988;252:199–206.

[17] Kienesberger PC, Oberer M, Lass A, Zechner R. Mammalian patatin domain containing proteins: a family with diverse lipolytic activities involved in multiple biological functions. J Lipid Res 2009;50:63–8.

[18] Banerji S, Flieger A. Patatin-like proteins: a new family of lipolytic enzymes present in bacteria? Microbiology 2004;150:522–5.

[19] Gacser A, Stehr F, Kroeger C, Kredics L, et al. Lipase 8 affects the pathogenicity of *Candida albicans*. Infect Immun 2007;75:4710–8.

[20] Jaeger KE, Ransac S, Dijkstra BW, Colson C, et al. Bacterial lipases. FEMS Microbiol Rev 1994;15:29–63.

[21] Kamil JP, Tischer BK, Trapp S, Nair VK, et al. vLIP, a viral lipase homologue, is a virulence factor of Marek's disease virus. J Virol 2005;79:6984–96.

[22] Stehr F, Felk A, Gakser A, Krestschmar M, Mahn B, et al. Expression analysis of the *Candida albicans* lipase gene family during experimental enfections and in patient samples. FEMS Yeast Res 2004;4:401–8.

[23] Hu C, Xiong N, Zhang Y, Rayner S, Chen S. Functional characterization of lipase in the pathogenesis of *Staphylococcus aureus*. Biochem Biophys Res Commun 2012;419:617–20.

[24] Lowe AM, Beattie DT, Deresiewicz RL. Identification of novel staphylococcal virulence genes by in vivo expression technology. Mol Microbiol 1998;27:967–76.

[25] Gupta R, Gupta N, Rathi P. Bacterial lipases: an overview of production, purification and biochemical properties. Appl Microbiol Biotechnol 2004;64:763–81.

[26] Lazniewski M, Steczkiewicz K, Knizewski L, Wawer I, Ginalski K. Novel transmembrane lipases of alpha/beta hydrolase fold. FEBS Lett 2011;585:870–4.

[27] Jaeger KE, Dijkstra BW, Reetz MT. Bacterial biocatalists: molecular biology, three dimensional structure and biotechnological applications of lipases. Annu Rev Microbiol 1999;53:315–51.

[28] Ryding U, Renneberg J, Rollof J, Christensson B. Antibody response to *Staphylococcus aureus* whole cell, lipase and staphylolysin in patients with *S. aureus* infections. FEMS Microbiol Immunol 1992;4:105–10.

[29] Rollof E, Vinge P, Nilsson EP, Braconier JH. Aggregation of humans granulocytes by *Staphylococcus aureus* lipase. J Med Microbiol 1992;36:52–5.

[30] Rollof J, Braconier JH, Soderstrom C, Nilsson Ehle P. Interference of *Staphylococcus aureus* lipase with human granulocyte function. Eur J Clin Microbiol Infect Dis 1988;7:505–10.

[31] Priatkin RG, Kuzmenko OM. Secreted proteins of *Staphylococcus aureus*. Zh Mikrobiol Epidemiol Immunobiol 2010;4:118–24.

[32] Xie W, Khosasih V, Suwanto A, Kim HK. Characterization of lipases from *Staphylococcus aureus* and *Staphylococcus epidermidis* isolated from human facial sebaceous skin. J Microbiol Biotechnol 2012;22:84–91.

[33] Long JP, Hart J, Albers W, Kapral FA. The production of fatty acid modifying enzyme (FAME) and lipase by various staphylococcal species. J Med Microbiol 1992;37:232–4.

[34] Kapral FA, Smith S, Lal D. The esterification of fatty acids by *Staphylococcus aureus* fatty acid modifying enzyme (FAME) and its inhibition by glycerides. J Med Microbiol 1992;37:235–7.

[35] Lu T, Park JY, Parnell K, Fox LK, McGuire MA. Characterization of fatty acid modifying enzyme activity in staphylococcal mastitis isolates and other bacteria. BMC Res Notes 2012;22:323.

[36] Savini V, Passeri C, Mancini G, Iuliani O, Marrollo R, Fazii P, et al. Coagulase-positive staphylococci: my pet's two faces. Res Microbiol 2013;164:371–4.

[37] Saising J, Singdam S, Ongsakul M, Voravuthikunchai SP. Lipase, protease, and biofilm as the major virulence factors in staphylococci isolated from acne lesions. Biosci Trends 2012;6:160–4.

[38] Bowden MG, Visai L, Longshaw CM. Is the GehD lipase from *Staphylococcus epidermidis* a collagen binding adhesin? J Biol Chem 2002;277:43017–23.

[39] Xiong N, Hu C, Zhang Y. Interaction of sortase and lipase 2 in the inhibition of *Staphylococcus aureus* biofilm formation. Arch Microbiol 2009;191:879–84.

[40] Sakinc T, Woznowski M, Ebsen M, Gatermann SG. The surface associated protein of *Staphylococcus saprophyticus* is a lipase. Infect Immun 2005;73:6419–28.

[41] Gatermann S, Kreft B, Marre R, Wanner G. Identification and characterization of a surface associated protein (Ssp) of *Staphylococcus saprophyticus*. Infect Immun 1992;60:1055–60.

[42] Schmid A, Dordick JS, Hauer B, Kiener A, Wubbolts M, Witholt B. Industrial biocatalysis today and tomorrow. Nature 2001;409:258–68.

[43] Davranov K. Microbial lipases in biotechnology, a review. Appl Biochem Microbiol 1994;30:527–34.

[44] Jaeger KE, Reetz MT. Microbial lipases form versatile tools for biotechnology. Trends Biotechnol 1998;16:396–403.

[45] Pandey A, Benjamin S, Soccol CR, Nigam P, Krieger N. The realm of microbial lipases in biotechnology. Biotechnol Appl Biochem 1999;29:119–31.

[46] Sharma R, Chisti Y, Banerjee UC. Production, purification, characterization and applications of lipase. Biotechnol Adv 2001;19:627–62.

[47] Svendsen A. Lipase protein engineering. Biochim Biophys Acta 2000;1543:223–38.

[48] Keskin SD, Sumnu G, Sahin S. Usage of enzymes in a novel baking process. Mol Nutr Food Res 2004;48:156–60.

[49] Hasan F, Shah AA, Hameed A. Industrial applications of microbial lipases. Enzym Microb Technol 2006;39:235–51.

[50] Shah S, Sharma S, Gupta MN. Biodiesel preparation by lipase catalysed transesterification of Jatropha oil. Energy Fuel 2004;18:154–9.

[51] Aravindan R, Anbumathi P, Viruthagiri T. Lipase application in food industry. Indian J Biotechnol 2007;6:141–58.

[52] Sumner C, Krause S, Sabot A, Turner K, McNeil CJ. Biosensor based on enzyme-catalysed degradation of thin polymer films. Biosens Bioelectron 2001;16:709–14.

[53] Linko YY, Lamsa M, Wu X, Uosukainen E, Seppala J, Linko P. Biodegradable products by lipase biocatalysis. J Biotechnol 1998;66:41–50.

[54] Ray A. Application of Lipase in Industry. Asian J Pharm Technol 2012;2:33–7.

[55] Osborn HT, Akoh CC. Structured lipids—novel fats with medical, nutraceutical, and food applications. Compr Rev Food Sci Food Saf 2002;1:93–103.

[56] Reimerdes EH, Franke K, Sell M. In: Influencing functional properties of egg yolk by using phospholipases. Conf food structure quality, Cork 3–7 October; 2004.

[57] Horchani H, Ben Salem N, Chaari A, Sayari A, et al. Staphylococcal lipases stereoselectively hydrolyse the sn-2 position of monomolecular films of diglyceride analogs. Application to sn-2 hydrolysis of triolein. J Colloid Interface Sci 2010;347:301–8.

[58] Chang RC, Chou SJ, Shaw JF. Site-directed mutagenesis of a highly active *Staphylococcus epidermidis* lipase fragment identifies residues essential for catalysis. J Am Oil Chem Soc 2000;77:1021–6.

[59] Chang RC, Chou SJ, Shaw JF. Site-directed mutagenesis of a highly active *Staphylococcus epidermidis* lipase fragment identifies residues essential for catalysis. J Am Oil Chem Soc 2000;77:1021–6.

[60] Liebeton K, Zonta A, Schimossek K, Nardini M, et al. Directed evolution of an enantioselective lipase. Chem Biol 2000;7:709–18.

[61] Sangeetha R, Arulpandi I, Geetha A. Bacterial lipases as potential industrial biocatalysts: an overview. Res J Microbiol 2011;6:1–24.

[62] Chakravorty D, Parameswaran S, Dubey VK, Patra S, et al. In silico characterization of thermostable lipases. Extremophiles 2011;15:89–103.

[63] Jiang X, Xu X, Huo Y, Wu Y, et al. Identification and characterization of novel esterases from a deep-sea sediment metagenome. Arch Microbiol 2012;194:207–14.

[64] Chauhan M, Chauhan RS, Garlapati VK. Evaluation of a new lipase from *Staphylococcus* sp. for detergent additive capability. Biomed Res Int 2013, https://doi.org/10.1155/2013/374967.

[65] Horchani H, Mosbah H, Ben Salem N, Gargouri Y, Sayari A. Biochemical and molecular characterisation of a thermoactive, alkaline and detergent-stable lipase from a newly isolated *Staphylococcus aureus* strain. J Mol Catal B Enzym 2009;56:237–45.

[66] Arora PK. *Staphylococcus lipolyticus* sp. nov., a new cold-adapted lipase producing marine species. Ann Microbiol 2013;63:913–22.

[67] Horchani H, Aissa I, Ouertani S, Zarai Z, et al. Staphylococcal lipases: biotechnological applications. J Mol Catal B 2012;76:125–32.

[68] Volpato G, Filice M, Rodrigues RC, Heck JX, et al. Modulation of a lipase from *Staphylococcus warneri* EX17 using immobilization techniques. J Mol Catal B Enzym 2009;60:125–32.

FURTHER READING

Gotz F, Verheij HM, Rosenstein R. Staphylococcal lipases: molecular characterization, secretion and processing. Chem Phys Lipids 1998;93:15–25.

Hirshberg HJ, Simons J, Dekker N, Egmond MR. Cloning, expression, purification and characterization of patatin, a novel phospholipase A. Eur J Biochem 2001;268:5037–44.

Jaeger KE. Extracellular enzymes of *Pseudomonas aeruginosa* as virulence factors. Immun Infekt 1994;22:177–80.

Nardini M, Dijstra BW. α/β hydrolase fold enzymes: the family keeps growing. Curr Opin Struct Biol 1999;9:732–7.

Simons JW, Adams H, Cox RC, Dekker N, et al. The lipase from *Staphylococcus aureus*: expression in *Escherichia coli*, large-scale purification and comparison of substrate specificity to *Staphylococcus hyicus* lipase. Eur J Biochem 1996;242:760–9.

Chapter 13

Staphylococcal Bacteriocins

Paweł Mak
Department of Analytical Biochemistry, Faculty of Biochemistry, Biophysics and Biotechnology, Jagiellonian University, Krakow, Poland

13.1 INTRODUCTION

Bacteriocins are ribosomally synthesized peptides or proteins produced by bacteria and are able to kill phylogenetically related strains at very low concentrations. Three general features—formation in ribosomes, activity at low concentrations (in the nanomolar range), and a narrow spectrum of action—distinguish bacteriocins from antibiotics. The latter, in fact, are synthesized by multienzymatic synthetase complexes, usually act at higher concentrations (in the micromolar range), and are effective against a broad spectrum of microorganisms.

It is estimated that 99% of bacterial strains produce at least one bacteriocin [1]. The sequences and physicochemical parameters of these molecules, along with information about their structure, taxonomy, bibliography, and biological activity have been integrated in a convenient database, called BACTIBASE [2]. At present, BACTIBASE contains 219 entries, from both Gram-negative and Gram-positive bacteria, describing diverse peptides and proteins of various sizes, structures, mechanisms of action, as well as gene organization. Also, such diversity greatly complicates bacteriocins' classification, which is endlessly under debate. Concise presentation of it, however, as well as of the aforementioned structural variety, is essential to understand the present chapter, which focuses on staphylococcal bacteriocins and the role they are presumed to play in zoonoses.

Bacteriocins from Gram-positive bacteria comprise four groups of peptides and proteins: class I, including posttranslationally modified peptides; class II, containing unmodified, heat-stable, small peptides below 10 kDa; class III, comprising large, over 10 kDa, heat-labile proteins; and class IV, which is formed by poorly characterized complex proteins, containing carbohydrate or lipid moieties [3–8]. A more detailed structural classification together with major criteria on which their organization is based are presented in Fig. 13.1.

The majority of bacteriocins are synthesized as inactive biologically proforms, containing an N-terminal leader sequence. Such a sequence maintains

Pet-to-Man Travelling Staphylococci: A World in Progress. https://doi.org/10.1016/B978-0-12-813547-1.00013-3

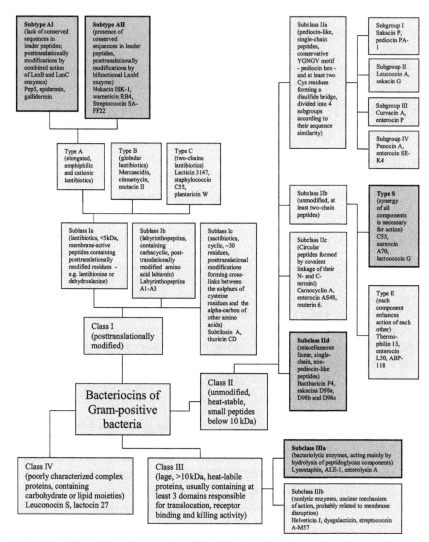

FIG. 13.1 Scheme illustrating criteria and classification of bacteriocins from Gram-positive bacteria. Selected representative compounds are also presented. Categories containing staphylococcins have been marked in bold.

the molecule in inactive form, allows for interaction with exporting machinery, and, in the case of posttranslationally modified bacteriocins, is recognized by enzymes responsible for modifications [9]. The leader sequence is cleaved during secretion from the cell, and this process is realized by a particular bacteriocin exporting system or, in some cases, by a general secretory (Sec) pathway. Moreover, expression of bacteriocin must be accompanied by production of specific immunity proteins, which cause the bacterial cell to resist the action of the bacteriocin. Immunity proteins may serve as an antagonistic receptor for particular bacteriocin, and/or act as specialized ATP-binding cassette (ABC) transporter systems, able to pump bacteriocin from the cell of the producer [10,11]. Expression of immunity proteins, as well as their high specificity, is critical because they allow for distinction between cells of producer and target cells—which usually are closely related and exist in the same ecological niche. Considering the data, it seems obvious that the genes of bacteriocins are usually complex and code pro-form of bacteriocin, immunity proteins, and transporter proteins, as well as enzymes involved in posttranslational modifications [12]. Expression of bacteriocin as well as bacteriocin immunity proteins is often a part of bacterial quorum-sensing mechanisms [13] and additionally may be induced by stress factors (DNA damage, UV light, antibiotics) acting via SOS promoters [14]. According to some recent studies [15], expression of bacteriocins may be also interpreted as a factor promoting stable genetic transmission of plasmids that encode bacteriocin—analogically as the similar, widespread among bacteria, toxin-antitoxin (TA) systems [16]. All these features prove that the biological role of bacteriocins is complex and is not simply limited to inhibition of the growth of other strains in the same ecological niche during overcrowding and depletion of nutrients.

13.2 STAPHYLOCOCCAL BACTERIOCINS

Staphylococcal bacteriocins are called "staphylococcins" and, so far, they form a relatively narrow group of compounds. They are produced by both coagulase-negative (CNS) and coagulase-positive (CPS) species/strains and belong to four groups: subclass Ia, lantibiotics (nine members, all from CNS strains), subclasses IIb and IId (two and three staphylococcins per subclass, respectively), and class III (comprising two bacteriocins). Many staphylococci also produce the so-called BLIS (bacteriocin-like inhibitory substances) which, so far, have been neither fully characterized nor classified. All the factors mentioned herein are presented in the following sections.

13.2.1 Subclass Ia Staphylococcins

The most numerous group of staphylococcins is formed by peptides belonging to class I (single-chain, posttranslationally modified bacteriocins), subclass Ia (lantibiotics—small peptides of molecular weight below 5 kDa, containing thioeter or dehydrated residues—lanthionine, 3-methyl-lanthionine, dehydroalanine or dehydrobutyrine), type A (lantibiotics forming elongated, amphipathic, screw-shaped molecules). Genes for these peptides (denoted as lanA) also encode, together with the peptide chain, a C-terminal propeptide and an N-terminal leader sequence. After synthesis, the unmodified linear prepeptide undergoes subsequent dehydration, formation of intramolecular thioether bonds, cleavage of the leader sequence and, finally, secretion [17]. For some staphylococcins (epicidin

280, epilancin K7 and epilancin 15X), enzymatic modification of N-terminal amino acid occurs as well [18]. Each stage of these complex processes is carried out by enzymes that are encoded by separate genes: lanB encodes proteins responsible for dehydration; lanC encodes enzymes catalyzing thioether bond formation; lanD encodes decarboxylases of the C-terminal cysteine residue found in some lantibiotics; lanP encodes a peptidase that removes the leader sequence; lanT encodes a transporter protein; lanK and lanR encode proteins involved in expression regulation; finally, lanI, lanH as well as lanEFG encode immunity proteins. Lantibiotics that are encoded by the separate genes lanB and lanC have been included into subtype AI, whereas those encoded by the single gene lanM (instead of lanB and lanC) form the subtype AII. Most staphylococcins belong to subtype AI, and the majority of them are produced by CNS. This group comprises seven distinct 21-34 amino acid-long peptides: Pep5 [19] and the very similar epicidin 280 [20], both from *Staphylococcus epidermidis* (strains 5 and BN 280, respectively); epilancin K7 and the closely related epilancin 15X, produced by *S. epidermidis* strains K7 and 15X154, respectively [21,22]; three similar peptides: epidermin, from *S. epidermidis* Tü3298 [19] (the peptide is identical to the previously described *S. epidermidis* staphylococcin 1580 [23]), gallidermin, isolated from *Staphylococcus gallinarum* F16/P57 Tü3298 [24] and identical to the previously described staphylococcin T from *Staphylococcus cohnii* T [25]; as well as BacCH91, from *Staphylococcus aureus* CH91/DSM26258 [26].

On the other hand, subtype AII staphylococcins are represented by two peptides, produced by CNS strains as well: nukacin ISK-1 [27] and the very similar warnericin RB4 (isolated from the *Staphylococcus warneri* strains Nukadoko and RB4, respectively) [28]. Recent studies have shown that nukacin ISK-1 is identical to the previously described peptide simulancin 3299 isolated from *Staphylococcus simulans* strain 3299 [29].

All the mentioned staphylococcins exert their killing properties toward sensitive bacteria through disruption of their cytoplasmic membranes, and consequent depolarization, cell content efflux, and death. Membrane destabilization is initiated by electrostatically driven interactions of peptides with the membrane surface and following formation of transmembrane channels [18]. Pep5 is also able to bind negatively charged teichoic and lipoteichoic acids, causing release and activation of cell wall hydrolyzing enzymes, and finally autolysis [18]. It was also demonstrated that gallidermin and epidermin specifically bind the cell wall precursor membrane-bound lipid II. Such a phenomenon facilitates pore formation in the membrane and simultaneously inhibits peptidoglycan biosynthesis [30].

13.2.2   Subclass IIb Staphylococcins

Staphylococcins that contain two or more peptide chains belong to class II bacteriocins (unmodified, heat-stable peptides with molecular weight below

10 kDa), subclass IIb (two or more chain-peptides), type S (synergistic: presence of all subunits is necessary to exert activity) and are represented by two peptides staphylococcin C55 (from *S. aureus* C55) [31] and aureocin A70 (from *S. aureus* A70) [32]. Staphylococcin C55 appeared to be identical to bacteriocin BacR1 from *S. aureus* UT0007 [33] and is composed of the two peptide chains C55α and C55β (30 and 32 amino acid residues, respectively), that exclusively exert bacterial killing when acting synergistically, in equimolar proportions. C55 is plasmid-encoded, and computer analysis of its sequence revealed the presence of the two genes sacαA, sacβB encoding peptide chains, sacM1, encoding a protein of unknown functions and sharing homology with a product of the lantibiotic lanM gene, and sacT, which encodes a protein similar to ABC transporters [33]. Very little is known about the C55 specific mechanism of action, but studies on similar two-peptide bacteriocins [34] suggest that both chains/halves of this staphylococcin enhance their own helicity, which is necessary for aggregation, interaction with the membrane, and pore formation through the "barrel-stave" or "wormhole" mechanisms. One can also speculate that two-component bacteriocins are capable of chiral-specific recognition of certain membrane receptors.

Aureocin A70, on the other hand, is composed of four 30-31 amino acid-long, strongly cationic peptides that, as in C55, are all are necessary to display full bactericidal activity. Aureocin A70 is encoded by a plasmid organized by two operons. The first one consists of the four genes aurABCD, encoding peptide chains, while the second contains the single gene aurT that encodes an ABC-transporter-like enzyme [32]. The A70-related plasmid also contains the putative operons orfA, orfB, and orfM, encoding proteins similar to those involved in regulation of lantibiotics expression, as well as those similar to immunity proteins. Unfortunately, nothing is known, currently, about the aureocin A70 mechanism of action. However, a recent study has demonstrated that this bacteriocin significantly potentiates its activity in the presence of aureocin A53 [35].

13.2.3   Subclass IId Staphylococcins

So far, three staphylococcal bacteriocins have been classified as belonging to subclass IId (unmodified, heat-stable peptides with molecular weight below 10 kDa); that is, aureocin A53, isolated from *S. aureus* A53 [36]; epidermicin NI01, from *S. epidermidis* 224 [37]; and BacSp222, from *Staphylococcus pseudintermedius* 222 [38]. Many authors consider these peptides as atypical members of subclass IIa (or IIa-like), which contains pediocin-like single-chain peptides. However, such an assignation is highly imperfect and the mentioned peptides should be better classified into subclass IId, which includes linear, single-chain, nonpediocin-like peptides [39].

All such staphylococcins are made up of 50–51 amino acids, are rich in tryptophan as well as lysine residues, show a high net cationic charge, and are both plasmid-encoded. They are synthesized without a leader sequence or signal-peptide, and encoding genes are surrounded by several Open Reading Frames (ORFs), that, in their turn, encode ABC-transporter-like proteins involved in peptide secretion. Again, a few ORFs encode immunity proteins, while others harbor the code for proteins of unknown functions. Aureocin A53, epidermicin NI01, and BacSp222 mechanisms of action have not yet been elucidated, but studies focusing on the similar peptide lacticin Q [40] suggest they might bind bacterial membranes without penetrating them. Afterward, they would induce rapid, so-called "flip-flop" trans-bilayer movements of lipids, which are responsible for pore formation, cell content efflux, then bacterial cell death. On the other hand, BacSp222 interacts with the phospholipid bilayer through the barrel-stave pore formation [41]. Moreover, BacSp222 seems to be one of the first examples of the multifunctional peptides that possess features characteristic for both bacteriocins and virulence factors. Besides its bactericidal activity, BacSp222 also demonstrates significant cytotoxic activities toward various eukaryotic cells. Moreover, at nanomolar concentrations, it also possesses modulatory properties toward host immune cells, efficiently enhancing interferon gamma-induced nitric oxide release in murine macrophage-like cell lines [38].

13.2.4 Subclass IIIa Staphylococcins

Class III bacteriocins comprise large, heat-labile proteins. This class is further subdivided into subclass IIIa, including bacteriolytic enzymes, and subclass IIIb, formed by nonlytic enzymes. Among staphylococcins, subclass IIIa is made up of two members; that is, lysostaphin, isolated from *S. simulans* biovar staphylolyticus ATCC1362 [42,43], and endopeptidase ALE-1, produced by *Staphylococcus capitis* EPK1 [44,45]. Lysostaphin is a 25-kDa, zinc-containing metallopeptidase encoded by a plasmid that also contains the life gene (responsible for immunity). However, in contrast to other typical bacteriocins, resistance to lysostaphin also depends on peptidoglycan interpeptide bridges modification. Lysostaphin is synthesized as a preproenzyme; the N-terminal leader sequence is removed during secretion, while 15 tandem repeats of propeptides, also at the N-terminus, are released by extracellular cysteine peptidase, yielding a fully active enzyme. Lysostaphin is a two-domain enzyme, consisting of an N-terminal peptidase and a C-terminal domain responsible for peptidoglycan binding. The catalytic domain exerts three distinct activities: glycylglycine endopeptidase, endo-β-N-acetyl glucosamidase, and N-acetyl-muramyl-L-alanine amidase; the first, particularly, is essential for the lytic properties of the enzyme, which primarily hydrolyzes glycine-glycine bonds (characteristic of pentaglycine motifs of peptidoglycan in the cell wall of *S. aureus*, *S. simulans*, *Staphylococcus carnosus*, and other organisms).

Endopeptidase ALE-1 is a homologue of lysostaphin, with which it shares similar molecular, structural, and functional features.

13.3 BACTERIOCIN-LIKE INHIBITORY SUBSTANCES

The term "BLIS" refers to diverse bactericidal factors that act as bacteriocins, but are neither obtained in pure form nor fully characterized. In spite of the controversial and unclear characteristics of this category, the acronym "BLIS" has been present in the literature since 1991 [46], and BLIS production has represented a useful tool for typing bacterial strains, both in the clinical and in the ecology context [18]. Currently, this group comprises a high-molecular weight lipoprotein-carbohydrate complex described in *S. aureus* 414 and named staphylococcin 414 [47]; a c.9 kDa-peptide staphylococcin 462, found in *S. aureus* 462 [48]; a c.5 kDa-peptide staphylococcin IYS2 isolated from *S. aureus* IYS2 [49]; a c.6 kDa-peptide Bac1829 obtained from culture medium of *S. aureus* KSI1829 [50]; the 41 kDa heat-stable bactericidal protein Bac201, isolated from *S. aureus* AB201 [51]; and the 4 kDa-peptide staphylococcin 188, obtained from *S. aureus* AB188 [52]. A separate presentation is necessary for staphylococcin Au-26, isolated and preliminarily characterized from the vaginal *S. aureus* strain 26 [53]. Seventeen years after investigations on Au-26, this compound was ascertained in community-acquired methicillin-resistant *S. aureus* (MRSA) strains from infection cases [54], and named as Bsa (Bacteriocin of *S. aureus*) [54]. Very detailed analyses of the putative Au-26/Bsa operons facilitated, in those MRSA isolates, the identification of genes encoding the very similar peptides BsaA1 and BsaA2, which share a high degree of homology with epidermin and gallidermin. Mass spectrometry studies confirmed their presence among the studied strains; however, neither of these peptides have been obtained in pure form, nor has the bactericidal action of homogenous peptides been confirmed. Hence, such molecules should be better classified as BLIS; also, the hypothesis according to which their production confers a competitive ecological advantage versus community-acquired strains should be confirmed.

It is also worth noting that the unclear chemical structure of many BLIS compounds often does not limit even advanced applicative studies on these substances. Staphylococcin 188 is the best example of this situation, and it shows a pronounced antibacterial, antidermatophytic, and antimycobacterial potential that may play a therapeutic role [55].

13.4 CONCLUSIONS

The microbicidal potential of bacteriocins is, among their features, the one that is mostly related to zoonoses and, more generally, to infectious diseases. This is primarily due both to the widespread clinical use of antimicrobials as well as to food and environment pollution by residual antibiotics, which select bacterial resistances [56]. Use of bacteriocins or bacteriocin-producing commensal strains with the aim to treat or limit the spread of infectious diseases seems to be cost-effective, nontoxic for humans and animals, and safe for the natural environment. Bacteriocins are effective toward microbes causing both human and animal diseases. One of the best examples is lysostaphin, clinical studies on which historically date back to 1965 [57]. So far, the effectiveness of this bacteriocin has been ascertained against organisms responsible for nasal and systemic infections, bacteremia and endocarditis, and the enzyme, moreover, also showed a pronounced protective effect against biofilm formation on intravascular catheter surfaces [18,58–60]. Additionally, and of diagnostic interest, lysostaphin may be used to presumptively discriminate between *Staphylococcus pasteuri* (lysostaphin-resistant) and *S. warneri* (lysostaphin-susceptible) [61].

Numerous studies also focus on the effectiveness of staphylococcins such as C55, Pep5, epidermin, aureocin A53, and the BLISS Bac188, for treatment of other human diseases, even caused by multidrug-resistant strains [18]. Likewise, C55, epidermin, aureocin 53, and Bac188 demonstrated a high efficiency in inhibiting pathogens involved in several veterinary diseases, such as bovine mastitis [62], bovine keratoconjuctivitis [63], poultry New Castle Disease [64], swine rhinitis [50] and various zoodermatoses [55].

However, little has been discovered, so far, about the role of bacteriocins in the field of zoonoses, which is related to animal-to-man transmission of pathogens and microbial pathogenic determinants, and vice versa. Results of a recent study on BacCH91, a staphylococcin originating from a poultry-associated *S. aureus* strain, showed that the encoding gene may be found in two allelic variants [26]. Variant I has been identified only in poultry-related isolates, while strains from humans and other animals possessed the allelic variant II. These results suggest that bacteriocins or their genes may be related to pathogenicity or virulence of particular strains. Such a hypothesis is also supported by the fact that, as mentioned herein, bacteriocin expression may be interpreted as a factor promoting stable genetic transmission of particular plasmids [15].

REFERENCES

[1] Klaenhammer TR. Genetics of bacteriocins produced by lactic acid bacteria. FEMS Microbiol Rev 1993;12:39–85.

[2] Hammami R, Zouhir A, Le Lay C, Ben Hamida J, Fliss I. BACTIBASE second release: a database and tool platform for bacteriocin characterization. BMC Microbiol 2010;10:22.

[3] Riley MA, Chavan MA, editors. Bacteriocins ecology and evolution. Berlin: Springer; 2007.

[4] Drider D, Rebuffat S, editors. Prokaryotic antimicrobial peptides. New York: Springer; 2011.

[5] Smarda J, Benada O. Phage tail-like (high-molecular-weight) bacteriocins of *Budvicia aquatica* and *Pragia fontium* (Enterobacteriaceae). Appl Environ Microbiol 2005;71:8970–3.

[6] Gebhart D, Williams SR, Bishop-Lilly KA, Govoni GR, Willner KM, Butani A, et al. Novel high-molecular-weight, R-type bacteriocins of *Clostridium difficile*. J Bacteriol 2012;194:6240–7.

[7] Michel-Briand Y, Baysse C. The pyocins of *Pseudomonas aeruginosa*. Biochimie 2002;84:499–510.

[8] O'Connor EB, Cotter PD, O'Connor P, O'Sullivan O, Tagg JR, Ross RP, et al. Relatedness between the two-component lantibiotics lacticin 3147 and staphylococcin C55 based on structure, genetics and biological activity. BMC Microbiol 2007;7:24.

[9] Cotter PD, Hill C, Ross RP. Bacteriocins: developing innate immunity for food. Nat Rev Microbiol 2005;3:777–88.

[10] Peschel A, Gotz F. Analysis of the *Staphylococcus epidermidis* genes epiF, -E, and -G involved in epidermin immunity. J Bacteriol 1996;178:531–6.

[11] Havarstein LS, Diep DB, Nes IF. A family of bacteriocin ABC transporters carry out proteolytic processing of their substrates concomitant with export. Mol Microbiol 1995;16:229–40.

[12] Venema K, Kok J, Marugg JD, Toonen MY, Ledeboer AM, Venema G, et al. Functional analysis of the pediocin operon of *Pediococcus acidilactici* PAC1.0: PedB is the immunity protein and PedD is the precursor processing enzyme. Mol Microbiol 1995;17:515–22.

[13] Nes IF, Eijsink VGH. Regulation of group II peptide bacteriocin synthesis by quorumsensing mechanisms. In: Dunny GM, Winans SC, editors. Cell-cell signalling in bacteria. Washington: ASM (American Society for Microbiology); 1999.

[14] Gillor O, Vriezen JA, Riley MA. The role of SOS boxes in enteric bacteriocin regulation. Microbiology 2008;154:1783–92.

[15] Inglis RF, Bayramoglu B, Gillor O, Ackermann M. The role of bacteriocins as selfish genetic elements. Biol Lett 2013;9:20121173.

[16] Van Melderen L, Saavedra De Bast M. Bacterial toxin-antitoxin systems: more than selfish entities? PLoS Genet 2009;5:e1000437.

[17] Chatterjee C, Paul M, Xie L, van der Donk WA. Biosynthesis and mode of action of lantibiotics. Chem Rev 2005;105:633–84.

[18] Bastos MC, Ceotto H, Coelho ML, Nascimento JS. Staphylococcal antimicrobial peptides: relevant properties and potential biotechnological applications. Curr Pharm Biotechnol 2009;10:38–61.

[19] Fontana MB, de Bastos Mdo C, Brandelli A. Bacteriocins Pep5 and epidermin inhibit *Staphylococcus epidermidis* adhesion to catheters. Curr Microbiol 2006;52:350–3.

[20] Heidrich C, Pag U, Josten M, Metzger J, Jack RW, Bierbaum G, et al. Isolation, characterization, and heterologous expression of the novel lantibiotic epicidin 280 and analysis of its biosynthetic gene cluster. Appl Environ Microbiol 1998;64:3140–6.

[21] van de Kamp M, van den Hooven HW, Konings RN, Bierbaum G, Sahl HG, Kuipers OP, et al. Elucidation of the primary structure of the lantibiotic epilancin K7 from *Staphylococcus epidermidis* K7. Cloning and characterisation of the epilancin-K7-encoding gene and NMR analysis of mature epilancin K7. Eur J Biochem 1995;230:587–600.

[22] Ekkelenkamp MB, Hanssen M, Danny Hsu ST, de Jong A, Milatovic D, Verhoef J, et al. Isolation and structural characterization of epilancin 15X, a novel lantibiotic from a clinical strain of *Staphylococcus epidermidis*. FEBS Lett 2005;579:1917–22.

[23] Sahl HG. Staphylococcin 1580 is identical to the lantibiotic epidermin: implications for the nature of bacteriocins from gram-positive bacteria. Appl Environ Microbiol 1994;60:752–5.

[24] Kellner R, Jung G, Horner T, Zahner H, Schnell N, Entian KD, et al. Gallidermin: a new lanthionine-containing polypeptide antibiotic. Eur J Biochem 1988;177:53–9.

[25] Furmanek B, Kaczorowski T, Bugalski R, Bielawski K, Bohdanowicz J, Podhajska AJ. Identification, characterization and purification of the lantibiotic staphylococcin T, a natural gallidermin variant. J Appl Microbiol 1999;87:856–66.

[26] Wladyka B, Wielebska K, Wloka M, Bochenska O, Dubin G, Dubin A, et al. Isolation, biochemical characterization, and cloning of a bacteriocin from the poultry-associated *Staphylococcus aureus* strain CH-91. Appl Microbiol Biotechnol 2013;97:7229–39.

[27] Islam MR, Nishie M, Nagao J, Zendo T, Keller S, Nakayama J, et al. Ring A of nukacin ISK-1: a lipid II-binding motif for type-A(II) lantibiotic. J Am Chem Soc 2012;134:3687–90.

[28] Minamikawa M, Kawai Y, Inoue N, Yamazaki K. Purification and characterization of warnericin RB4, anti-Alicyclobacillus bacteriocin, produced by *Staphylococcus warneri* RB4. Curr Microbiol 2005;51:22–6.

[29] Ceotto H, Holo H, da Costa KF, Nascimento Jdos S, Salehian Z, Nes IF, et al. Nukacin 3299, a lantibiotic produced by *Staphylococcus simulans* 3299 identical to nukacin ISK-1. Vet Microbiol 2010;146:124–31.

[30] Bonelli RR, Schneider T, Sahl HG, Wiedemann I. Insights into in vivo activities of lantibiotics from gallidermin and epidermin mode-of-action studies. Antimicrob Agents Chemother 2006;50:1449–57.

[31] Navaratna MA, Sahl HG, Tagg JR. Two-component anti-Staphylococcus aureus lantibiotic activity produced by *Staphylococcus aureus* C55. Appl Environ Microbiol 1998;64:4803–8.

[32] Netz DJ, Sahl HG, Marcelino R, dos Santos Nascimento J, de Oliveira SS, Soares MB, et al. Molecular characterisation of aureocin A70, a multi-peptide bacteriocin isolated from *Staphylococcus aureus*. J Mol Biol 2001;311:939–49.

[33] Navaratna MA, Sahl HG, Tagg JR. Identification of genes encoding two-component lantibiotic production in *Staphylococcus aureus* C55 and other phage group II S. aureus strains and demonstration of an association with the exfoliative toxin B gene. Infect Immun 1999;67:4268–71.

[34] Garneau S, Martin NI, Vederas JC. Two-peptide bacteriocins produced by lactic acid bacteria. Biochimie 2002;84:577–92.

[35] Varella Coelho ML, Santos Nascimento JD, Fagundes PC, Madureira DJ, Oliveira SS, Vasconcelos de Paiva Brito MA, et al. Activity of staphylococcal bacteriocins against *Staphylococcus aureus* and *Streptococcus agalactiae* involved in bovine mastitis. Res Microbiol 2007;158:625–30.

[36] Netz DJ, Pohl R, Beck-Sickinger AG, Selmer T, Pierik AJ, Bastos Mdo C, et al. Biochemical characterisation and genetic analysis of aureocin A53, a new, atypical bacteriocin from *Staphylococcus aureus*. J Mol Biol 2002;319:745–56.

[37] Sandiford S, Upton M. Identification, characterization, and recombinant expression of epidermicin NI01, a novel unmodified bacteriocin produced by *Staphylococcus epidermidis* that displays potent activity against staphylococci. Antimicrob Agents Chemother 2012;56:1539–47.

[38] Wladyka B, Piejko M, Bzowska M, Pieta P, Krzysik M, Mazurek L, et al. A peptide factor secreted by *Staphylococcus pseudintermedius* exhibits properties of both bacteriocins and virulence factors. Sci Rep 2015;5:14569.

[39] Sawa N, Koga S, Okamura K, Ishibashi N, Zendo T, Sonomoto K. Identification and characterization of novel multiple bacteriocins produced by *Lactobacillus sakei* D98. J Appl Microbiol 2013;115:61–9.

[40] Yoneyama F, Ohno K, Imura Y, Li M, Zendo T, Nakayama J, et al. Lacticin Q-mediated selective toxicity depending on physicochemical features of membrane components. Antimicrob Agents Chemother 2011;55:2446–50.

[41] Pieta P, Majewska M, Su Z, Grossutti M, Wladyka B, Piejko M, et al. Physicochemical studies on orientation and conformation of a new bacteriocin BacSp222 in a planar phospholipid bilayer. Langmuir 2016;32:5653–62.

[42] Schindler CA, Schuhardt VT. Lysostaphin: a new bacteriolytic agent for the *Staphylococcus*. Proc Natl Acad Sci U S A 1964;51:414–21.

[43] Kumar JK. Lysostaphin: an antistaphylococcal agent. Appl Microbiol Biotechnol 2008;80:555–61.

[44] Sugai M, Fujiwara T, Akiyama T, Ohara M, Komatsuzawa H, Inoue S, et al. Purification and molecular characterization of glycylglycine endopeptidase produced by *Staphylococcus capitis* EPK1. J Bacteriol 1997;179:1193–202.

[45] Hirakawa H, Akita H, Fujiwara T, Sugai M, Kuhara S. Structural insight into the binding mode between the targeting domain of ALE-1 (92AA) and pentaglycine of peptidoglycan. Protein Eng Des Sel 2009;22:385–91.

[46] James SM, Tagg JR. The prevention of dental caries by BLIS-mediated inhibition of mutans streptococci. N Z Dent J 1991;87:80–3.

[47] Gagliano VJ, Hinsdill RD. Characterization of a *Staphylococcus aureus* bacteriocin. J Bacteriol 1970;104:117–25.

[48] Hale EM, Hinsdill RD. Characterization of a bacteriocin from *Staphylococcus aureus* strain 462. Antimicrob Agents Chemother 1973;4:634–40.

[49] Nakamura T, Yamazaki N, Taniguchi H, Fujimura S. Production, purification, and properties of a bacteriocin from *Staphylococcus aureus* isolated from saliva. Infect Immun 1983;39:609–14.

[50] Crupper SS, Iandolo JJ. Purification and partial characterization of a novel antibacterial agent (Bac1829) produced by *Staphylococcus aureus* KSI1829. Appl Environ Microbiol 1996;62:3171–5.

[51] Iqbal A, Ahmed S, Ali SA, Rasool SA. Isolation and partial characterization of Bac201: a plasmid-associated bacteriocin-like inhibitory substance from *Staphylococcus aureus* AB201. J Basic Microbiol 1999;39:325–36.

[52] Saeed S, Ahmad S, Rasool SA. Antimicrobial spectrum, production and mode of action of staphylococcin 188 produced by *Staphylococcus aureus* 188. Pak J Pharm Sci 2004;17:1–8.

[53] Scott JC, Sahl HG, Carne A, Tagg JR. Lantibiotic-mediated anti-lactobacillus activity of a vaginal *Staphylococcus aureus* isolate. FEMS Microbiol Lett 1992;72:97–102.

[54] Daly KM, Upton M, Sandiford SK, Draper LA, Wescombe PA, Jack RW, et al. Production of the Bsa lantibiotic by community-acquired *Staphylococcus aureus* strains. J Bacteriol 2010;192:1131–42.

[55] Saeed S, Rasool SA, Ahmed S, Khanum T, Khan MB, Abbasi A, et al. New insight in staphylococcin research: bacteriocin and/or bacteriocin-like inhibitory substance(s) produced *by S. aureus* AB188. World J Microbiol Biotechnol 2006;22:713–22.

[56] Marshall BM, Levy SB. Food animals and antimicrobials: impacts on human health. Clin Microbiol Rev 2011;24:718–33.

[57] Harrison EF, Cropp CB. Comparative in vitro activities of lysostaphin and other antistaphylococcal antibiotics on clinical isolates of *Staphylococcus aureus*. Appl Microbiol 1965;13:212–5.

[58] von Eiff C, Kokai-Kun JF, Becker K, Peters G. vitro activity of recombinant lysostaphin against *Staphylococcus aureus* isolates from anterior nares and blood. Antimicrob Agents Chemother 2003;47:3613–5.

[59] Climo MW, Patron RL, Goldstein BP, Archer GL. Lysostaphin treatment of experimental methicillin-resistant *Staphylococcus aureus* aortic valve endocarditis. Antimicrob Agents Chemother 1998;42:1355–60.

[60] Oluola O, Kong L, Fein M, Weisman LE. Lysostaphin in treatment of neonatal *Staphylococcus aureus* infection. Antimicrob Agents Chemother 2007;51:2198–200.

[61] Savini V, Catavitello C, Bianco A, Balbinot A, D'Antonio D. Epidemiology, pathogenicity and emerging resistances in *Staphylococcus pasteuri*: from mammals and lampreys, to man. Recent Pat Antiinfect Drug Discov 2009;4:123–9.

[62] Varella Coelho ML, Santos Nascimento JD, Fagundes PC, Madureira DJ, Oliveira SS, Vasconcelos de Paiva Brito MA, et al. Activity of staphylococcal bacteriocins against *Staphylococcus aureus* and *Streptococcus agalactiae* involved in bovine mastitis. Res Microbiol 2007;158:625–30.

[63] Senturk S, Cetin C, Temizel M, Ozel E. Evaluation of the clinical efficacy of subconjunctival injection of clindamycin in the treatment of naturally occurring infectious bovine keratoconjunctivitis. Vet Ophthalmol 2007;10:186–9.

[64] Saeed S, Rasool SA, Ahmad S, Zaidi SZ, Rehmani S. Antiviral activity of staphylococcin 188: a purified bacteriocin-like inhibitory substance isolated from *Staphylococcus aureus* AB188. Res J Microbiol 2007;2:796–806.

Chapter 14

Phage-Associated Virulence Determinants of *Staphylococcus aureus*

Weronika M. Ilczyszyn, Maja Kosecka-Strojek, Jacek Międzobrodzki
Department of Microbiology, Faculty of Biochemistry, Biophysics and Biotechnology, Jagiellonian University, Krakow, Poland

14.1 INTRODUCTION

Bacterial genomes, due to their relatively high plasticity, are characterized by a rapid pace of evolution unprecedented in eukaryotic organisms. DNA rearrangements and the accumulation of point mutations through the generations are heavily responsible for the observed variability of prokaryotes. However, the main driving force of bacterial evolution is the ability to acquire multiple genes in a single recombination event called horizontal gene transfer (HGT). The three main mechanisms of lateral gene transfer among bacteria are transformation, conjugation, and bacteriophage transduction.

Transformation was the first mechanism to be discovered. During this process, DNA is taken up directly from the environment, integrated into the genome, and expressed by a prokaryotic cell, referred to as the "recipient" or "transformant." Some bacterial species are in a permanent physiological competent state while others are prone to transformation only under certain conditions, or never. Conjugation is the direct transfer of DNA from the donor to the recipient cell by direct cell-to-cell contact established through a channel called "conjugative bridge." "Transduction" refers to the transfer of DNA mediated by viral particles. This mechanism, however, is a complex phenomenon that occurs in many forms. If a group of bacterial viruses, known as bacteriophages, always carry the same genetic sequences from the host cell to other cells, the mechanism is referred to as "specific transduction." However, general transduction takes place when phages carry random fragments of DNA, cleaved and mis-packed during the viral replication process [1].

HGT has been particularly significant during the evolution of Gram-positive bacteria belonging to the *Staphylococcus aureus* species. Its complex genetic background and phenotypic plasticity makes this organism both a commensal

Pet-to-Man Travelling Staphylococci: A World in Progress. https://doi.org/10.1016/B978-0-12-813547-1.00014-5

and a potential human pathogen [2,3]. Although ubiquitous in the biosphere, its reservoirs are animals and humans, as staphylococci, in general, can proliferate only in higher, warm-blooded organisms. *S. aureus* is normally a harmless inhabitant of the human skin and mucous membranes; nonetheless, it can be responsible for many serious systemic or local infections, especially in immunocompromised patients. Currently, *S. aureus* is one among the main etiological agents of nosocomial, community-acquired or farm-acquired infections. This fact is strongly related to the emergence of multidrug-resistant pathogens. In addition, being that *S. aureus* is the cause of serious, acute, or chronic human and animal diseases, studies conducted on staphylococci gained significant economic importance. The phenomenon mentioned herein stems from the fact that *S. aureus* strains have acquired a veritable arsenal of diverse virulence factors, superantigens, and resistance determinants that are responsible for the clinical disease. The biggest advantage of staphylococci is their adaptive power, which allows them to survive under extreme conditions and colonize even hostile and diverse environments [4–6]. Genomic analysis shows that HGT is of paramount importance for the differentiation of *S. aureus*: between 22% and 25% of its genomic DNA consists of "accessory sequences" variable among strains and constituted of mobile or once mobile genetic elements (MGEs) [4]. About 2% of the entire staphylococcal DNA was obtained as a result of phage infection, primarily by prophage integration [7]. Such a high impact of transduction on staphylococci evolution may have its origin in the low competence of these Gram-positive bacteria to acquire DNA directly from the environment [8]. It should also be emphasized that the importance of bacteriophages is not only limited to their role in the transfer of large blocks of genetic material, but it has been proven that prophages may also actively participate in pathogenesis of bacterial infections [3,9].

14.2 BACTERIOPHAGES AND *STAPHYLOCOCCUS AUREUS* VIRULENCE

All viruses specifically attacking bacteria, known as "bacteriophages," can be divided into three distinctive groups: lytic, lysogenic, and chronic. For each of these groups, the penetration of phages into bacterial cells can have different outcomes. During the lytic infection, phage particles replicate until depletion of metabolic resources, which is followed by host cell lysis, and release of a large number of new viral particles into the environment. Lytic phages are classified mainly into the *Myoviridae* family and, due to their high specificity, they are used in experimental phage therapy [7,10]. Bacteria, infected by phages of a chronic nature, retain the ability to grow and divide in spite of the continuous viral replication and new viruses" release.

In comparison with their lytic counterparts, lysogenic phages of the *Siphoviridae* family do not only influence bacterial evolution due to the intensive "arms race" between "predator and prey" (that cause lysogenic infection resulting in propagation of phages and lysis of the host) but, during a lysogenic infection, the viral genome integrates into the bacterial chromosome in a latent state called "prophage" [2,11]. Microbial genomics clearly shows that lysogenization is more the rule than the exception, and many bacterial species contain more than one prophage in their genome [9,12]. The *S. aureus* strain NCTC 8325 was one of the first staphylococci to undergo whole-genome sequencing. Its genome contains three bacteriophage sequences in comparison with other members of the *Staphylococcus* genus, which typically have only up to two such insertions. Genomic comparison between the NCTC 8325 strain and strain MSSA-476 revealed that the major cause of differences is sequences of bacteriophage origin. However, the diversity of strains N315 and NW2 is due to the presence of different transposons, insertional elements, and pathogenicity islands [13]. The reader should, however, by no means come to a conclusion that bacteriophage sequences are present in each staphylococcal genome. The *Staphylococcus epidermidis* ATCC 12228 strain, for instance, does not contain a single prophage in its genetic material [14].

Some lysogenic phages carry additional genes called "morons" or lysogenic conversion genes (LCG). After prophage integration into the host chromosome, gene expression causes the bacterial phenotype to change. [1]. The lysogenic conversion phenomenon has been associated with the production of some endotoxins by *S. aureus* when more than 40 years ago it was discovered that phage infections can convert nonhemolytic strains into alpha-hemolysin-producing strains. Since then, it has been proved that numerous bacterial virulence determinants are, in fact, encoded by phage genomes. During positive lysogenic conversion, the LGC-type gene expression can significantly influence the host's adaptive powers or, in the case of pathogenic bacteria, hold responsibility for the specificity and occurrence of certain clinical symptoms during the course of an infection [3]. Staphylococcal virulence factors of major clinical significance obtained through phage transmission are the chemotaxis inhibitory protein (CHIPS), the staphylococcal complement inhibitor (SCIN), the Panton-Valentine leukocidin (PVL), the staphylokinase, the exfoliative toxin A, and enterotoxins (Table 14.1).

Nevertheless, lysogenization may result in loss of expression of functional proteins if the bacteriophage integrates within the coding sequences. An example of negative lysogenic conversion is phenotypes that are deprived of lipase or of β-toxin production after having integrated phages L54a and φ13, respectively [26]. The CHIPS protein (*chp*)-transducing phage also contains staphylokinase (*sak*) and enterotoxin A (*sea*) gene-encoding sequences in its genome, but eliminates β-hemolysin expression by insertional inactivation of the *hlb* gene [23]. The phenomenon of simultaneous positive and negative lysogenic conversion has also been associated with a *sak*+, *sea*+, and p-lysine, absent phenotype [27]. In conclusion, the majority of virulence factors propagate within the *Staphylococcus* genus by means of specialized transduction. Only recently has the direct link between generalized and specific transduction of chromosomal pathogenicity determinants been discovered [3].

14.3 STAPHYLOCOCCAL PATHOGENICITY ISLANDS SaPIs

Generally, pathogenicity islands (PI) are often flanked by repetitive sequences such as insertion elements (IS), and may also contain the *int* gene, that encodes a phage-like integrase. Particularly because of the close proximity of repeatable sequences and the presence of integrase homologues, PI were assumed to be examples of MGEs. Until recently, however, mobilization has been fully assigned to only a few staphylococcal islands whose horizontal transfer occurs by means of bacteriophage transduction. Staphylococcal PIs (SaPIs) constitute a large and relatively coherent family of genetic elements ranging from 15 to 27 kb [2,15]. Within their sequence, SaPIs carry genes encoding several staphylococcal virulence factors such as toxic shock syndrome toxin-1 (TSST-1), enterotoxins B, C, L, K, P, and Q, the biofilm- associated protein (Bap), the homologue of the penicillin binding protein (PBP), and the genomic variant of the von Willebrand binding protein (vWBp$_s$) (Table 14.1). SaPIs were primarily reported during whole-genome or comprehensive nucleotide fragment analyses of *S. aureus* strains; however, other representatives of these elements were detected in the genomes of other Gram-positive bacteria of the *Staphylococcus* and *Lactococcus* genera [15]. This year, new genomic islands named SaPIivm10, SaPIishikawa11, SaPIivm60, SaPIhhms2, SaPIj11, SaPIno10, and SaPIhirosaki4 have been detected in the staphylococcal genome using a comprehensive LA-PCR system, targeting all SaPI and SaPI-related insertion sites: 8′, 9′, 18′, 19′, 44′, and 49′ [20]. Until now, SaPI1 and SaPIbov1, found in the *S. aureus* RN4282 and RF122 strains, respectively, are the best described and most representative SaPIs [28]. The mechanism of SaPIs excision from the chromosome by means of recombination between flanking repetitive sequences is analogous to the process in which a prophage is released from its host, and is dependent on the Xis protein coded by the island itself [29]. Replication and mobilization of SaPIs can be a consequence of either a SOS-induced excision of the helper prophage present in the same strain, or of the infection by a helper phage or of the joint entry of SaPI and the helper phage [4,15]. Several helper phages for different SaPIs have been described, including f11, 53, 80, 80α, and ΦNM1 and, in their absence, the islands are maintained in a repressed state by the master repressor Stl. The Stl proteins of different SaPIs are widely divergent, thus the ability of a particular helper phage to derepress a given SaPI appears to be one of the determinants of their specificity. SaPI1 is derepressed by Sri, the product of 80α ORF22; SaPIbov1 is derepressed by Dut, the product of 80α ORF32; SaPIbov2, finally, is derepressed by the product of 80α ORF15 [30].

After induction, SaPIs redirect replication and the DNA packaging machinery of their helpers, which gives them the name "molecular pirates. There are at least three known strategies for how SaPIs can interfere with their helper

TABLE 14.1 Selected SaPI- and Phage-Related Virulence Factors of *Staphylococcus aureus*

Toxin/Virulence Determinant (*Gene*)	MGE	Description/Mechanisms/ Symptoms	Ref
Biofilm associated-protein Bap (*bap*)	SaPIbov2	Associated with the formation of a biofilm inside the cow's udder, involved in mastitis	[2,15]
Host-specific variant of von Willebrand binding protein, vWBp$_s$ (*vwb$_{ps}$*)	SaPIbov2 SaPIbov4, SaPIbov5, SaPIeq1 SaPIbov2	Genomic variant of von Willebrand binding protein, vWBp$_s$, its presence allows *S. aureus* cells to coagulate bovine or equine plasma in accordance with the nature of the host, from which the SaPI-bearing strain was isolated	[16,17]
CHIPS, chemotaxis inhibitory protein (*chip*)	φ13, φtp310-3, φ252B, φMu3A, φN315, φNM3, φSa3JH1, φSa3mw, φSa3 ms, φSa3JH9, φSa3USA300, φβCUSA300_TCH1516	CHIPS binds to neutrophils receptors C5AR1 and FPR1 reducing their activity toward complement C5a and formylated peptides, respectively	[2,18]
Exfoliative toxin A (*eta*)	ΦETA	The toxin responsible for the symptoms associated with staphylococcal scalded skin syndrome (SSSS)	[2,19]
Penicillin-binding protein (*ear*)	SaPI1, SaPIivm10, SaPIishikawa11, SaPIivm60, SaPIno10, SaPIhirosaki4	Homolog of penicillin-binding protein	[15,20]
Enterotoxin A (*sea*)	ΦSa3ms,ΦSa3mw, Φ252B, ΦNM3, ΦMu50a	Staphylococcal food poisoning	[21]
Enterotoxin B (*seb*)	SaPI3, SaPIivm10, SaPIishikawa11, SaPIivm60, SaPIno10, SaPIhirosaki4	Staphylococcal food poisoning	[20,21]
Enterotoxin C (*sec*)	SaPIn1,SaPIm1, SaPImw2, SaPIbov1	Staphylococcal food poisoning	[21]

Continued

TABLE 14.1 Selected SaPI- and Phage-Related Virulence Factors of Staphylococcus aureus—cont'd

Toxin/Virulence Determinant (Gene)	MGE	Description/Mechanisms/ Symptoms	Ref
Staphylococcal enterotoxin-like protein K (selk)	ΦSa3ms, ΦSa3mw, SaPI1, SaPI3, SaPIbov1, SaPI5, SaPIJ11	Staphylococcal food poisoning	[20,21]
Staphylococcal enterotoxin-like protein L (sell)	SaPIn1, SaPIm1, SaPImw2, SaPIbov1	Staphylococcal food poisoning	[21]
Staphylococcal enterotoxin-like protein P (selp)	ΦN315, ΦMu3A, SaPIJ11	Staphylococcal food poisoning	[21]
Staphylococcal enterotoxin-like protein Q (selq)	ΦSa3ms, ΦSa3mw, SaPI1, SaPI3, SaPI5	Staphylococcal food poisoning	[21]
PVL, Panton-Valentine leukocidin, two-subunit toxin (lukS, lukF)	ΦSA2pvl ΦSLT, ΦPVL, ΦSA2MW, ΦSA2usa	The toxin is combined of two subunits, LukS and LukF, encoded on the bacteriophage PVL. During infection, PVL causes lysis of mammalian leucocytes. Its expression stimulates neutrophils to excessively produce inflammatory factors. Such an acute inflammatory response ultimately leads to tissue necrosis. The PVL toxin is considered to be one of the most dangerous staphylococcal virulence factors associated with a raised risk of transmission, complications, and hospitalization	[22,23]
SCIN, Staphylococcal complement inhibitor (scn)	φ13, φN315, φ252B, φNM3, φMu50A, φSa3JH1, φSa3 ms, φSa3mw, φSa3JH9, φMu3A, φtp310-3, φSa3USA300, φβCUSA300_TCH1516	SCIN specifically binds to and inhibits the activity of the complement C3bBb convertase. Therefore, SCIN interferes with the process of phagocytosis following opsonization of S. aureus by C3b deposition on its surface	[2,18]

TABLE 14.1 Selected SaPI- and Phage-Related Virulence Factors
of Staphylococcus aureus — cont'd

Toxin/Virulence Determinant (*Gene*)	MGE	Description/Mechanisms/ Symptoms	Ref
Staphylokinase (*sak*)	phl3, ph42D, phφC, φN315, φMu50A	Staphylokinase is responsible for the removal of IgG and complement C3b by proteolytic cleavage from the bacterial cell surface, thus interfering with the process of phagocytosis. Complex binding between staphylokinase and plasminogen results in the formation of active plasmin, a proteolytic enzyme with a broad spectrum of activity, facilitating the penetration of *S. aureus* to surrounding tissues and their destruction during infection	[2,24]
Toxic shock syndrome toxin 1 (*tst-1*)	SaPI1, SaPI2, SaPIbov1, SaPI3	Main toxin responsible for the onset of toxic shock syndrome characterized by high fever, dysfunction of internal organs, hypotension and exfoliation in the convalescence phase. TSST-1 stimulates the epithelial secretion of proinflammatory cytokines	[15,25]

MGE, mobile genetic element.

phages" growth, packing, and spread, but they are not used altogether in the interaction between a particular island and a given phage:

- production of capsids that are smaller than the normal 80α capsid, which matches with the smaller size of the SaPI1 genome, thus the chance to assemble viable viral particles is negligible. The presence of CpmA (*gp7*) and CpmB (*gp6*) proteins, encoded by SaPI1, is required in the formation of small capsids [31];
- inhibition of phage DNA packaging by the SaPI-encoded *ppi* genes: the Ppi protein binds to a small terminase subunit TerS$_P$, specifically blocking the

packaging of phage DNA into assembled capsids, and does not respond to SaPI-encoded TerS$_S$ which ensures preferential packing of SaPI particles [32];
- inhibition of phage 80 growth by SaPI2-encoded product of ORF17. This mechanism, however, still is to be elucidated [32].

Furthermore, a high proportion of the newly replicated SaPI particles is never packed into capsids and, with the cell lysates, it is released into the environment, from where DNA, through transformation, may be incorporated into new bacterial cells [15]. Horizontal transfer of the SaPIs occurring by modified transduction and transformation can be associated with the progressive spread of toxic shock toxin genes, such as that encoding TSST-1 and superantigens within the *S. aureus* species [33]. Some SaPIs, such as SaPI2, and SaPI5 are found in clinically important methicillin-resistant *S. aureus* (MRSA) strains, including USA200 and USA300 [30]. Such observations followed by the recent report that *S. aureus* Newman (animal infection model strain, isolated from a case of "secondary tubercular osteomyelitis") prophages ΦNM1 and ΦNM2 can mobilize SaPI1 and SaPIbov1 suggest that prophages of clinical isolates may participate in spread and establishment of pathogenicity islands in the hospital environment [28].

Although recent screening of coagulase-negative staphylococci (CoNS) as *Staphylococcus caprae*, *S. epidermidis*, *Staphylococcus warneri*, *Staphyloccus simulans*, or *Staphylococcus saprophyticus* for SaPI-associated superantigens showed no positive results, it is known that SaPIbov2 can be transduced to *Staphyloccus xylosus* or *S. epidermidis* [34]. It was also reported that transfer of SaPI1 and SaPIbov1 to *Listeria monocytogenes*, a nonstaphylococcal Gram-positive species, occurred spontaneously after prophages induction in raw cow's milk [35]. Thus, the ability of SaPIs to shift hosts is not yet commonly reported, but this phenomenon could gradually lead to the spread of virulence factors among CoNS strains and organisms belonging to other genera that do not posses SaPIs. Additionally, treatment of bacterial infections can also contribute, in their turn, to the spread of phage- and bacteria-encoded genes.

14.4 ANTIBIOTIC DEPENDENT SOS RESPONSE AND THE INDUCTION OF PROPHAGES

It was mentioned herein that staphylococcal pathogenicity islands can be mobilized after SOS-induced prophage propagation in the host strain. The SOS-system is a global response to DNA damage that leads to the increased expression of genes involved in DNA repair and survival of the bacterial cell; it is regulated by two proteins, LexA and RecA. LexA is a repressor that binds to the promoter regions of SOS-response regulatory genes. The presence of DNA lesions activates the RecA protein, which stimulates the autocatalytic digestion of a specific Ala-Gly bond in the LexA protein. This process prevents the further binding of LexA repressors to DNA, and thereby the SOS genes can be transcribed and translated. After the genetic material is repaired, newly synthesized

LexA and RecA proteins prevent further SOS-response [36]. Fluoroquinolone antibiotics inhibit two key DNA replication enzymes: gyrase and topoisomerase IV, which are responsible for introducing negative supercoils into DNA and parting mother and daughter DNA strands after replication [37]. Lack of this mechanism results in damage of genetic material, cell cycle arrest and, if repair systems fail, cell death. By targeting transmembrane PBSs, β-lactams exert a bactericidal effect, which is caused by the inhibition of bacterial cell wall synthesis. Furthermore, β-lactams up-regulate expression of the DpiA protein (a subunit of the DpiAB transcription system) which, if overproduced, introduces mutations, stops DNA replication, and thereby induces the SOS system [38]. β-Lactams and flouroquinolones, through SOS response, promote prophage induction, propagation, as well as amplification and expression of the LCG [39]. A similar phenomenon was observed in *S. aureus* strains bearing the SaPI1, SaPIbov1, and SaPIbov3 islands, which were subjected to subinhibitory concentrations of ampicillin and penicillin (both β-lactams), ciprofloxacin (a fluoroquinolone), or mitomycin C [37]. Moreover, the usage of fluoroquinolone antibiotics in the treatment of hemolytic uremic syndrome caused by the Shiga toxin-producing *Escherichia coli* strain H57:O157 leads to intensification of the clinical disease and, possibly, to the patient's death [39]. So far, however, there have been no reports of patients with toxic shock syndrome showing health deterioration in response to antibiotic treatment, but such a possibility cannot be ruled out.

14.5   CONCLUSIONS

In conclusion, bacteriophages participate in the spread of pathogenicity islands indirectly and directly, by lysogenic conversion, and so contribute to the emergence of diverse and highly virulent *S. aureus* strains. The recent report of a TSST-1- and PVL-producing strain may serve as confirmation of this statement [40]. Beside the pathogenicity factors, bacteriophages and other MGEs strongly influence the *S. aureus* adaptation to both human and animal hosts, providing new genetic information and facilitating the loss of unnecessary DNA in new ecological niches [5]. Tracking the dynamics of staphylococci and phage co-evolution will remain one of the major tasks of advanced microbiology and epidemiology for a long time.

ACKNOWLEDGMENTS

This work has been supported in part by the grant no. N N401 017740 from the National Science Centre, Poland.

CONFLICT OF INTEREST

None declared.

REFERENCES

[1] Brüssow H, Canchaya C, Hardt WD. Phages and the evolution of bacterial pathogens: from genomic rearrangements to lysogenic conversion. Microbiol Mol Biol Rev 2004;68(3):560–602.

[2] Malachowa N, DeLeo FR. Mobile genetic elements of *Staphylococcus aureus*. Cell Mol Life Sci 2010;67(18):3057–71.

[3] Wagner PL, Waldor MK. Bacteriophage control of bacterial virulence. Infect Immun 2002;70(8):3985–93.

[4] Deghorain M, Van Melderen L. The staphylococci phages family: an overview. Viruses 2012;4(12):3316–35.

[5] Fitzgerald JR. Human origin for livestock-associated methicillin-resistant *Staphylococcus aureus*. MBio 2012;3(2):e00082–12.

[6] Miedzobrodzki J, Malachowa N, Markiewski T, et al. Differentiation of *Staphylococcus aureus* isolates based on phenotypical characters. Postepy Hig Med Dosw 2008;62:322–7.

[7] Kuroda M, Ohta T, Uchiyama I, et al. Whole genome sequencing of meticillin-resistant *Staphylococcus aureus*. Lancet 2001;357(9264):1225–40.

[8] Canchaya C, Fournous G, Brüssow H. The impact of prophages on bacterial chromosomes. Mol Microbiol 2004;53(1):9–18.

[9] Łoś M, Węgrzyn G. Chapter 9: Pseudolysogeny. In: Szybalski MŁ, editor. Bacteriophages, Part A: Advances in Virus Research. 82. Academic Press; 2012. p. 339–49.

[10] Miedzybrodzki R, Fortuna W, Weber-Dabrowska B, et al. Phage therapy of staphylococcal infections (including MRSA) may be less expensive than antibiotic treatment. Postepy Hig Med Dosw (Online) 2007;61:461–5.

[11] Blair JE, Carr M. Lysogeny in staphylococci. J Bacteriol 1961;82(6):984–93.

[12] Canchaya C, Fournous G, Chibani-Chennoufi S, et al. Phage as agents of lateral gene transfer. Curr Opin Microbiol 2003;6(4):417–24.

[13] Baba T, Takeuchi F, Kuroda M, et al. Genome and virulence determinants of high virulence community-acquired MRSA. Lancet 2002;359(9320):1819–27.

[14] Canchaya C, Proux C, Fournous G, et al. Prophage genomics. Microbiol Mol Biol Rev 2003;67(2):238–76.

[15] Novick RP, Christie GE, Penadés JR. The phage-related chromosomal islands of Gram-positive bacteria. Nat Rev Microbiol 2010;8(8):541–51.

[16] Guinane CM, Ben Zakour NL, Tormo-Mas MA, et al. Evolutionary genomics of *Staphylococcus aureus* reveals insights into the origin and molecular basis of ruminant host adaptation. Genome Biol Evol 2010;2:454–66.

[17] Viana D, Blanco J, Tormo-Más MA, et al. Adaptation of *Staphylococcus aureus* to ruminant and equine hosts involves SaPI-carried variants of von Willebrand factor-binding protein. Mol Microbiol 2010;77(6):1583–94.

[18] Serruto D, Rappuoli R, Scarselli M, et al. Molecular mechanisms of complement evasion: learning from staphylococci and meningococci. Nat Rev Microbiol 2010;8(6):393–9.

[19] Bukowski M, Wladyka B, Dubin G. Exfoliative toxins of *Staphylococcus aureus*. Toxins (Basel) 2010;2(5):1148–65.

[20] Sato'o Y, Omoe K, Ono HK, et al. A novel comprehensive analysis method for *Staphylococcus aureus* pathogenicity islands. Microbiol Immunol 2013;57(2):91–9.

[21] Argudín MA, Mendoza MC, Rodicio MR. Food poisoning and *Staphylococcus aureus* enterotoxins. Toxins 2010;2(7):1751–73.

[22] Finck-Barbançon V, Duportail G, Meunier O, et al. Pore formation by a two-component leukocidin from *Staphylococcus aureus* within the membrane of human polymorphonuclear leukocytes. Biochim Biophys Acta 1993;1182(3):275–82.

[23] Veldkamp KE, Heezius HC, Verhoef J, et al. Modulation of neutrophil chemokine receptors by *Staphylococcus aureus* supernate. Infect Immun 2000;68(10):5908–13.

[24] Bokarewa MI, Jin T, Tarkowski A. *Staphylococcus aureus*: staphylokinase. Int J Biochem Cell Biol 2006;38(4):504–9.

[25] Schaefers MM, Breshears LM, Anderson MJ, et al. Epithelial proinflammatory response and curcumin-mediated protection from staphylococcal toxic shock syndrome toxin-1. PLoS One 2012;7(3):e32813.

[26] Coleman D, Knights J, Russell R, et al. Insertional inactivation of the *Staphylococcus aureus* beta-toxin by bacteriophage phi 13 occurs by site- and orientation-specific integration of the phi 13 genome. Mol Microbiol 1991;5(4):933–9.

[27] Coleman DC, Sullivan DJ, Russell RJ, et al. *Staphylococcus aureus* bacteriophages mediating the simultaneous lysogenic conversion of beta-lysin, staphylokinase and enterotoxin A: molecular mechanism of triple conversion. J Gen Microbiol 1989;135(6):1679–97.

[28] Dearborn AD, Dokland T. Mobilization of pathogenicity islands by *Staphylococcus aureus* strain Newman bacteriophages. Bacteriophage 2012;2(2):70–8.

[29] Mir-Sanchis I, Martínez-Rubio R, Martí M, et al. Control of *Staphylococcus aureus* pathogenicity island excision. Mol Microbiol 2012;85(5):833–45.

[30] Christie GE, Dokland T. Pirates of the Caudovirales. Virology 2012;434(2):210–21.

[31] Spilman MS, Damle PK, Dearborn AD, et al. Assembly of bacteriophage 80α capsids in a *Staphylococcus aureus* expression system. Virology 2012;434(2):242–50.

[32] Ram G, Chen J, Kumar K, et al. Staphylococcal pathogenicity island interference with helper phage reproduction is a paradigm of molecular parasitism. Proc Natl Acad Sci USA 2012;109(40):16300–5.

[33] Subedi A, Ubeda C, Adhikari RP, et al. Sequence analysis reveals genetic exchanges and intraspecific spread of SaPI2, a pathogenicity island involved in menstrual toxic shock. Microbiology 2007;153(Pt 10):3235–45.

[34] Maiques E, Ubeda C, Tormo MA, et al. Role of staphylococcal phage and SaPI integrase in intra- and interspecies SaPI transfer. J Bacteriol 2007;189(15):5608–16.

[35] Chen J, Novick RP. Phage-mediated intergeneric transfer of toxin genes. Science 2009;323(5910):139–41.

[36] Maiques E, Ubeda C, Campoy S, et al. Beta-lactam antibiotics induce the SOS response and horizontal transfer of virulence factors in *Staphylococcus aureus*. J Bacteriol 2006;188(7):2726–9.

[37] Ubeda C, Maiques E, Knecht E, et al. Antibiotic-induced SOS response promotes horizontal dissemination of pathogenicity island-encoded virulence factors in staphylococci. Mol Microbiol 2005;56(3):836–44.

[38] Miller C, Thomsen LE, Gaggero C, et al. SOS response induction by beta-lactams and bacterial defense against antibiotic lethality. Science 2004;305(5690):1629–31.

[39] Panos GZ, Betsi GI, Falagas ME. Systematic review: are antibiotics detrimental or beneficial for the treatment of patients with *Escherichia coli* O157:H7 infection? Aliment Pharmacol Ther 2006;24(5):731–42.

[40] Li Z, Stevens DL, Hamilton SM, et al. Fatal *Staphylococcus aureus* hemorrhagic pneumonia: genetic analysis of a unique clinical isolate producing both PVL and TSST-1. PLoS One 2011;6(11):e27246.

Chapter 15

Diagnostics: Routine Identification on Standard and Chromogenic Media, and Advanced Automated Methods

Andrzej Kasprowicz, Anna Białecka, Joanna Białecka
Center of Microbiological Research and Autovaccines of Dr. Jan Bóbr Ltd., Krakow, Poland

15.1   INTRODUCTION

The first phase of the diagnostic cycle is to properly collect the appropriate clinical specimen for culture (i.e., a representative sample taken from an ongoing infection). The material must then be adequately handled and transported, under appropriate conditions, to a microbiology laboratory. Specimens are to be transported as quickly as possible using adequate transport media that preserve potential etiological factors, allow the preliminary assessment of the bacterial species present, especially in mixed cultures, and that reduce the ability of concomitant flora to multiply.

The next stage of the diagnostic cycle is the analytical phase of laboratory testing. This phase begins with conducting direct Gram stains of the clinical material (e.g., pus, blood, etc.), which helps determine the presence of microorganisms. The direct specimen has a high diagnostic value in samples that are physiologically sterile. Gram-positive cocci arranged in irregular grape-like clusters may suggest a staphylococcal infection. These bacteria may also form chains, or be arranged in pairs or as single cells (Figs. 15.1 and 15.2).

The clinical specimen is inoculated onto agar plates, for example, Tryptic Soy Agar supplemented with 5% sheep blood or onto selective media (Baird-Parker Agar, Chapman Agar). Inoculation must always be carried out by streaking, a technique that isolates individual bacterial colonies. If performed correctly, preliminary identification of staphylococci in the clinical specimen is possible. Bacteria of the *Staphylococcus* genus inoculated onto agar plates with 5% sheep blood, after an incubation period from 18 to 24h at a temperature of 35–37°C in

Pet-to-Man Travelling Staphylococci: A World in Progress. https://doi.org/10.1016/B978-0-12-813547-1.00015-7

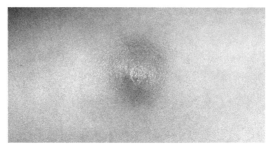

FIG. 15.1 Boil—infiltration of the skin.

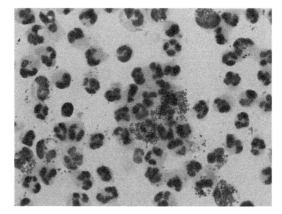

FIG. 15.2 Pus—a Gram stained sample (visible clusters of Gram+ cocci).

aerobic conditions, form round, raised, smooth, and shiny colonies, which range from 1 to 3 mm in diameter. Colonies may display a white pigment, different shades of yellow to golden pigments, and even orange coloring. The intensity of the color depends on the culture conditions (i.e., medium composition, light intensity, temperature). The distinctive coloring of colonies is not a feature that can help assign a strain to a certain species, even to the *Staphylococcus* genus. Most staphylococcal strains grown on agar, supplemented with 5% sheep blood, are beta-haemolytic (causing complete lysis of sheep erythrocytes) (Figs. 15.3 and 15.4).

15.2 STANDARD AND CHROMOGENIC CULTURE MEDIA

Selective media are used in the routine diagnosis of staphylococcal infections. They support the distinctive growth of staphylococci and inhibit or partially inhibit the growth of other bacteria present in the tested clinical specimen. The most commonly used selective media are

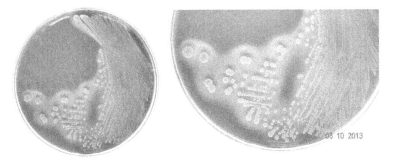

FIG. 15.3   Streak plate technique, growth of *Staphylococcus aureus.*

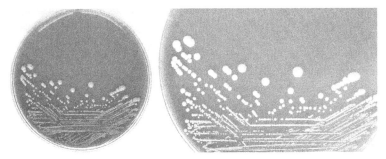

FIG. 15.4   Streak plate technique, growth of *Staphylococcus epidermidis.*

Chapman agar (Mannitol salt agar, MSA-F), a selective and differential medium that is recommended for differentiating and isolating staphylococci from clinical specimens. Strains that have the ability to ferment mannitol, produce acid products, and so form yellow colonies. The presence of a high salt concentration (7.5% NaCl) inhibits the growth of concomitant flora, including Gram-negative bacteria (Fig. 15.5).

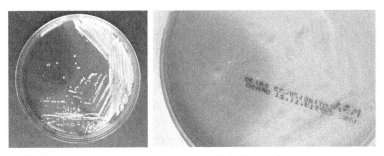

FIG. 15.5   *Staphylococcus aureus* streaked on a mannitol salt agar plate.

Baird-Parker Agar is a selective and diagnostic medium for the enumeration and identification of coagulase-positive staphylococci colonies found in food products intended for human and animal consumption and in clinical specimens. This method applies the ability of *Staphylococcus aureus* to reduce potassium tellurite, as well as their ability to proteolyse egg yolk, which is a substrate added to detect lipase activity.

The selective agents, lithium chloride and tellurite, are responsible for suppressing the growth of most bacteria without inhibiting coagulase-positive staphylococci; whereas pyruvate and glycine stimulate the growth of staphylococci. After the incubation period, *S. aureus* reduces the potassium tellurite forming distinctive black colonies. Clear zones may also form around colonies caused by the proteolysis of egg yolk. Additionally, an opaque zone of precipitation may form due to lipase activity (Fig. 15.6).

SASelect (BIO-RAD) is selective medium for the isolation and direct identification of *S. aureus*. Detection of phosphatase activity allows for the direct identification of *S. aureus* colonies, which appear pink to orange after an incubation period of 18–24 h at a temperature of 35°C. Glycosidic activity allows for the differentiation of other staphylococci, which appear blue (Fig. 15.7A and B).

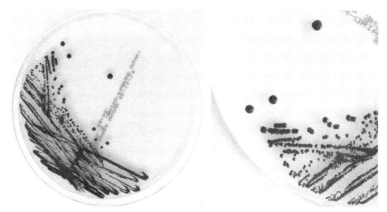

FIG. 15.6 *Staphylococcus aureus* streaked on a Baird-Parker plate.

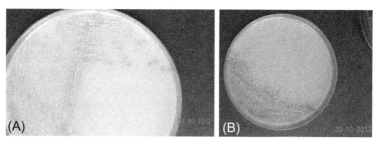

FIG. 15.7 (A) *Staphylococcus aureus* streaked on a SASelect plate. (B) A mixed culture on a SASelect plate.

SAID (BioMerieux) is a chromogenic media for the direct identification of *S. aureus* and the selective isolation of staphylococci. Direct identification of *S. aureus* is based on the spontaneous green coloration of colonies, which produce α-glucosidase. Differentiation of *S. aureus* is possible due to the appropriate nutrient content of the medium, detection sensitivity, and coloration specificity. The optimal differentiation of mixed cultures is possible due to the presence of a second chromogenic substrate (white, pink, or mauve colonies are formed), which simultaneously inhibits the growth of other bacteria (Gram + and Gram −) and yeasts (Fig. 15.8).

The next stage of the diagnostic cycle is to perform a Gram stain of a colony, which, based on its morphology, may suggest a staphyloccocal infection.

15.3 IDENTIFICATION TESTS

15.3.1 Gram Staining of Culture Samples

After having prepared a smear of the single colony and staining it by performing a Gram strain, Gram-positive cocci arranged in grape-like clusters (from the Greek word *staphyle* meaning bunch of grapes) may be observed. These spherical cells with a diameter of 0.8–1.0 μm grow in clusters because they divide in different planes (Fig. 15.9).

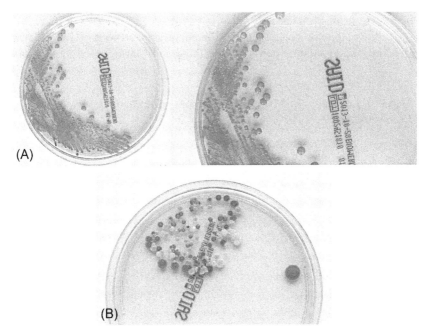

FIG. 15.8   (A) *Staphylococcus aureus* streaked on a SAID plate. (B) A mixed culture of staphylococci on a SAID plate.

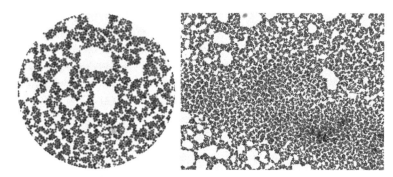

FIG. 15.9 Gram stain of *Staphylococcus*.

Performing a catalase test in order to differentiate staphylococci (catalase-positive) from streptococci (catalase negative) is the following stage of the diagnostic cycle. Another method used for differentiating staphylococci (sensitive) from other Gram-positive cocci (resistant) is the furazolidone susceptibility test (Table 15.1).

15.3.2 The Catalase Test

Catalase is an enzyme that is produced by most aerobic, facultative anaerobic, and some anaerobic bacteria. This enzyme decomposes hydrogen peroxide, which is toxic for bacteria, into water and oxygen. The catalase test is essential for differentiating staphylococci from streptococci. The test is conducted by picking the inoculum from a plate culture and placing it on a slide and

TABLE 15.1 Features That Differentiate Staphylococci From Other Gram-Positive Cocci

Genus	Morphology	Catalase	Furazolidone	Colonies
Staphylococcus	Clusters	+	Sensitive	Large, creamy white
Micrococcus	Large cells	+	Resistant	Medium sized, yellow to red
Streptococcus	Chains	−	−	Small to large, gray, alfa, beta and gamma hemolysis
Enterococcus	Chains	−	−	Large, gray, gamma hemolysis

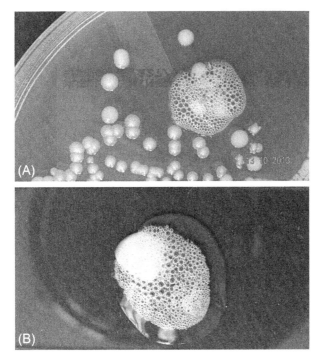

FIG. 15.10   (A) A positive catalase reaction on a Mueller-Hinton Agar plate. (B) Slide catalase test.

submerging it in a drop of 3% hydrogen peroxide. A positive reaction is evident when oxygen bubbles are observed (Fig. 15.10). It is not recommended to add the hydrogen peroxide into the culture itself. This may result in a nonspecific reaction caused by the presence of the different media components. This method may not be applied if bacterial cultures are grown on agar containing blood cells.

15.3.3   The Furazolidone Susceptibility Test

This method is used for the differentiation of staphylococci and micrococci (genera *Staphylococcus* and *Micrococcus*). Micrococci are resistant to furazolidone.

Susceptibility to furazolidone can be determined by using the filter paper-disk diffusion method. A plate of Mueller-Hinton agar is inoculated and a filter paper disk, impregnated with 100 µg of furazolidone, is placed on the plate. The plate is then incubated for 16–18 h at 35°C. Zones of inhibition having a diameter >15 mm show that the tested specimen is susceptible to furazolidone, indicating that further tests need to be run that will lead to the identification of *Staphylococcus* (Fig. 15.11).

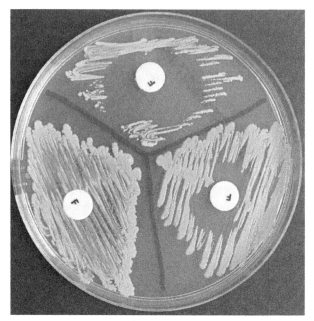

FIG. 15.11   Furazolidone susceptibility testing performed by a disk diffusion test.

The next stage of the diagnostic cycle is to perform tests that will determine the presence of coagulase (i.e., bound coagulase), free coagulase, or routine kits that simultaneously mark the presence of bound coagulase, protein A, and/or capsular polysaccharides.

15.3.4   Bound Coagulase

Bound coagulase or clumping factors react directly with fibrinogen-causing cell agglutination, without the presence of an activator. This is why rapid results are achieved by performing this test. The test is referred to as the slide coagulase test in which rabbit plasma is stirred into the bacterial suspension. If *S. aureus* is present in the tested suspension, clumps of bacteria will start to form within 30 s during the stirring process (Fig. 15.12).

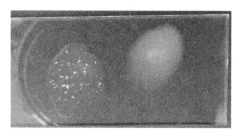

FIG. 15.12   Clumping factor. A positive test result (shown on the left), a negative test result (shown on the right).

15.3.5   Free Coagulase

Coagulase clots human and rabbit plasma in the presence of its activator. As a result of the coagulase-activator interaction, a substance similar to thrombin is produced, which converts fibrinogen to fibrin. To perform the test, a homogeneous bacterial suspension is mixed with sterile rabbit plasma in a citrate buffer.

If a clot forms after 4 h of incubation at 35°C, a positive result has been achieved, indicating the presence of *S. aureus*. If, however, the test result is negative after 4 h of incubation, it is required to continue incubation and examine the sample after 12, 18, and 24 h (Fig. 15.13).

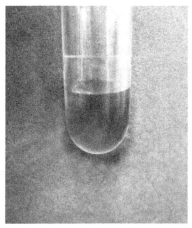

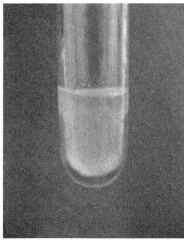

FIG. 15.13   Coagulase test. A negative test result (top test tube), a positive test result (bottom test tube).

15.3.6 The Detection of Clumping Factor, Protein A, and Capsular Polysaccharides by Using the Pastorex Staph-Plus Test (Bio-Rad)

The Pastorex Staph-Plus test is a rapid agglutination test used for detecting clumping factor, protein A, and capsular polysaccharides of *S. aureus*. To run the test it is essential to confirm that the tested specimen is a Gram positive and catalase-positive staphylococci.

The kit consists of latex particles that are sensitized with fibrinogen, immunoglobulin IgG, and monoclonal antibodies directed against capsular polysaccharides of *S. aureus*. A combination of these three factors allows for the effective identification of encapsulated *S. aureus* strains and nonencapsulated strains. A test is considered positive if there is visible agglutination of the latex particles within 30 s of card rotation, after having come into contact with the bacterial suspension on the agglutination card. A negative reaction is detected when no aggregates are produced in either of the reaction circles and the suspension, in both the tested circle and the negative control circle, retains their milky color.

The next stage of the diagnostic cycle is to identify staphylococcal strains at a species-level using a diverse range of biochemical tests.

15.4 IDENTIFICATION-AUTOMATED TESTS

Staphylococcal species can be relatively quickly identified with the help of a wide range of systems based on biochemical reactions (i.e., tube tests, miniaturized biochemical tests, and computerized identification systems). An example of such an identification system is the ID 32 STAPH test (BioMerieux). This system consists of 32 dehydrated substrates of specific reactions, for example, for the fermentation of glucose, fructose, lactose, trehalose, mannitol, raffinose, ribose, cellobiose, sucrose, methyl D-glucopyranoside, *N*-acetylglucosamine, for the reduction of nitrates to nitrites, and for the production of alkaline phosphatase, arginine dihydrolase, urea, and more. Other similar systems are RAPIDEC Staph, API STAPH VITEK, Vitek 2 (BioMerieux), MicroScan Pos ID panel, MicroScan Rapid Pos ID panel (Baxter Diagnostics Inc.), Crystal gram-positive identification system, Crystal rapid gram-positive identification system, BD Phoenix (Becton, Dickinson and Company), Pasco MIC/ID gram-positive panel (MIDI Inc.), and the GP MicroPlate test panel. Apart from using biochemical tests to identify staphylococci, chromatographic methods such as the MIDI Sherlock identification system or molecular methods such as the AccuProbe culture identification test for *S. aureus* (Gen-Probe) may also be applied (Fig. 15.14).

Different diagnostic systems can be used in the diagnosis of staphylococcal infections, including those that focus on biochemical reactions as well those that are based on other distinctive features such as sensitivity to novobiocin or

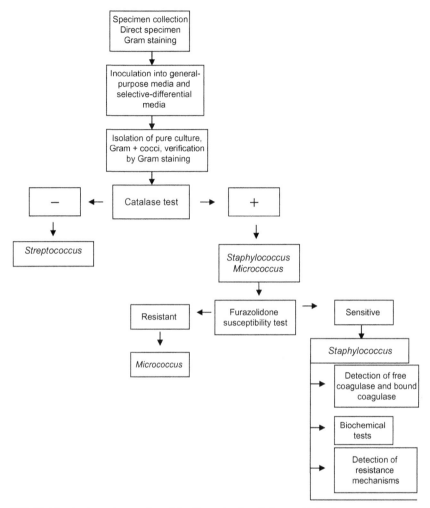

FIG. 15.14   A schematic overview of the diagnosis of staphylococcal infections.

the ability of strains to produce DNases. Table 15.2 provides a detailed presentation of clinically significant *Staphylococcus* species.

15.5   CONCLUSIONS

Accurate and punctual staphylococcal diagnostics are a fundamental basis for future effective therapy. An effective identification of staphylococcal basis on bacteria culture on selective specific growth media is necessary. Personal experience of diagnosticians and good practice are required, despite advanced instrumental or automated methods.

TABLE 15.2 Key Tests for Identification of the Most Clinically Significant *Staphylococcus* Species

	Species	*S. aureus* subsp. *aureus*	*S. epidermidis*	*S. haemolyticus*	*S. hyicus* (veterinary)	*S. intermedius* (veterinary)	*S. lugdunensis*	*S. schleiferi* subsp. *schleiferi*	*S. saprophyticus* subsp. *saprophyticus*
Result of test for	Colony pigmentation	+	-	d	-	-	d	-	d
	Staphylo-coagulase	+	-	-	d	+	-	-	-
	Clumping factor	+	-	-	-	d	(+)	+	-
	Heat stable nuclease	+	-	-	+	+	-	+	-
	Alkaline phosphatase	+	+	-	+	+	-	+	-
	Pyrrolidonyl arylamidase	-	-	+	-	+	+	+	-
	Ornithine decarboxylase	-	(d)	-	-	-	+	-	-
	Urease	d	+	-	d	+	d	-	+
	β-Galactosidase	-	-	-	-	+	-	(+)	+

Acetoin production	+	+	+	−	−	+	+	+
Novobiocin resistance	+	−	−	−	−	−	−	−
Polymyxin B resistance	−	−	d	−	+	−	+	+
Acid production (aerobically) from								
D-Trehaloze	+	d	+	+	+	+	−	+
D-Mannitol	d	−	−	(d)	−	d	−	+
D-Mannose	−	+	+	+	+	−	(+)	+
D-Turanose	+	−	(d)	d	−	(d)	(d)	+
D-Xylose	−	−	−	−	−	−	−	−
D-Cellobiose	−	−	−	−	−	−	−	−
Maltose	+	−	+	(±)	−	+	+	+
Sucrose	+	−	+	+	+	+	+	+

Data from Murray PR. Manual of clinical microbiology, vol. 22. 6th ed.; 2005. p. 284.

CONFLICT OF INTEREST

None declared.

FURTHER READING

Szewczyk EM. Diagnostyka Bakteriologiczna. Warszawa: PWN; 2013. p. 27–30, 431–5 [in Polish].

Murray PE. Manual of clinical microbiology. 9th ed. Washington, DC: ASM Press; 2007. p. 285–91.

Sellenriek P, Holmes J, Ferrett R, et al. Comparison of MicroScan Walk-Away®, Phoenix™ and VITEK-TWO® Microbiology systems used in the identification and susceptibility testing of bacteria. Abstr 105th General Meeting of the American Society for Microbiology (ASM), 2005.

Kloos WE, George CG. Identification of *Staphylococcus* species and subspecies with the MicroScan Pos ID and Rapid Pos ID panel systems. J Clin Microbiol 1991;29(4):738–44.

Eigner U, Cagnic A, Amburst M, et al. Evaluation of the New Automated Phoenix™ Microbiology System for Gram-Positive Cocci As presented at the 11th European Congress of Clinical Microbiology and Infectious Diseases (ECCMID), 2001.

Miedzobrodzki J, Małachowa N, Markiewski T, et al. Differentiation of *Staphylococcus aureus* isolates based on phenotypical characters. Postepy Hig Med Dosw 2008;62:322–7.

Gotz F. The genera *Staphylococcus* and *Macrococcus*. Prokaryotes 2006;4:5–75.

D'Souza HA, Baron EJ. BBL CHROMagar *Staph aureus* is superior to mannitol salt for detection of *Staphylococcus aureus* in complex mixed infections. Am J Clin Pathol 2005;123:806–8.

Gaillot OM, Wetch N, Fortineau N, et al. Evaluation of CHROMagar *Staph aureus*, a new chromogenic medium, for isolation, and presumptive identification of *Staphylococcus aureus* from human clinical specimens. J Clin Microbiol 2000;38:1587–91.

Kloos WE, Schleifer KH. Isolation and characterization of staphylococci from human skin. II Description of four new species: *Staphylococcus warneri*, *Staphylococcus capitis*, *Staphylococcus hominis*, and *Staphylococcus simulans*. Int J Syst Bacteriol 1975;25:62–79.

Perry JD, Rennison C, Butterworth LA, et al. Evaluation of *S. aureus* ID, a new chromogenic agar medium for detection of *Staphylococcus aureus*. J Clin Microbiol 2003;41:5695–8.

Kasprowicz A, Heczko PB. Evaluation of rapid tests used in clinical laboratory to differentiate staphylococci from micrococci. In: Jeljaszewicz J, editor. The staphylococci, Zbl BaktSuppl, vol. 14. Stuttgart: Gustaw Fischer Verlag; 1985. p. 177–9.

Kloss WE, Bannerman TL. Update on clinical significance of coagulase-negative staphylococci. Clin Microbiol Rev 1994;7:117–40.

Błaszczyk-Kostanecka M. Dermatologia w Praktyce. Warszawa: PZWL; 2005. p. 17–8 [in Polish].

Chapter 16

Molecular Identification and Genotyping of Staphylococci: Genus, Species, Strains, Clones, Lineages, and Interspecies Exchanges

Beata Krawczyk, Józef Kur
Department of Molecular Biotechnology and Microbiology, Faculty of Chemistry, Gdańsk University of Technology, Gdańsk, Poland

16.1   INTRODUCTION

Staphylococcal species are widely distributed in various environments, including skin and mucous membranes of humans and animals, as well as soil, sand, and water. Only half of the 45 species and 21 subspecies described thus far have been cultured from human specimens. *Staphylococcus aureus* is the most clinically relevant species of the genus, but coagulase-negative staphylococci (CoNS) are increasingly recognized as etiologic agents of human disease as well. CoNS have been identified as a major cause of hospital-acquired infections that typically affect immunocompromised patients and those with implanted medical devices. Identification to the species level is necessary to provide a better understanding of CoNS pathogenic potential and may support therapeutic decisions. Nevertheless, the emerging clinical importance of both *S. aureus* and CoNS makes it mandatory to obtain accurate identification to the subspecies level, too.

Again, bacterial typing at the strain level is particularly important for diagnosis, treatment, and epidemiological surveillance of bacterial diseases, particularly in the case of highly antibiotic resistant organisms and those involved in nosocomial or pandemic infections. Strain typing, additionally, finds application in the study of bacterial population dynamics. Over the past two decades, molecular methods have progressively replaced phenotypic assays in the field of bacterial identification, and current genotyping methods can be classified

Pet-to-Man Travelling Staphylococci: A World in Progress. https://doi.org/10.1016/B978-0-12-813547-1.00016-9

into three main categories: (1) DNA banding pattern-based methods, which classify bacteria according to the size of the fragments generated by amplification and/or enzymatic digestion of genomic DNA; (2) DNA sequencing-based methods, studying polymorphism of DNA sequences; (3) DNA hybridization-based methods that use nucleotidic probes.

Here, we describe and compare the applications of the mentioned methods in staphylococcal diagnostics.

16.2 MOLECULAR IDENTIFICATION METHODS

Several polymerase chain reaction (PCR) sequencing-based methods have been developed for the identification and phylogenetic study of staphylococci. A number of genes have been evaluated as molecular targets, including, for instance, the 16S rDNA [1–4], *hsp60* [5–9], *sodA* [10,11], *rpoB* [12–14], *tuf* [15–20], *gap* [19,21–23], *dnaJ* [24,25], *nuc* [26–28], and the 16S–23S spacer sequence [29–34]. Numerous studies have demonstrated that genotyping methods are superior to phenotypic methods. However, sequences of some genes are not sufficiently discriminative to differentiate closely related *Staphylococcus* species, and the databases only include a limited number of species. Methods used for staphylococcal identification are summarized in Table 16.1.

16.2.1   Species Identification Based on the 16S rDNA Sequence

As a tool for classification of microbial organisms, comparative DNA sequence analysis of the genes of conserved macromolecules has become commonplace in microbiology. The most useful taxonomic marker molecules, and the most extensively investigated, are the larger rDNAs and their coding genes, particularly 16S rDNAs and, to a lesser extent, 23S rDNAs. The considerable quantity of 16S rDNA sequence data available in public databases means that this gene has been the favorite choice in many studies. In fact, its usefulness has been broadly examined for identification of staphylococci [1–4] and it has been shown that, as it is a very conservative gene within this genus, the sequences used are not discriminative enough to differentiate all subspecies [1]. Conversely, it was determined that the 5′ end of 16S rDNA contains sufficient information for identification of almost every staphylococcal species [2]. As a part of the Ribosomal Differentiation of Microorganisms (RIDOM) project, a new ribosomal gene sequence database was created by sequencing both strands of the 5′ end of the 16S rDNA for the 81 type- and reference strains comprising all the validly described staphylococcal (sub)species [1]. Analyses of the newly determined sequences of the 16S rDNA 5′ end revealed, however, that they contain information that is suitable for identification of almost all staphylococci species, and with a few exceptions, subspecies. The increasing feasibility of

TABLE 16.1 Comparison of Chosen Molecular Methods for Identification of Staphylococci

Method Based on	Possibilities	References
16S rDNA sequence	PCR-RFLP of 16S rDNA PCR followed by direct sequencing	[1–4]
hsp60 gene sequence	PCR followed by direct sequencing PCR-RFLP of *hsp60* gene	[5–9]
sodA gene sequence	PCR followed by direct sequencing	[10,11]
rpoB gene sequence	PCR followed by direct sequencing	[12–14]
tuf gene sequence	PCR assay in combination with species-specific probes PCR-RFLP of *tuf* gene PCR followed by direct sequencing	[15–20]
gap gene sequence	PCR-RFLP of *gap* gene Terminal restriction fragment length polymorphism (T-RFLP) PCR followed by direct sequencing	[19–23]
dnaJ gene sequence	PCR-RFLP of *dnaJ* gene PCR followed by direct sequencing	[24,25]
nuc multiplex PCR		[26–28]
Internal transcribed spacer PCR (ITS-PCR)	Single PCR with conserved primers PCR-RFLP	[29–35]
Real-time PCR	Melt curve analysis With probes for *tuf* gene With probes for *sodA* gene	[36–40]

high-throughput sequencing, accompanied by falling costs, suggests that 16S rDNA sequencing is a promising, rapid alternative to biochemical and phenotypic procedures in general in the field of bacterial typing.

16.2.2 Species Identification and Phylogenetic Relationships Based on Different Partial *hsp*60 Gene Sequences and PCR-Restriction Fragment Length Polymorphism of the *hsp*60 Gene

The *hsp60* (*groEL*) gene, which encodes a 60-kDa polypeptide known as GroEL, 60-kDa chaperonin, or Hsp60 for heat shock protein 60, has the potential to serve as a general phylogenic marker. This gene has been proven to be an ideal universal DNA target for identification to the species level because it has well-conserved DNA sequences within a given species, but with sufficient sequence variations to allow species-specific identification [5]. Kwok and co-workers [6] have determined that the interspecies level sequence similarity among 28 distinct species of staphylococci ranges from 74% to 93%. Importantly, phylogenetic analysis based on the neighbor-joining distance method has revealed a remarkable concordance between the tree derived from partial *hsp60* gene sequences and the one based on genomic DNA-DNA hybridization, while 16S rRNA gene sequences correlate less well.

The *groEL* gene has been used successfully as a tool for the identification of the main *Staphylococcus* species involved in human infections by PCR restriction fragment length polymorphism (RFLP) [7]. Using the same approach, namely the PCR-RFLP of the *groEL* gene, the study was extended to the most common staphylococcal strains isolated from cows with mastitis [8] and to the CoNS strains causing mastitis in goats [9].

The results obtained to date indicate that DNA sequence analysis of the highly conserved and ubiquitous *hsp60* genes may offer advantages over both DNA-DNA hybridization and 16S rRNA sequencing in defining the taxonomy and phylogenetic relationships within the genus *Staphylococcus*. In addition, sequence data provide direct evidence for the presence of species specific *hsp60* gene sequences, which can be employed as a genotypic method for the species identification of staphylococci, and CoNS in particular.

16.2.3 The *sod*A Gene Polymorphism in *Staphylococcus* spp. Identification

Poyart et al. [10] proposed staphylococci identification based on the *sodA* gene sequences encoding the manganese-dependent superoxide dismutase. They have determined the *sodA* internal (429 bp) sequences of 40 types of CoNS strains and demonstrated the usefulness of this database for species-level identification of staphylococcal isolates. This method consists of a PCR carried out with a single pair of degenerate oligonucleotides and based on amplification of a staphylococcal *sodA* fragment, then direct sequencing of the PCR product with the same degenerate primers. The topology of the phylogenetic tree obtained in this way was in general agreement with the one inferred from an analysis of their 16S rDNA or *hsp60* gene sequences. Sequence analysis revealed that the staphylococcal *sodA* genes exhibit a higher divergence than the corresponding 16S ribosomal DNA does. These results confirm that the *sodA* gene constitutes a highly discriminative target sequence for differentiating closely related bacterial species. However, it was concluded that, like the *hsp60* gene, the *sodA* gene does not allow discrimination at the subspecies level.

Sivadon et al. used the *sodA* sequencing successfully for the identification of CoNS clinical isolates [11] and for genotypic identification in a prospective study aiming to examine the distribution of CoNS strains recovered from septic orthopedic surgery.

16.2.4   Identification of Staphylococci Species Based on *rpo*B Gene Sequence

The *rpoB* gene encoding the highly conserved β sub-unit of the bacterial RNA polymerase has been demonstrated to be a suitable target for identifying bacteria. It has been also shown to be more discriminative than the 16S rDNA. A staphylococcal *rpoB* sequence database was established and made it possible to develop a molecular method for identifying *Staphylococcus* isolates by PCR, followed by direct sequencing of the 751-bp amplicon [12]. The *rpoB* sequence similarity of 29 staphylococcal *rpoB* partial sequences showed interspecific similarity values of 71.6%–93.6%, with the most closely related sequences being those of *S. aureus* subsp. *aureus* and *S. aureus* subsp. *anaerobius*. The phylogenetic tree inferred from neighbor joining showed nine clusters characterized by a sequence similarity level of >85% within the cluster. The nine clusters were *Staphylococcus epidermidis*, *Staphylococcus saprophyticus*, *Staphylococcus simulans*, *Staphylococcus intermedius*, *Staphylococcus hyicus*, *Staphylococcus sciuri*, *Staphylococcus auricularis*, *Staphylococcus aureus*, and *Staphylococcus caseolyticus*.

In 2009, Sampimon and co-workers [14] successfully used *rpoB* sequencing for the identification of 172 CoNS isolated from bovine milk. They were then classified into 17 species. The most frequently isolated species were *Staphylococcus chromogenes* and *S. epidermidis*, followed by *Staphylococcus xylosus*, *Staphylococcus warneri*, and *Staphylococcus equorum*, at 37%, 13%, 9%, 8%, and 6% of the isolates, respectively. For comparison, the API Staph ID 32 and the Staph-Zym correctly identified 41% and 31% of the isolates, respectively. The best agreement with the *rpoB* sequence-based species identification was found for *S. epidermidis*, *S. hyicus*, and *S. xylosus*, at 100%, 89%, and 87%, respectively. Good sensitivity was found for *S. epidermidis*, *S. simulans,* and *S. xyloxus*, at 100%, 78%, and 73%, respectively, while sensitivity was poor with *S. chromogenes*, *S. warneri*, and *S. equorum* (0%, 54%, and 0%, respectively). Both phenotypic tests misidentified a large proportion of the CoNS isolates and were thus unsuitable for the identification of CoNS species from bovine milk samples.

16.2.5   Identification of Staphylococci Species Based on *tuf* Gene Sequences

The *tuf* gene, which encodes the elongation factor Tu (EF-Tu), is involved in peptide chain formation and is an essential constituent of the bacterial genome. This fact makes it a target of choice for diagnostic purposes. Martineau and co-workers [16] have developed a *tuf*-based PCR assay in combination with five species-specific probes targeting *S. aureus*, *S. epidermidis*, *S. haemolyticus*, *S. hominis*, and *S. saprophyticus*. On the basis of the published nucleotide sequence of the *tuf* gene from 11 different staphylococcal species, including the most common staphylococcal pathogens, Kontos et al. [17] performed

PCR-RFLP for a part of the *tuf* gene (370 bp), showing this method to be reliable and reproducible for identification of the staphylococcal species in question. The 16S rRNA and *tuf* sequence-based relationships obtained by Heikens and co-workers [15] accorded with the phylogenetic trees published previously [3,16]. The trees derived from both the 16S rRNA and the *tuf* sequence indicate that *S. sciuri* is the most distant *Staphylococcus* species [15]. Furthermore, the *tuf* sequence-derived tree indicates that *S. warneri* is associated with the *S. epidermidis* group, which includes *S. capitis*. This concurs with results previously described in PCR-sequencing assays targeting the *sodA* gene [10]. The *tuf*-derived data showed intraspecies sequence divergence more often than the 16S rRNA-derived data. Apparently, the 16S rDNA gene is more highly conserved than the *tuf* gene. Pairwise comparison of the *tuf* sequences revealed that, at 92.6%, their mean identity is lower than the mean identity of 16S rDNA sequences, at 95.9 [15]. These results indicate that, in the differentiation of closely related *Staphylococcus* species, *tuf* constitutes a more discriminatory target gene than the 16S rDNA gene. This represents a major point of interest in the use of *tuf* results from the small size of the amplicon (660 bp) required, along with the fact that nondegenerate oligonucleotide primers can be used; these two conditions have not been achieved simultaneously by most other targets [18]. The *tuf* gene has thus emerged as a reliable molecular tool for an accurate identification of *Staphylococcus* species [15,19,20].

16.2.6 Species Identification of Staphylococci Based on Different Partial Gap Gene Sequences and PCR-RFLP of the Gap Gene

The *gap* gene encodes a 42-kDa transferring-binding protein (Tpn) located within the staphylococcal cell wall. Tpn is a member of the newly emerging family of multifunctional cell wall-associated glyceraldehyde-3-phosphate dehydrogenases, well known for its glycolytic function of converting D-glyceraldeyde-3-phosphate to 1,3-bisphosphoglycerate. The *gap* has proved to be a very well-conserved gene that may be a useful tool in an RFLP-PCR assay for differentiating staphylococcal species. Genetic uniformity was found in *S. aureus* strains analyzed using this procedure. Yugueros and co-workers [21] described a PCR-based DNA amplification method combined with RFLP for the identification of 12 staphylococcal species relevant for humans. The RFLP-PCR of the *gap* gene enabled the detection of intraspecies polymorphism among the *S. epidermidis*, *S. hominis*, and *S. simulans* strains. In their following study [22], they showed that *Alu*I digestion of PCR-generated *gap* gene products rendered distinctive RFLP patterns that allowed 24 *Staphylococcus* spp. strains to be identified with high specificity.

In 2007, Layer and co-workers [23] described a terminal restriction fragment length polymorphism (T-RFLP) analysis that facilitated the identification of 28 *Staphylococcus* species when amplification of the partial *gap* gene sequence was performed, followed by fingerprinting of the digest fragments of *Dde*I, *Bsp*HI, and *Taq*I. These results demonstrated that the T-RFLP method based on partial *gap* gene sequences is feasible and applicable for rapid and accurate staphylococci species identification.

In 2008, Ghebremedhin et al. [19] partially sequenced the *gap* gene (931 bp) from 27 *Staphylococcus* species. The partial sequences had an interspecies homology of 24.3%–96% and were useful for staphylococcal identification. The DNA sequence similarities of the partial staphylococcal *gap* sequences were found to be lower than those for 16S rRNA (97%), *rpoB* (86%), *hsp60* (82%), and *sodA* (78%). The phylogenetically-derived trees revealed four statistically supported groups: *S. hyicus/S. intermedius*, *S. sciuri*, *S. haemolyticus/S. simulans*, and *S. aureus/epidermidis*. The phylogenetic analysis based on the *gap* gene sequences revealed similarities between the dendrograms based on other gene sequences. For example, the *gap* gene sequences enabled the differentiation of the closely related staphylococci species, *S. caprae* and *S. capitis*, which was impossible with 16S rDNA sequencing. Based on these results, they have proposed the partial sequencing of the *gap* gene as an alternative molecular tool for taxonomical analysis of *Staphylococcus* species and for decreasing the possibility of misidentification.

16.2.7   *dnaJ* Gene Polymorphism in *Staphylococcus* spp. Identification

The *dnaJ* gene encodes the DnaJ protein, which is also known as Hsp40, and it is a member of the heat shock protein (Hsp) family. In 2007, Shah and co-workers [24] proposed a *dnaJ* gene sequence-based assay for species identification and phylogenetic grouping in the genus *Staphylococcus*. A portion of approximately 883 bp of the *dnaJ* gene sequence from 45 staphylococcal type strains was compared with 16S rDNA and other conserved gene sequences available in public databases, namely *hsp60*, *sodA*, and *rpoB*. It was also determined that the *dnaJ* gene sequence is more discriminative than the sequences of other conserved, housekeeping genes used in *Staphylococcus* taxonomy. In addition, the evolutionary rate of substitution of the *dnaJ* sequence is much faster than that of the 16S rDNA gene sequence [24]. The *dnaJ* gene sequence is thus potentially useful for the identification of genetically related species and even subspecies. Nucleotide sequence comparisons revealed that, at a mean similarity of 77.6%, the staphylococcal *dnaJ* gene showed higher discrimination than the 16S rDNA, with a mean similarity of 97.4%, the *rpoB*, with a mean similarity of 86%, the *hsp60*, with a mean similarity of 82%, and the *sodA*, with a mean similarity of 81.5%. Analysis of the *dnaJ* gene sequence from 20 *Staphylococcus* isolates representing two clinically important species, *S. aureus* and *S. epidermidis*, showed <1% sequence divergence.

In 2008, Hauschild and Stepanovic [25] described a method based on PCR-RFLP analysis of the *dnaJ* gene with *Xap*I or *Bsp*143I digestion and showed that this method is an adequate tool for correct identification of almost all prevalent *Staphylococcus* species and subspecies, irrespective of their phenotypic characterization. It is worth mentioning that this method also has the capability of discriminating subspecies of the species *S. capitis*, *S. carnosus*, *S. cohnii*, and *S. hominis*. However, the paucity of data on the divergence of the *dnaJ* sequence within staphylococcal species makes it difficult to state whether this technique would suffer the same criticism over the accurate identification of staphylococci to the species level, similar to phylogenetic studies based on 16S rDNA gene sequence analysis.

16.2.8 Human-Associated Staphylococci Species Identification by *nuc* Multiplex PCR

Hirotaki and co-workers [26] developed a rapid and accurate multiplex-PCR (M-PCR) assay for species identification of human-associated staphylococci. It encompassed *S. aureus*, *S. capitis*, *S. caprae*, *S. epidermidis*, *S. haemolyticus*, *S. hominis*, *S. lugdunensis*, *S. saprophyticus*, and *S. warneri*. The method was based on nucleotide sequences of the thermonuclease (*nuc*) genes that had been used for staphylococcal identification ([26–28], for example). During the validation process, 361 staphylococcal strains that had been identified to the species level by sequence analysis of the *hsp60* genes were studied. The outcome was that the M-PCR demonstrated both sensitivity and a specificity of 100%. This method might be useful in clinical research by virtue of its simplicity and accuracy.

16.2.9 The Use of Internal Transcribed Spacer-PCR

In prokaryotes, the genes for 16S, 23S, and 5S rRNAs are separated by spacer regions that show a high degree of variability in both sequence and size at the genus and species level. The diversity of the intergenic spacer regions comes about, in part, owing to variations in the number and type of tRNA sequences found among these spacers. In staphylococci, there are several copies of the *rrn* operon. The highly polymorphic nature of the 16S–23S spacer sequences may be analyzed by PCR, using conserved sequences from the adjacent 16S and 23S genes as primers. This method is known as internal transcribed spacer-PCR (ITS-PCR). Identification of staphylococci by ITS-PCR was first studied by Jensen et al. [29], who successfully applied this technique in order to differentiate *S. aureus*, *S. epidermidis*, *S. saprophyticus*, and *S. warneri* strains. Some authors tested ITS-PCR for the identification of staphylococci from diverse origins, using different protocols [30,31], whereas others designed PCR primers based on species-specific sequences of 16S–23S spacers for the detection of particular staphylococcal species [32,33]. In a study presented by Couto and co-workers [34], the ITS-PCR was used to identify a collection of 617 clinical staphylococcal isolates and 29 staphylococcal species control strains. Each of the 29 control strains tested showed distinct amplification patterns, which subsequently facilitated the identification of nearly 600 clinical isolates, corresponding to 95.95% of the samples tested. It was also shown that the restriction of the amplicons with restriction enzymes as proposed by other authors [29,31] is not necessary for the identification of staphylococci of human origin. The majority of these species were easily identified by direct comparison with the control strain profile, and this additional resolution step may therefore not be necessary. ITS-PCR proved to be a valuable alternative for the identification of staphylococci, offering higher reliability than the currently available commercial systems, within the same response time and at a lower cost.

16.2.10 Real-Time PCR

For the past few years, a variety of real-time PCR chemistries and detection instruments have appeared on the market, and many of them lend themselves to applications for detection and quantitation of microorganisms. These approaches may amplify DNA, as well as detect and confirm target sequence identity, in a closed-tube format with the use of a variety of fluorophores, labeled probes, or both, and without the need to run gels. All the studies using real-time PCR demonstrated that this technique is a powerful tool for the detection of microorganisms, including staphylococci. Edwards and colleagues [36] used melt curves generated by the combination of a fluorescent intercalating probe with a labeled hybridization one to discriminate 15 staphylococcal species. Skow and co-workers [37] developed a rapid-cycle, real-time PCR assay that would distinguish various clinically relevant *Staphylococcus* strains via the unique melting curve profiles of each species. The assay was based on the ability of this approach to detect sequence variation present within the 16S rDNA amplicons of each species. Nine common staphylococcal species, namely *S. aureus*, *S. capitis*, *S. epidermidis*, *S. haemolyticus*, *S. hominis*, *S. lugdunensis*, *S. schleiferi*, *S. simulans*, and *S. warneri*, were examined. This PCR/melt curve analysis achieved an accuracy of nearly 100% and performed better than biochemical testing. Kobayashi et al. [38] developed a real-time PCR with primers and probes for the *tuf* gene that could easily differentiate *S. aureus* from *S. epidermidis*. Iwase and co-workers [39] have described a species-specific quantitative detection method involving 5′ nuclease real-time PCR and using a minor groove binder probe designed from the *sodA* gene for the purpose of *S. epidermidis* identification. This method distinguished *S. epidermidis* from other staphylococci and specifically quantified the bacterium. The same group of authors [40] showed that real-time PCR based on the *sodA* is useful for the identification and quantitative analysis of CoNS such as *S. capitis*, *S. haemolyticus*, and *S. warneri*.

16.2.11 The Use of Mass Spectrometry in Staphylococci Identification

Matrix-assisted laser desorption/ionization time-of-flight mass spectrometry (MALDI-TOF-MS) enables rapid identification of organisms through molecular mass profiling of protein biomarkers. For a given bacterial strain, MALDI-TOF-MS yields, within minutes, a reproducible spectrum consisting of a series of peaks corresponding to the mass-to-charge (*m/z*) ratios of ions released from bacterial proteins during laser desorption. The commercial application of this technology in pathogens' identification has only begun in the past 6 years, and new, potentially life-saving applications of these technologies are already rapidly identifying organisms that have proven to be notoriously difficult to identify in the past. MALDI-TOF-MS has been used successfully for identification of almost all staphylococcal species.

Bergeron et al. [18] used MALDI-TOF-MS technology in conjunction with techniques based on partial nucleotide sequences of the *tuf* or *gap* genes for comparison purposes in order to evaluate their suitability, practicability, and discriminatory power for staphylococcal identification. MALDI-TOF-MS correctly identified 138 isolates (74.2%). Four strains were misidentified, 39 were unidentified, five were identified at the group (*hominis/warneri*) level, and one strain was identified at the genus level. One hundred and eighty-four strains (98.9%) were correctly identified by *tuf* gene sequencing. Only one strain was misidentified and one was unidentified. The authors found an overall superiority for the molecular method, even though the MALDI-TOF-MS-based method is faster and more cost-effective. The slight inferiority of the MALDI-TOF-MS versus the *tuf*-based method was rather unexpected given the number of enthusiastic reports on the performance of this technology for species identification.

A study by Dupont and co-workers [41] was performed in order to compare MALDI-TOF-MS linked to a recently engineered microbial identification database, and two rapid identification automated systems, BD Phoenix (Becton Dickinson Diagnostic Systems, France) and VITEK-2 (BioMérieux, Marcy L'Etoile, France), in respect to CoNS identification. All isolates were also characterized by *sodA* gene sequencing, which facilitated the interpretation of the identification results obtained using the respective database of each apparatus. Overall, correct identification results were obtained in 93.2%, 75.6%, and 75.2% of the cases with the MALDI-TOF-MS, Phoenix and VITEK-2 systems, respectively. This study demonstrated the robustness and high sensitivity of MALDI-TOF-MS microbial identification database and confirmed that this approach represents a powerful tool for the fast identification of clinical CoNS isolates. Dubois et al. [42] also showed a very high percentage of staphylococcal strains' correct identifications obtained using the MALDI-TOF-MS. A total of 151 out of 152 strains (99.3%) were correctly identified to the species level. Only one strain was identified to the genus level. The MALDI-TOF-MS method also revealed different *S. epidermidis* clonal lineages that were of either human or environmental origin, suggesting that the MALDI-TOF-MS method could be useful in the profiling of staphylococcal strains [43,44]. Based on the results they obtained, the authors indicated that the MALDI-TOF-MS technology, linked with a broad-spectrum reference database, is an effective tool for the swift and reliable identification of staphylococci and a powerful method for the rapid identification of clonal strains of *S. aureus*, which might be useful for tracking nosocomial outbreaks of MRSA and for epidemiologic studies of infectious diseases in general.

16.3   MOLECULAR METHODS FOR STAPHYLOCOCCI GENOTYPING

Molecular typing is widely employed in the differentiation of staphylococcal strains, and has provided an excellent basis for understanding the molecular evolution of strains and establishing the major clades, or clonal clusters and dominant subclades, both in the global and in local population. Genotyping techniques have been described for the monitoring of staphylococcal infections, especially during a suspected outbreak of MRSA, as well as for the control of the track spread of strains such as MRSA among units or hospitals. Genotyping may also be used for patients with recurrent staphylococcal infections in order to indicate whether a second episode has occurred owing to a relapse or a recurrence. The identification of *S. aureus* with an identical genotype during independent episodes of infection in a given patient has a high risk of involvement in relapse.

The genotyping methods used for *Staphylococcus* strains isolated from individuals who are persistently carriers, although silent, of staphylococci in their nasopharynx and/or other body sites provide the possibility of documenting whether effective colonization and infection is limited to certain individual bacterial isolates or to major clusters of strains [45,46].

Genotyping methods have been also used to compare genetic differences between *S. aureus* strains from infected animals and colonized or infected humans [47]; in this context, in fact, *S. aureus* cross-colonization and/or cross-infection between domestic animals and their owners has been observed by genotyping [48].

For genotyping to be effective, an appropriate method must be used and the suitability of its discriminatory power should be taken into account. A choice of method also depends on whether there are long- or short-term epidemiological studies of local outbreaks. Furthermore, given that genomic instability depends on the region within a given chromosome, even regions for genetic characterization should be chosen with care [49]. Thus the analysis can involve the whole genomic DNA, the selected fragment, the gene or a gene fragment.

16.3.1   Methods Based on Whole Genome Analysis

This group of methods, often called *genetic fingerprinting,* includes random amplified polymorphic DNA (RAPD), pulsed-field gel electrophoresis (PFGE), and ligation-mediated PCR (LM PCR). These methods mainly use restriction analysis, or the amplification of DNA fragments in vitro using PCR, or both approaches to a particular strain specific DNA bands profile. Detection is performed by means of electrophoretic separation. For all the methods based on the analysis of a whole genome, the use of purified DNA from a single colony or pure culture of bacteria is crucial.

16.3.1.1 Techniques Based on Restriction Analysis

REA-PFGE

PFGE is the reference technique for typing of staphylococcal isolates recovered from patients and the environment in epidemiological studies. The macrorestriction analysis of DNA-fragments (REA-PFGE) is done by using rare-cutting restriction endonuclease, such as *Sma*I, *Xba*I, or *Cfr*9I, for example, followed by separation on an agarose gel using PFGE. With the aid of the computerized gel scanning and analysis software, it is possible to create data banks of PFGE patterns for all *S. aureus* strains, enabling the creation of reference databases to which any new strain can be compared for identification (i.e., the phylogenetic relationship to other apparently similar strains). PFGE is known to be highly discriminatory and is frequently used in investigations on outbreaks. However, its stability may be insufficient for reliable application in long-term epidemiological studies [50]. HARMONY, a multinational, European Union project was set up in order to establish a European database of representative epidemic MRSA (EMRSA) strains. A new "harmonized" PFGE protocol, which was developed by consensus, was applied. This approach demonstrated sufficient reproducibility to enable successful comparison of pulsed-field gels among laboratories and the tracking of strains around the EU [51]. The database thus developed contains profiles of DNA patterns for the 21 strains representing the most widespread epidemic clones of MRSA. It holds the exact characteristics of these strains, their drug resistance profiles, toxin production, phage type, and epidemiological data [51,52].

Despite efforts to standardize protocols and interpretation criteria [53,54] for PFGE data, comparison of interlaboratory results usually remains difficult. REA-PFGE is still limited by demanding protocols, low throughput, and high costs. PFGE is recommended for reference laboratories and research centers that are asked to provide epidemiologically useful information about the emergence and spread of MRSA clones.

16.3.1.2 Techniques Based on PCR

Random Amplified Polymorphic DNA

RAPD [55,56] typing, which is also known as arbitrarily primed polymerase chain reaction (AP-PCR), is a technique adapted for rapid detection of genomic polymorphism. The technique is based on the amplification of the genomic DNA with either a single or multiple short oligonucleotide primers of an arbitrary or random sequence. The number and location of these random primer sites varies for different strains of a bacterial species. What results from the separation of the amplification products by agarose gel electrophoresis is therefore a pattern of bands which, in theory, is characteristic of the bacterial strain in question. In most papers, the high discriminatory power of the RAPD pattern is described, though PFGE has been reported [57] as being more useful than RAPDs in distinguishing among *S. aureus* strains. In most cases, the sequences of the RAPD primers, which generate the best DNA pattern for differentiation, must be determined empirically. Another problem is both the inter- and intral-aboratory reproducibility of the method, and it has been demonstrated that, for typing *S. aureus* strains, the successful use of RAPD heavily depends on the optimal use of PCR protocols [58].

16.3.1.3   Techniques Based on Restriction Analysis and PCR
LM PCR Methods
The diagnostic methods based on DNA amplification by PCR preceded by the li-gation of adapters to the restriction fragments, which is known as ligation medi-ated PCR (LM PCR), include amplified fragment length polymorphism (AFLP) [59], amplification of DNA surrounding rare restriction sites (ADSRRS) [60], and PCR melting profiles (PCR MP) [61]. These methods have three common steps: (i) a genomic total DNA is digested with one or two restriction enzyme/s; (ii) the mixture of DNA fragments is ligated with synthetic adapters; either one or two types of adapter are used, depending on the restriction enzymes; (iii) the PCR. AFLP, ADSRRS, and PCR MP are image-based DNA fingerprint-ing techniques that compare DNA fragment patterns generated by restriction in combination with amplification. However, each of the methods differs in the technique for selecting the amplified PCR fragments.

In AFLP analysis, the PCR primers used for amplification contain DNA sequences homologous to the linker along with 1–3 selective bases at their 3′ ends. The selective nucleotides only allow the amplification of a subset of the genomic restriction fragments. The AFLP method has a high discrimination power. By using a combination of several primers, even single differences be-tween closely related organisms can be detected. Comparison of three *S. aureus*-strain genotyping methods, namely AFLP, multilocus sequence typing (MLST) and PFGE, revealed that AFLP is more reproducible than PFGE and probably more suitable for interlaboratory data exchange, but is still less reproducible than DNA sequence-based MLST [62].

The principle of the ADSRRS-fingerprinting method is amplification of the exclusive DNA fragments surrounding relatively rare nucleotide sequences, such as rare restriction sites, for example, and the PCR suppression (SP PCR) phenomenon for obtaining limited representation of the DNA fragments that form the bacterial genome.

The PCR MP technique is based on the use of low denaturation tempera-tures during LM PCR. The method allows the specific, gradual amplification of the genomic DNA, starting from the less stable DNA fragments, which are amplified at lower denaturation temperatures and moving on to the more stable ones, which are amplified at a higher temperature. Lowering the denaturation temperature during the PCR should decrease the number of fragments amplified because only single-stranded DNA molecules can serve as templates. The advantage of the PCR MP method lies in the possibility of verifying a questionable genetic pattern through the use of a short-denaturation temperature gradient.

ADSRRS-fingerprinting and PCR MP have been successfully used for the genotyping of *S. aureus* strains isolated from patients with a history of furunculosis [63]. ADSRRS-fingerprinting and PCR MP have been shown to be as effective in discriminating closely related strains as the PFGE method, which is considered to be the "gold standard" for epidemiological studies and can be used as alternatives for studies on the intraspecies genetic relatedness of *S. aureus* strains.

16.3.1.4   Whole Genome Sequencing

Whole genome sequencing (WGS) refers to the construction of the complete nucleotide sequence of a genome and allows clinical isolates of *S. aureus* to be compared with each other and with reference sequences across time and space, with an accuracy of a single nucleotide difference. The high resolution offered by WGS has the potential to revolutionize our understanding and management of *S. aureus* infection. WGS allows accurate characterization of transmission events and outbreaks, and provides information concerning the genetic basis of phenotypic characteristics, including antibiotic susceptibility and virulence [64]. The first complete *S. aureus* genome was published in 2001 [65]; only 12 complete genomic sequences of *S. aureus* were available in 2008, but, by 2012, there were 178 fully annotated *S. aureus* genomes publically available in the National Centre for Biotechnology Information (NCBI) Reference Sequences (RefSeq) database.

16.3.1.5   GeneChip-Based Technique

The *S. aureus* Affymetrix oligonucleotide array was constructed in order to determine the relatedness and genetic composition of a series of *S. aureus* strains. In 2004, Dunman et al. [66] indicated that the technology can make it possible to determine whether a group of similar strains under investigation are clonal or slightly divergent in genetic composition. This distinction is a critical aspect of monitoring strain outbreaks.

16.3.2   Methods Based on the Analysis of the Characteristic Regions of the Genome

16.3.2.1   Multiple-Locus Variable-Number Tandem Repeat Fingerprinting

The Multiple-Locus Variable-Number Tandem Repeat Fingerprinting (MLVF) method is applied for the analysis of a limited number of polymorphic genes with variable numbers of tandem repeats (VNTRs) and encoding surface proteins of *S. aureus*, such as *sdr*, *clfA*, *clfB*, *ssp*, or *spa*. These genes are characterized by the presence of a variable number of short tandem repeat sequences. Sabat et al. [67] have described a system for discrimination of *S. aureus* clinical isolates based on analysis of the VNTR by means of multiplex PCR. The sequence polymorphism of these loci makes it possible to obtain various electrophoretic patterns. Strains with electrophoretic patterns differing in 6 or more bands from a maximum of 14 are considered as unrelated [68]. Multilocus VNTR typing appears to be the most useful PCR-based method for achieving rapid genotyping of *S. aureus* strains with the purposes of long-term epidemiological studies.

16.3.2.2   Multilocus Sequence Typing

DNA sequence-based typing has become more popular owing to the progress in the development of a large-scale sequencing methodology, ease of data transfer, and the excellent comparability of the results. MLST [69] characterizes bacterial isolates on the basis of sequence polymorphism within internal fragments of seven housekeeping genes, *arcC*, *aroE*, *glp*, *gmk*, *pta*, *tpi*, and *yqiL*, which represent the stable "core" of the staphylococcal genome. PCR is used to amplify the DNA at each locus and PCR products with a length of 400 to ~500 bp are sequenced. *S. aureus* strains can then be compared using the same loci. Each gene fragment is translated into a distinct allele, and each isolate is classified as a sequence type (ST) by the combination of alleles of the seven housekeeping loci [70,71]. There are now more than 2500 STs on the *S. aureus* MLST database (https://pubmlst.org/). MLST has been used successfully for MRSA strain typing and has been proposed both for long-term global studies and for the assessment of evolutionary relationships among *S. aureus* strains [72]. The method is achieving global recognition, and a web-based database of strain types has been developed (http://www.mlst.net). The *S. aureus* MLST website currently contains information on isolates from humans and animals from 40 different countries and represents a useful global resource for the study of the epidemiology of this species and the surveillance of hypervirulent and/or antibiotic-resistant clones.

16.3.2.3   spa Typing

spa typing is a common genotyping tool for methicillin-resistant *S. aureus* (MRSA) in Europe and is based on the polymorphic direct repeats found in the staphylococcal protein A gene (*spa*). This region of the *spa* locus consists of a variable number of 24-bp short sequence repeats (SSR). The diversity of the SSR region seems to arise from deletion and duplication of the repetitive units as well as from point mutations [73]. The *spa* typing method is based on two techniques, that is PCR and sequencing of amplification products. The primers are designed to target the evolutionary conserved sequences flanking the polymorphic region of the *spa* gene. Correlation has been found between SSRs variability and pathogenicity.

The website www.seqnet.org provides information on identified *spa* types and sequences of the various regions of repetitive SSR. Tandem repeat sequences are numbered and processed using the appropriate software Ridom StaphType, which is available at http://spaserver.ridom.de. The result is given in the form of a numeric code that distinguishes the *spa* types and contains information concerning the order and changes in repetitive regions. *spa* types are grouped on the complex with the same SSR sequence (*spa*-clonal complexes). Specialized software for automated sequence analysis ensure a common typing nomenclature. *SeqNet.org* is an initiative of currently 60 laboratories from 29 European countries. To date, around 12,928 *spa* types have been identified and belong to 280,288 characterized strains. A wide geographical distribution has been found, with some prevalent in all of the European countries in question; *spa* types t067, t041, t032, and t003 were the four most significant regional clusters [74].

Reproducibility of *spa* sequencing has been evaluated in a multicenter study using specialized software for sequence analysis [75]. DNA sequence-based typing of *S. aureus* protein A gene (*spa* amplification product from 110 to 422 bp) showed 100% intra- and interlaboratory reproducibility, without extensive harmonization of protocols, for 30 blind-coded *S. aureus* DNA samples sent to 10 laboratories. *spa*-typing is now widely used for MRSA typing on account of its concordance with MLST. Unfortunately, though, it is limited in terms of resolving power as it poorly distinguishes outbreak isolates from background isolates, which is of relevance in a local epidemic situation [76].

MRSA can also be identified by means of a high-resolution melting real-time PCR analysis on protein A (*spa*) typing [77]. This simple and inexpensive assay can be applied in laboratories in various geographical regions by adopting the different *spa* types circulating in the diverse areas. It has great potential in terms of becoming a useful tool for MRSA epidemiological analysis and infection control.

16.3.2.4   Agr Typing

The accessory gene regulator (*agr*), a crucial regulatory *S. aureus* component, is involved in the control of bacterial virulence factor expression. *S. aureus agr* is a 3-kb locus showing highly conserved and hyper variable regions among *S. aureus* strains [78]. The sequence of this hyper variable segment can be used as a target of PCR amplification aiming to define *agr* types. Four allelic groups of *agr* have been identified within the species and a real-time multiplex quantitative PCR assay for rapid determination of *S. aureus agr* type has also been developed [79].

16.3.2.5   SCCmec Typing

The SCC (Staphylococcal Cassette Chromosome) *mec* element is a mobile genetic element carrying the central determinant for betalactam resistance, which is encoded by the *mecA* gene and is inserted into the chromosome of resistant strains

[80–84]. SCC*mec* elements are highly diverse in their structural organization and genetic content and have been classified into types and subtypes. The SCC*mec* type is determined by the *mec* complex and the *ccr* genes. The regions that are not part of the *mec* complexes and *ccr* genes are called J (junkyard) regions.

SCC*mec* typing is based on the analysis of the hyper-variable components of the SCC*mec* cassette. SCC*mec* types are defined by the combination of the *ccr* gene complex type, which is represented by the *ccr* gene allotype(s), and the class of the *mec* gene complex. The first SCC*mec* typing methods were based on the detection of the polymorphisms surrounding the *mec*A gene by means of hybridization with *mec*A and *Tn554* probes with *Cla*I digested genomic DNA (*mecA:Tn554 probe typing*) [85]. Nowadays, SCC*mec* typing methods can be divided into three types: (1) based on restriction enzyme digestion; (2) based on simplex PCR and multiplex PCR; or (3) based on real-time PCR. For example, a method based on PCR amplification of the *ccrB* gene in combination with RFLP employing endonucleases *Hinf*I and *Bsm*I was demonstrated in 2005 by van der Zee et al. in 2005 [86], and by Yang et al. in 2006 [87]. Multiplex PCR-based SCC*mec* typing methods focused on the analysis of *J* or *mec* class and *ccr* type regions [88–96] are more popular. Milheiriço et al. [91] have developed a multiplex PCR assay which is based on the amplification of six specific loci within the J1 region of SCC*mec* type IV variants. SCC*mec* typing methods were modified many times in order to allow detection of the most popular types of cassettes. In 2007 [88], Boye et al. designed a quick and easily interpreted method based on a single-tube multiplex PCR using primers for the specific detection of both the *mec* class and *ccr* type of SCC*mec* types I to V. In 2006, Lina et al. [94] described a multilocus typing method for SCC*mec* based on *ccr* gene complex polymorphisms. That study demonstrated a strong correlation between the *ccr* complex sequence and the type of SCC*mec*.

In parallel to classical PCR based methods, those based on real-time PCR for SCC*mec* typing have been developed. A multiplex scheme based on a real-time PCR targeting the *ccrB* regions of SCC*mec* types I to IV was published in 2004 by Francois et al. [97]. A multiplex PCR strategy for the rapid identification of structural types and variants of the *mec* element in methicillin-resistant *S. aureus* allows a better understanding of the evolutionary relationships among MRSA clones.

SCC*mec* has been identified not only in *S. aureus*, but also in other coagulase-positive staphylococci (CoPS) and CoNS. In particular, CoPS other than *S. aureus*, namely *S. pseudointermedius* and *S. schleiferi* subsp. *coagulans,* are agents of disease in dogs and cats, where they cause pyoderma and otitis externa, although they may alternatively behave as commensals [98–102]. SCC*mec* elements have been reported in several methicillin-resistant CoNS (MR-CoNS). CoNS species may serve as reservoirs of SCC*mec* that may be subsequently acquired by *S. aureus* (that therefore becomes MRSA) [103]. According to the available data, SCC*mec* elements are more diverse in MR-CoNS, with new variants of *ccr* genes continuing to be identified. The presence of very similar

SCC*mec* elements in MRSA and CoNS strongly suggests the horizontal transfer of SCC*mec* [104–109].

The International Working Group on the Classification of Staphylococcal Cassette Chromosome Elements (IWG-SCC) has established a webpage dedicated to the SCC*mec* element, providing updated information on SCC*mec* elements, a detailed classification system based on the differences in J regions, and currently available typing methods (http://www.staphylococcus.net and/or http://www.SCCmec.org).

16.4 CONCLUSION

In the genomic era, bacterial genotyping benefits increasingly from bioinformatics such as the computer-based selection of typing markers, the design of the typing strategy, the storage, exchange, and comparison of genotyping data (namely DNA banding patterns, DNA sequences and microarray profiles), the development of improved analysis tools, and phylogenetic analysis. Genotyping databases containing banding patterns, DNA sequences, or DNA microarray profiles are convenient for interlaboratory comparison, retrospective studies, and long-term epidemiological surveillance of bacterial infections. Locus-specific typing methods now use multiple loci rather than a single locus, and whole-genome analysis has allowed selection of typing markers to gradually become less empirical and more rational. Genome sequencing of one bacterial strain not only provides detailed information of intraspecies diversity, but also enables the selection of locus-specific typing markers and the rational design of genotyping strategies. Genome sequences also facilitate the rational choice of primers for PCR-based genotyping methods.

Each identification and genotyping method offers advantages and features disadvantages. Consequently, choosing an appropriate molecular method is not an easy task and selection may depend on the objective of the study, as well as on a number of variables, including typeability, reproducibility, resolution, cost, simplicity of execution, ease of interpretation, along with time required to obtain results (Box 16.1).

BOX 16.1

Bacterial strain identification and typing methods have been evolving rapidly in recent years. Genotyping methods are of paramount importance to define whether a local outbreak is due to a single or diverse strains, to investigate characteristics of unrelated bacterial isolates, as well as to compare isolates from a certain region or area with those collected worldwide. Finally, in phylogenetic research, genome-based comparisons are necessary to study relationship among strains.

REFERENCES

[1] Becker K, Harmsen D, Mellmann A, Meier C, Schumann P, Peters G, et al. Development and evaluation of a quality controlled ribosomal sequence database for 16S ribosomal DNA based identification of *Staphylococcus* species. J Clin Microbiol 2004;42:4988–95.

[2] Gribaldo S, Cookson B, Saunders N, Marples R, Stanley J. Rapid identification by specific PCR of coagulase-negative staphylococcal species important in hospital infection. J Med Microbiol 1997;46:45–53.

[3] Takahashi T, Satoh I, Kikuchi N. Phylogenetic relationships of 38 taxa of the genus *Staphylococcus* based on 16S rRNA gene sequence analysis. Int J Syst Bacteriol 1999;49:725–8.

[4] Onni T, Sanna G, Cubeddu GP, Marogna G, Lollai S, Leori G, et al. Identification of coagulase-negative staphylococci isolated from ovine milk samples by PCR-RFLP of 16S rRNA and *gap* genes. Vet Microbiol 2011;144:347–52.

[5] Goh SH, Potter S, Wood JO, Hemmingsen SM, Reynolds RP, Chow AW. HSP60 gene sequences as universal targets for microbial species identification: studies with coagulase-negative staphylococci. J Clin Microbiol 1996;34:818–23.

[6] Kwok AY, Su SC, Reynolds RP, et al. Species identification and phylogenetic relationships based on partial HSP60 gene sequences within the genus *Staphylococcus*. Int J Syst Bacteriol 1999;49:1181–92.

[7] Barros EM, Iório NLP, de Freire Bastos MC, Santos KRN, Giambiagi-deMarval M. Species-level identification of clinical staphylococcal isolates based on polymerase chain reaction restriction fragment length polymorphism analysis of a partial *groEL* gene sequence. Diagn Microbiol Infect Dis 2007;59:251–7.

[8] Santos OCS, Barros EM, Brito MAVP, de Freire Bastos MC, dos Santos KN, Giambiagi-deMarval M. Identification of coagulase-negative staphylococci from bovine mastitis using RFLP-PCR of the *groEL* gene. Vet Microbiol 2008;130:134–40.

[9] Onni T, Vidili A, Bandino E, Marogna G, Schianchi S, Tola S. Identification of coagulase-negative staphylococci isolated from caprine milk samples by PCR-RFLP of *groEL* gene. Small Ruminant Res 2012;104:185–90.

[10] Poyart C, Quesne G, Boumaila C, Trieu-Cuot P. Rapid and accurate species-level identification of coagulase-negative staphylococci by using the *sodA* gene as a target. J Clin Microbiol 2001;39:4296–301.

[11] Sivadon V, Rottman M, Chaverot S, Quincampoix JC, Avettand V, de Mazancourt P, et al. Use of genotypic identification by *sodA* sequencing in a prospective study to examine the distribution of coagulase-negative *Staphylococcus* species among strains recovered during septic orthopedic surgery and evaluate their significance. J Clin Microbiol 2005;43:2952–4.

[12] Drancourt M, Raoult D. *rpoB* gene sequence-based identification of *Staphylococcus* species. J Clin Microbiol 2002;40:1333–8.

[13] Mellmann A, Becker K, von Eiff C, Keckevoet U, Schumann P, Harmsen D. Sequencing and staphylococci identification. Emerg Infect Dis 2006;12:333–6.

[14] Sampimon OC, Zadoks RN, De Vliegher S, Supré K, Haesebrouck F, Barkema HW, et al. Performance of API Staph ID 32 and Staph-Zym for identification of coagulase-negative staphylococci isolated from bovine milk samples. Vet Microbiol 2009;136:300–5.

[15] Heikens E, Fleer A, Paauw A, Florijn A, Fluit AC. Comparison of genotypic and phenotypic methods for species level identification of clinical isolates of coagulase-negative staphylococci. J Clin Microbiol 2005;43:2286–90.

[16] Martineau F, Picard FJ, Ke D, Paradis S, Roy PH, Ouellette M, et al. Development of a PCR assay for identification of staphylococci at genus and species levels. J Clin Microbiol 2001;39:2541–7.

[17] Kontos F, Petinakia E, Spiliopouloub I, Maniatia M, Maniatisa AN. Evaluation of a novel method based on PCR restriction fragment length polymorphism analysis of the *tuf* gene for the identification of *Staphylococcus* species. J Microbiol Methods 2003;55:465–9.

[18] Bergeron M, Dauwalder O, Gouy M, Freydiere AM, Bes M, Meugnier H, et al. Species identification of staphylococci by amplification and sequencing of the *tuf* gene compared to the *gap* gene and by matrix-assisted laser desorption ionization time-of-flight mass spectrometry. Eur J Clin Microbiol Infect Dis 2011;30:343–54.

[19] Ghebremedhin B, Layer F, König W, König B. Genetic classification and distinguishing of *Staphylococcus* species based on different partial *gap*, 16S rRNA, *hsp60*, *rpoB*, sodA, and *tuf* gene sequences. J Clin Microbiol 2008;46:1019–25.

[20] Hwang SM, Kim MS, Park KU, Song J, Kim EC. *tuf* gene sequence analysis has greater discriminatory power than 16S rRNA sequence analysis in identification of clinical isolates of coagulase-negative staphylococci. J Clin Microbiol 2011;49:4142–9.

[21] Yugueros J, Temprano A, Berzal B, Sánchez M, Hernanz C, Luengo JM, et al. Glyceraldehyde-3-phosphate dehydrogenase-encoding gene as a useful taxonomic tool for *Staphylococcus* spp. J Clin Microbiol 2000;38:4351–5.

[22] Yugueros J, Temprano A, Sánchez M, Luengo JM, Naharro G. Identification of *Staphylococcus* spp. by PCR-restriction fragment length polymorphism of *gap* gene. J Clin Microbiol 2001;39:3693–5.

[23] Layer F, Ghebremedhin B, König W, König B. Differentiation of *Staphylococcus* spp. by terminal-restriction fragment length polymorphism analysis of glyceraldehyde-3-phosphate dehydrogenase-encoding gene. J Microbiol Methods 2007;70:542–9.

[24] Shah MM, Iihara H, Noda M, et al. *dnaJ* gene sequence-based assay for species identification and phylogenetic grouping in the genus *Staphylococcus*. Int J Syst Evol Microbiol 2007;57:25–30.

[25] Hauschild T, Stepanovic S. Identification of *Staphylococcus* spp. by PCR-restriction fragment length polymorphism analysis of *dnaJ* gene. J Clin Microbiol 2008;46:3875–9.

[26] Hirotaki S, Sasaki T, Kuwahara-Arai K, Hiramatsu K. Rapid and accurate identification of human-associated staphylococci by use of multiplex PCR. J Clin Microbiol 2011;49:3627–31.

[27] Brakstad OG, Aasbakk K, Maeland JA. Detection of *Staphylococcus aureus* by polymerase chain reaction amplification of the *nuc* gene. J Clin Microbiol 1992;30:1654–60.

[28] Barski P, Piechowicz L, Galiński J, Kur J. Rapid assay for detection of methicillin-resistant *Staphylococcus aureus* using multiplex PCR. Mol Cell Probes 1996;10:471–5.

[29] Jensen MA, Webster JA, Straus N. Rapid identification of bacteria on the basis of polymerase chain reaction-amplified ribosomal DNA spacer polymorphisms. Appl Environ Microbiol 1993;59:945–52.

[30] Bes M, Guerin-Faublee V, Meugnier H, Etienne J, Freney J. Improvement of the identification of staphylococci isolated from bovine mammary infections using molecular methods. Vet Microbiol 2000;71:287–94.

[31] Mendoza M, Meugnier H, Bes M, Etienne J, Freney J. Identification of *Staphylococcus* species by 16S–23S rDNA intergenic spacer PCR analysis. Int J Syst Bacteriol 1998;48:1049–55.

[32] Forsman P, Tilsala-Timisjarvi A, Alatossava T. Identification of staphylococcal and streptococcal causes of bovine mastitis using 16S–23S rRNA spacer regions. Microbiology 1997;143:3491–500.

[33] Saruta K, Matsunaga T, Kono M, Hoshina S, Ikawa S, Sakai O, et al. Rapid identification and typing of *Staphylococcus aureus* by nested PCR amplified ribosomal DNA spacer. FEMS Microbiol Lett 1997;146:271–8.

[34] Couto I, Pereira S, Miragala M, Sanches IS, Lencastre H. Identification of clinical staph-ylococcal isolates from humans by internal transcribed spacer PCR. J Clin Microbiol 2001;39:3099–103.

[35] Layer F, Ghebremedhin B, Moder KA, König W, König B. Comparative study using various methods for identification of *Staphylococcus* species in clinical specimens. J Clin Microbiol 2006;44:2824–30.

[36] Edwards KJ, Kaufmann ME, Saunders NA. Rapid and accurate identification of coagulase-negative staphylococci by real-time PCR. J Clin Microbiol 2001;39:3047–51.

[37] Skow A, Mangold KA, Tajuddin M, Huntington A, Fritz B, Thomson Jr. RB, et al. Species-level identification of staphylococcal isolates by real-time PCR and melt curve analysis. J Clin Microbiol 2005;43:2876–80.

[38] Kobayashi N, Bauer TW, Sakai H, Daisuke Togawa D, Lieberman IH, Fujishiro T, et al. The use of newly developed real-time PCR for the rapid identification of bacteria in culture-negative osteomyelitis. J Bone Spine 2006;73:745–7.

[39] Iwase T, Hoshina S, Seki K, Shinji H, Masuda S, Mizunoe Y. Rapid identification and spe-cific quantification of *Staphylococcus epidermidis* by 5' nuclease real-time polymerase chain reaction with a minor groove binder probe. Diag Microbiol Infect Dis 2008;60:217–9.

[40] Iwase T, Seki K, Shinji H, Mizunoe Y, Masuda S. Development of a real-time PCR assay for the detection and identification of *Staphylococcus capitis*, *Staphylococcus haemolyticus* and *Staphylococcus warneri*. J Med Microbiol 2007;56:1346–9.

[41] Dupont C, Sivadon-Tardy V, Bille E, Dauphin B, Beretti JL, Alvarez AS, et al. Identification of clinical coagulase-negative staphylococci, isolated in microbiology laboratories, by matrix-assisted laser desorption/ionization-time of flight mass spectrometry and two automated sys-tems. Clin Microbiol Infect 2010;16:998–1004.

[42] Dubois D, Leyssene D, Chacornac JP, Kostrzewa M, Schmit PO, Talon R, et al. Identification of a variety of *Staphylococcus* species by matrix-assisted laser desorption ionization-time of flight mass spectrometry. J Clin Microbiol 2010;48:941–5.

[43] Hillenkamp F, Karas M, Beavis RC, Chait BT. Matrix-assisted laser desorption/ionization mass spectrometry of biopolymers. Anal Chem 1991;63:1193A–203A.

[44] Bernardo K, Pakulat N. Identification and discrimination of *Staphylococcus aureus* strains using matrix-assisted laser desorption/ionization-time of flight mass spectrometry. Pro-teomics 2002;2:747–53.

[45] Kluytmans J, van Belkum A, Verbrugh H. Nasal carriage of *Staphylococcus aureus*: epidemi-ology, underlying mechanisms, and associated risks. Clin Microbiol Rev 1997;10:505–20.

[46] Peacock SJ, de Silva I, Lowy FD. What determines nasal carriage of *Staphylococcus aureus*? Trends Microbiol 2001;9:605–10.

[47] van Leeuwen WB, Melles DC, Alaidan A, Al-Ahdal M, Boelens HA, Snijders SV, et al. Host- and tissue-specific pathogenic traits of *Staphylococcus aureus*. J Bacteriol 2005;187:4584–91.

[48] Simoons-Smit AM, Savelkoul PH, Stoof J, Starink TM, Vandenbroucke-Grauls CM. Trans-mission of *Staphylococcus aureus* between humans and domestic animals in a household. Eur J Clin Microbiol Infect Dis 2000;19:150–2.

[49] van Belkum A, Struelens M, de Visser A, Verbrugh H, Tibayrenc M. Role of genomic typ-ing in taxonomy, evolutionary genetics, and microbial epidemiology. Clin Microbiol Rev 2001;14:547–60.

[50] Hallin M, Deplano A, Denis O, De Mendonça R, De Ryck R, Struelens MJ. Validation of pulsed-field gel electrophoresis and spa typing for long-term, nationwide epidemiological surveillance studies of *Staphylococcus aureus* infections. J Clin Microbiol 2007;45:127–33.

[51] Murchan S, Kaufmann ME, Deplano A, de Ryck R, Struelens M, Zinn CE, et al. Harmonization of pulsed-field gel electrophoresis protocols for epidemiological typing of strains of methicillin-resistant *Staphylococcus aureus*: a single approach developed by consensus in 10 European laboratories and its application for tracing the spread of related strains. J Clin Microbiol 2003;41:1574–85.

[52] Cookson BD, Robinson DA, Monk AB, Murchan S, Deplano A, de Ryck R, et al. Evaluation of molecular typing methods in characterizing a European collection of epidemic methicillin-resistant *Staphylococcus aureus* strains: the HARMONY collection. J Clin Microbiol 2007;45:1830–7.

[53] Tenover FC, Arbeit RD, Goering RV, Mickelsen PA, Murray BE, Persing DH, et al. Interpreting chromosomal DNA restriction patterns produced by pulsed-field gel electrophoresis: criteria for bacterial strain typing. J Clin Microbiol 1995;33:2233–9.

[54] Tenover FC, Arbeit RD, Goering RV. How to select and interpret molecular strain typing methods for epidemiological studies of bacterial infections: a review for healthcare epidemiologists. Molecular Typing Working Group of the Society for Healthcare Epidemiology of America. Infect Control Hosp Epidemiol 1997;18:426–39.

[55] Williams JG, Kubelik AR, Livak KJ, Rafalski JA, Tingey SV. DNA polymorphisms amplified by arbitrary primers are useful as genetic markers. Nucleic Acids Res 1990;18:6531–5.

[56] Welsh J, McClelland M. Fingerprinting genomes using PCR with arbitrary primers. Nucleic Acids Res 1990;18:7213–8.

[57] Saulnier P, Bourneix C, Prévost G, Andremont A. Random amplified polymorphic DNA assay is less discriminant than pulsed-field gel electrophoresis for typing strains of methicillin-resistant *Staphylococcus aureus*. J Clin Microbiol 1993;31:982–5.

[58] van Belkum A, Kluytmans J, van Leeuwen W, Bax R, Quint W, Peters E, et al. Multicenter evaluation of arbitrarily primed PCR for typing of *Staphylococcus aureus* strains. J Clin Microbiol 1995;33:1537–47.

[59] Vos P, Hogers R, Bleeker M, Reijans M, van de Lee T, Hornes M, et al. AFLP: a new technique for DNA fingerprinting. Nucleic Acids Res 1995;23:4407–14.

[60] Masny A, Plucienniczak A. Fingerprinting of bacterial genomes by amplification of DNA fragments surrounding rare restriction sites. Biotechniques 2001;31:930–6.

[61] Masny A, Plucienniczak A. Ligation mediated PCR performed at low denaturation temperatures—PCR melting profiles. Nucleic Acids Res 2003;31:e114.

[62] Melles DC, van Leeuwen WB, Snijders SV, Horst-Kreft D, Peeters JK, Verbrugh HA, et al. Comparison of multilocus sequence typing (MLST), pulsed-field gel electrophoresis (PFGE), and amplified fragment length polymorphism (AFLP) for genetic typing of *Staphylococcus aureus*. J Microbiol Methods 2007;69:371–5.

[63] Krawczyk B, Leibner J, Barańska-Rybak W, Samet A, Nowicki R, Kur J. ADSRRS-fingerprinting and PCR MP techniques for studies of intraspecies genetic relatedness in *Staphylococcus aureus*. J Microbiol Methods 2007;71:114–22.

[64] Price JR, Didelot X, Crook DW, Llewelyn MJ, Paul J. Whole genome sequencing in the prevention and control of *Staphylococcus aureus* infection. J Hosp Infect 2013;83:14–21.

[65] Kuroda M, Ohta T, Uchiyama I, et al. Whole genome sequencing of methicillin-resistant *Staphylococcus aureus*. Lancet 2001;357:1225–40.

[66] Dunman PM, Mounts W, McAleese F, Immermann F, Macapagal D, Marsilio E, et al. Uses of *Staphylococcus aureus* GeneChips in genotyping and genetic composition analysis. J Clin Microbiol 2004;42:4275–83.

[67] Sabat A, Krzyszton-Russjan J, Strzalka W, Filipek R, Kosowska K, Hryniewicz W, et al. New method for typing *Staphylococcus aureus* strains: multiple-locus variable-number tandem repeat analysis of polymorphism and genetic relationships of clinical isolates. J Clin Microbiol 2003;41:1801–4.

[68] Sabat A, Malachowa N, Miedzobrodzki J, Hryniewicz W. Comparison of PCR-based methods for typing *Staphylococcus aureus* isolates. J Clin Microbiol 2006;44:3804–7.

[69] Maiden MC, Bygraves JA, Feil E, Morelli G, Russell JE, Urwin R, et al. Multilocus sequence typing: a portable approach to the identification of clones within populations of pathogenic microorganisms. Proc Natl Acad Sci USA 1998;95:3140–5.

[70] Enright MC, Day NP, Davies CE, Peacock SJ, Spratt BG. Multilocus sequence typing for characterization of methicillin-resistant and methicillin-susceptible clones of *Staphylococcus aureus*. J Clin Microbiol 2000;38:1008–15.

[71] Enright MC, Robinson DA, Randle G, Feil EJ, Grundmann H, Spratt BG. The evolutionary history of methicillin-resistant *Staphylococcus aureus* (MRSA). Proc Natl Acad Sci USA 2002;99:7687–92.

[72] Hallin M, Denis O, Deplano A, De Ryck R, Crèvecoeur S, Rottiers S, et al. Evolutionary relationships between sporadic and epidemic strains of healthcare-associated methicillin-resistant *Staphylococcus aureus*. Clin Microbiol Infect 2008;14:659–69.

[73] Strommenger B, Braulke C, Heuck D, Schmidt C, Pasemann B, Nübel U, et al. *spa* typing of *Staphylococcus aureus* as a frontline tool in epidemiological typing. J Clin Microbiol 2008;46:574–81.

[74] Grundmann H, Aanensen DM, van den Wijngaard CC, Spratt BG, Harmsen D, Friedrich AW, et al. Geographic distribution of *Staphylococcus aureus* causing invasive infections in Europe: a molecular-epidemiological analysis. PLoS Med 2010;7:e1000215.

[75] Aires-de-Sousa M, Boye K, de Lencastre H, Deplano A, Enright MC, Etienne J, et al. High interlaboratory reproducibility of DNA sequence-based typing of bacteria in a multicenter study. J Clin Microbiol 2006;44:619–21.

[76] Frénay HM, Bunschoten AE, Schouls LM, van Leeuwen WJ, Vandenbroucke-Grauls CM, Verhoef J, et al. Molecular typing of methicillin-resistant *Staphylococcus aureus* on the basis of protein A gene polymorphism. Eur J Clin Microbiol Infect Dis 1996;15:60–4.

[77] Chen JH, Cheng VC, Chan JF, She KK, Yan MK, Yau MC, et al. The use of high-resolution melting analysis for rapid spa typing on methicillin-resistant *Staphylococcus aureus* clinical isolates. J Microbiol Methods 2013;92:99–102.

[78] Jarraud S, Mougel C, Thioulouse J, Lina G, Meugnier H, Forey F, et al. Relationships between *Staphylococcus aureus* genetic background, virulence factors, agr groups (alleles), and human disease. Infect Immun 2002;70:631–41.

[79] Francois P, Koessler T, Huyghe A, Harbarth S, Bento M, Lew D, et al. Rapid *Staphylococcus aureus* agr type determination by a novel multiplex real-time quantitative PCR assay. J Clin Microbiol 2006;44:1892–5.

[80] Hiramatsu K, Cui L, Kuroda M, Ito T. The emergence and evolution of methicillin-resistant *Staphylococcus aureus*. Trends Microbiol 2001;9:486–93.

[81] Katayama Y, Ito T, Hiramatsu K. Genetic organization of the chromosome region surrounding *mecA* in clinical staphylococcal strains: role of IS431-mediated *mecI* deletion in expression of resistance in *mecA*-carrying, low-level methicillin-resistant *Staphylococcus haemolyticus*. Antimicrob Agents Chemother 2001;45:1955–63.

[82] Oliveira DC, Tomasz A, de Lencastre H. The evolution of pandemic clones of methicillin-resistant *Staphylococcus aureus*: identification of two ancestral genetic backgrounds and the associated *mec* elements. Microb Drug Resist 2001;7:349–61.

[83] Oliveira DC, Milheirico C, de Lencastre H. Redefining a structural variant of staphylococcal cassette chromosome *mec*, SCC*mec* type VI. Antimicrob Agents Chemother 2006;50:3457–9.

[84] Ito T, Ma XX, Takeuchi F, Okuma K, Yuzawa H, Hiramatsu K. Novel type V staphylococcal cassette chromosome *mec* driven by a novel cassette chromosome recombinase, *ccrC*. Antimicrob Agents Chemother 2004;48:2637–51.

[85] Leski T, Oliveira D, Trzcinski K, Sanches IS, Aires de Sousa M, Hryniewicz W, et al. Clonal distribution of methicillin-resistant *Staphylococcus aureus* in Poland. J Clin Microbiol 1998;36:3532–9.

[86] van der Zee A, Heck M, Sterks M, et al. Recognition of SCCmec types according to typing pattern determined by multienzyme multiplex PCR-amplified fragment length polymorphism analysis of methicillin-resistant *Staphylococcus aureus*. J Clin Microbiol 2005;43:6042–7.

[87] Yang JA, Park DW, Sohn JW, Kim MJ. Novel PCR-restriction fragment length polymorphism analysis for rapid typing of staphylococcal cassette chromosome mec elements. J Clin Microbiol 2006;44:236–8.

[88] Boye K, Bartels MD, Andersen IS, Møller JA, Westh H. A new multiplex PCR for easy screening of methicillin-resistant *Staphylococcus aureus* SCCmec types I-V. Clin Microbiol Infect 2007;13:725–7.

[89] Kondo Y, Ito T, Ma XX, et al. Combination of multiplex PCRs for staphylococcal cassette chromosome mec type assignment: rapid identification system for mec, ccr, and major differences in junkyard regions. Antimicrob Agents Chemother 2007;51:264–74.

[90] Milheiriço C, Oliveira DC, de Lencastre H. Update to the multiplex PCR strategy for assignment of mec element types in *Staphylococcus aureus*. Antimicrob Agents Chemother 2007;51:3374–7. [Erratum in: Antimicrob Agents Chemother 2007;51:4537].

[91] Milheiriço C, Oliveira DC, de Lencastre H. Multiplex PCR strategy for subtyping the staphylococcal cassette chromosome mec type IV in methicillin-resistant *Staphylococcus aureus*: 'SCCmec IV multiplex'. J Antimicrob Chemother 2007;60:42–8 [Erratum in: J Antimicrob Chemother 2007;60:708].

[92] Oliveira DC, de Lencastre H. Multiplex PCR strategy for rapid identification of structural types and variants of the mec element in methicillin-resistant Staphylococcus aureus. Antimicrob Agents Chemother 2002;46:2155–61.

[93] Zhang K, McClure JA, Elsayed S, Louie T, Conly JM. Novel multiplex PCR assay for characterization and concomitant subtyping of staphylococcal cassette chromosome mec types I to V in methicillin-resistant *Staphylococcus aureus*. J Clin Microbiol 2005;43:5026–33.

[94] Lina G, Durand G, Berchich C, Short B, Meugnier H, Vandenesch F, et al. Staphylococcal chromosome cassette evolution in *Staphylococcus aureus* inferred from *ccr* gene complex sequence typing analysis. Clin Microbiol Infect 2006;12:1175–84.

[95] Oliveira DC, Milheirico C, Vinga S, de Lencastre H. Assessment of allelic variation in the *ccr*AB locus in methicillin resistant *Staphylococcus aureus* clones. J Antimicrob Chemother 2006;58:23–30.

[96] Oliveira DC, Santos M, Milheiriço C, Carriço JA, Vinga S, Oliveira AL, et al. CcrB typing tool: an online resource for staphylococci *ccr*B sequence typing. J Antimicrob Chemother 2008;61:959–60.

[97] Francois P, Renzi G, Pittet D, Bento M, Lew D, Harbarth S, et al. A novel multiplex real-time PCR assay for rapid typing of major staphylococcal cassette chromosome mec elements. J Clin Microbiol 2004;42:3309–12.

[98] Morris DO, Rook KA, Shofer FS, Rankin SC. Screening of *Staphylococcus aureus*, *Staphylococcus intermedius*, and *Staphylococcus schleiferi* isolates obtained from small companion animals for antimicrobial resistance: a retrospective review of 749 isolates (2003–04). Vet Dermatol 2006;17:332–7.

[99] Sasaki T, Kikuchi K, Tanaka Y, Takahashi N, Kamata S, Hiramatsu K. Methicillin-resistant *Staphylococcus pseudintermedius* in a veterinary teaching hospital. J Clin Microbiol 2007;45:1118–25.

[100] Van Hoovels L, Vankeerberghen A, Boel A, Van Vaerenbergh K, De Beenhouwer H. First case of *Staphylococcus pseudintermedius* infection in a human. J Clin Microbiol 2006;44:4609–12.

[101] Savini V, Carretto E, Polilli E, Marrollo R, Santarone S, Fazii P, et al. Small colony variant of methicillin-resistant *Staphylococcus pseudintermedius* ST71 presenting as a sticky phenotype. J Clin Microbiol 2014;52(4):1225–7.

[102] Savini V, Passeri C, Mancini G, Iuliani O, Marrollo R, Argentieri AV, et al. Coagulase-positive staphylococci: my pet's two faces. Res Microbiol 2013;164:371–4.

[103] Hanssen AM, Ericson Sollid JU. SCC*mec* in staphylococci: genes on the move. FEMS Immunol Med Microbiol 2006;46:8–20.

[104] Ruppe E, Barbier F, Mesli Y, Maiga A, Cojocaru R, Benkhalfat M, et al. Diversity of staphylococcal cassette chromosome *mec* structures in methicillin-resistant *Staphylococcus epidermidis* and *Staphylococcus haemolyticus* strains among outpatients from four countries. Antimicrob Agents Chemother 2009;53:442–9.

[105] Jamaluddin TZ, Kuwahara-Arai K, Hisata K, Terasawa M, Cui L, Baba T, et al. Extreme genetic diversity of methicillin-resistant *Staphylococcus epidermidis* strains disseminated among healthy Japanese children. J Clin Microbiol 2008;46:3778–83.

[106] Kuroda M, Yamashita A, Hirakawa H, Kumano M, Morikawa K, Higashide M, et al. Whole genome sequence of *Staphylococcus saprophyticus* reveals the pathogenesis of uncomplicated urinary tract infection. Proc Natl Acad Sci USA 2005;102:13272–7.

[107] Higashide M, Kuroda M, Omura CT, Kumano M, Ohkawa S, Ichimura S, et al. Methicillin-resistant *Staphylococcus saprophyticus* isolates carrying staphylococcal cassette chromosome *mec* have emerged in urogenital tract infections. Antimicrob Agents Chemother 2008;52:2061–8.

[108] Mombach Pinheiro Machado AB, Reiter KC, Paiva RM, Barth AL. Distribution of staphylococcal cassette chromosome *mec* (SCC*mec*) types I, II, III and IV in coagulase-negative staphylococci from patients attending a tertiary hospital in southern Brazil. J Med Microbiol 2007;56:1328–33.

[109] Descloux S, Rossano A, Perreten V. Characterization of new staphylococcal cassette chromosome *mec* (SCC*mec*) and topoisomerase genes in fluoroquinolone- and methicillin-resistant *Staphylococcus pseudintermedius*. J Clin Microbiol 2008;46:1818–23.

Chapter 17

Methicillin Resistance in *Staphylococcus aureus*

Edoardo Carretto, Rosa Visiello, Paola Nardini
Clinical Microbiology Laboratory—IRCCS Arcispedale Santa Maria Nuova, Reggio Emilia, Italy

17.1 INTRODUCTION

The resistance of *Staphylococcus aureus* (SA) to methicillin was firstly described in 1961, a few years after the introduction of the β-lactamase resistant semisynthetic penicillins [1]. After the first report of outbreaks in European hospitals in the 1960s, methicillin resistance became a problem worldwide by the 1980s, and nowadays accounts for a large number of nosocomial infections [2]. The application of molecular methods to bacterial epidemiology has allowed researchers to establish that the circulation of methicillin-resistant *Staphylococcus aureus* (MRSA) was due to epidemic clones, each with a specific genetic background. A detailed discussion about the origin and the dissemination of these different clonal lineages is exhaustively summarized in the paper of Deurenberg and Stobbering [3] and will not furtherly discussed in the present chapter.

MRSA accounts for more than 60% of SA isolates in the ICUs of US hospitals and the number of deaths related to MRSA infections accounts for 19,000 cases/year (more than AIDS) [4]. According to ECDC, large intercountry variations in the occurrence of MRSA has been documented across Europe. In 2016, MRSA percentages were generally lower in northern Europe and higher in the southern and south-eastern parts (e.g., 1.2% in the Netherlands and 50.5% in Romania) [5].

The main mechanism of resistance to methicillin is due to the presence of a specific gene, called *mecA*, which promotes the synthesis of penicillin-binding proteins (PBP) with low affinity for β-lactam molecules. In detail, the group of enzymes known as PBPs, classified on the basis of their molecular weight and with chemical activity similar to serine proteases, catalyze the transpeptidation reactions that allow the peptidoglycan cross-linking of the bacterial cell wall. Both the penicillins and cephalosporins interfere with peptidoglycan formation by competitively inhibiting PBPs. The β-lactams covalently bind to PBPs, forming highly stable complexes. The bounded PBPs are no longer able to catalyze reactions necessary for cell wall synthesis.

Pet-to-Man Travelling Staphylococci: A World in Progress. https://doi.org/10.1016/B978-0-12-813547-1.00017-0

225

Staphylococcal strains can harbor the *mecA* gene (or its homologue, the *mecC* gene). This gene is carried on a mobile genetic element called staphylococcal chromosome cassette *mec* (SCC*mec*), which is described in detail herein. Its expression encodes for an additional 76-kDa penicillin binding protein (PBP2a), which has a decreased affinity for β-lactam antimicrobials [2]. PBP2a and native PBP2 work in concert to allow cell-wall synthesis despite the presence of β-lactam antibiotics, thus effectively conferring resistance to penicillins, cephalosporins, and the carbapenems. The SCC is integrated into the chromosome of *S. aureus* at a very specific location (*attBscc*) [6]. It is likely that MRSA originated from the stable transfer of SCC*mec* into lineages of methicillin susceptible SA, having a genetic background that made them fit for the integration of this mobile element.

Other types of methicillin resistance, significantly more infrequent compared with those encoded by the SCC*mec* complex, are due to overproduction of β-lactamaseses or, exceptionally, from spontaneous mutation of normal PBP genes [7,8].

The bacterial world is always amazing in its ability to adapt to the environment. The β-lactam resistance mediated by the SCC*mec* complex is a very multifaceted mechanism, which cannot be explained only on the basis of the antibiotic pressure. However, this resistance trait was identified only two years after the clinical use of β-lactamase resistant semisynthetic penicillins. A fascinating explanation of the origin of this kind of resistance has been proposed by de Lencastre et al., which postulated that SCC*mec* has originated from *Staphylococcus sciuri* with a *mecA* homologous gene (designated *pbpD,* that encodes a 84 kDa PBP by this microorganism). The finding was based on molecular studies that demonstrated that *mecA* of SA isolates gave a strong hybridization with more than 200 epidemiologically unrelated strains of *S. sciuri* [9]. It was supposed that the β-lactam resistance mechanism in *S. sciuri* initiated following the selective pressure of penicillin, which has been used prophylactically in veterinary practice since 1949 [6]. Because this microorganism does not have the ability to produce β-lactamases, lacking in proper gene complexes, it has been postulated that its way to survive was to change the drugs' target. Recently, a paper of Tsubakishita et al. demonstrated that the *mecA* locus of the *Staphylococcus fleurettii* genome has a sequence identical to that of the *mecA*-containing region of SCC*mec*. These results are in agreement with those previously published, because *S. sciuri*, *S. vitulinus*, and *S. fleurettii* evolved from a common ancestor, also demonstrating that *mecA* gene homologues of *S. fleurettii* descended from its ancestor and was not recently acquired [10].

17.2 DESCRIPTION OF THE SCC*mec* CASSETTE

This mobile genetic element, which can be easily transferred among staphylococcal species (not only SA), carries the determinants for the methicillin resistance (the *mec* gene complex) and other elements, such as cassette chromosome recombinases (*ccr*), integration site sequences (ISS), and flanking regions. Knowledge concerning the SCC*mec* is continuously evolving, and updated information can be found on the website www.sccmec.org.

The structure of SCC*mec* is quite characteristic, consisting essentially of

(1) The so-called *mec* complex, which contains the *mecA* gene and, its regulatory elements *mecI* (encoding the repressor protein MecI) and *mecR1* (encoding the signal transducer protein MecR1) and insertion sequences. At the moment of writing this chapter, six different *mec* gene complexes (defined as classes) are known. They vary in the position of the different genetic elements, as shown in Table 17.1.

Class A can be considered as the prototype complex. Class B is characterized by the absence of *mecI* and by a truncation of *mecR1* due to the insertion of IS1272. Class C has no *mecI* and two IS431s at the two extremities, which are oriented in the same direction in class C1, and in the opposite direction in class C2. The class D is similar to class B, but has no ISs downstream to *mecR1*, which is, however, truncated. Finally, class E is characteristic of the *mecC* gene, which will be discussed in detail as follows.

(2) The chromosomal cassette recombinase genes (*ccr genes*) mediate insertion and excision of SCC*mec* from the bacterial genome. At the moment, three different *ccr* genes (*ccrA*, *ccrB*, and *ccrC*) have been identified. For *ccrA* and *ccrB*, different allotypes are known, ranging from 60% to 82% of genome similarity between each other. Their combination allows for the classification of the seven different *ccr* gene complexes. The eighth is defined by the *ccrC* gene [11], which also has different variants that are not considered as allotypes having a mutual similarity of more than 87%. It should be noted that other *ccr* gene complexes have been defined for staphylococcal species other than SA.

(3) The J regions (or junkyard regions, more recently renamed as joining regions) are located beside the *mec* and *ccr* gene complexes. There are at least three joining regions, which have been described based on their position

TABLE 17.1 *mec* Gene Complexes

mec Gene Complexes	Elements Arrangement
Class A	IS431-mecA-mecR1-mecI
Class B	IS431-mecA-ΔmecR1-IS1272
Class C1-C2[a]	IS431-mecA-ΔmecR1-IS431
Class D	IS431-mecA-ΔmecR1
Class E	blaZ-mecA$_{LGA251}$-mecR1$_{LGA251}$-mecI$_{LGA251}$

[a] *Classes C1 and C2 differ for the arrangement of IS431: in the same direction for C1, in the opposite direction for C2.*

inside the SCC*mec* cassette: J1 is between the right chromosomal junction and the *ccr* gene complex, J2 is between the *ccr* gene complex and the *mec* gene complex, whereas J3 is located between the *mec* gene complex and the left chromosomal junction. In these elements are often present other genetic structures, such as transposons that can carry other resistant genes.

17.3 CLASSIFICATION OF THE SCC*mec* CASSETTES

The different combinations of *mecA* gene complex and *ccr* gene complexes allow to define the main SCC*mec* types, which are designated in roman numerals. Currently, 11 of these types have been defined [11–18]. However, because different genetic traits may be present in the different types, each SCC*mec* type has been divided in subtypes, essentially based on the polymorphisms or variation in J regions of the same type. To define SCC*mec* subtypes, two methods have been accepted to define generic names: (1) the J1 region variations characterize a subtype that is defined by a roman numeral plus small letters and (2) differences caused by the presence/absence of genetic element characterize a subtype that is defined by a roman numeral plus capital letters. The classification of SCC*mec* types and subtypes is summarized in Table 17.2.

TABLE 17.2 Classification of SCC*mec*

SCC*mec* Types	*ccr* Gene Complexes[a]	*mec* Gene Complexes	Subtypes[b]	References
I	1, A1B1	B	=	[12,13]
II	2, A2B2	A	A, b	[12,13]
III	3, A3B3	A	A	[12,13]
IV	2, A2B2	B	a, b, c, d, E, F, g, h	[15,18–20]
V	5, C1	C2	c	[11,18]
VI	4, A4B4	B	=	[16]
VII	5, C1	C1	=	[14]
VIII	4, A4B4	A	=	[17]
IX	1, A1B1	C2	=	[18]
X	7, A1B6	C1	=	[18]
XI	8, A1B3	E	=	[21]

[a] *ccr gene complex number, followed by the combination of the haplotypes.*
[b] *The most commonly described subtypes found in clinical isolates.*

17.4   THE *mecC* GENE

During a surveillance study performed in veterinary medicine, it was found a MRSA that did not have either the *mecA* gene, or the modified PBPs encoded by this gene. The strain was named SA LGA251 and molecular studies performed on its genome demonstrated that the isolate harbored a *mecA* homologue (*mecA*$_{LGA251}$), sharing a 70% similarity with this gene. The PBP encoded by *mecA*$_{LGA251}$ also has an aminoacidic sequence with a 63% similarity to PBP2a [19,20]. Retrospective studies on isolates stored in bacterial collections revealed that the SA harboring the *mecC* gene were already present in 1975 [20]. A *mecC* positive SA was first described in human microbiology in 2011, not being correlated with the studies previously cited. In this report it was proposed that to the newly SCC*mec* described could be attributed with the number XI [21].

The origin of the *mecC* gene is probably similar to those of *mecA*, with coagulase negative staphylococci playing a pivotal role in the selection and diffusion. The *mecC* gene was demonstrated in *Staphylococcus stepanovicii* [22] and its homologues in *S. sciuri* and in *S. xylosus* [23,24].

From the phenotypical point of view, SA strains possessing the *mecC* gene show susceptibility or low level-resistance to oxacillin, whereas the resistance to cefoxitin is higher. In this situation, the different substrate affinity of the PBP encoded by *mecC* can be a useful marker for this kind of resistance. Molecular confirmation is, however, always required. Different *home-made* methods are currently available to detect the *mecC* gene [19]. Clinicians should be aware that not all the commercially used systems that detect MRSA in different clinical specimens are able to reveal this kind of resistance.

17.5   MOLECULAR METHODS FOR SCC*mec* TYPING

Knowledge of the kind of SCC*mec* harbored in SA may be useful for epidemiological purposes, possibly also helping to define the origin of the staphylococci. As an example, SCC*mec* V is typically found in livestock-associated MRSA. To do that, a vast amount of molecular methods have been published. From a practical point of view, multiplex PCR (mPCR) is the molecular technique most widely used, because it is quick, easy to perform and, if properly set, accurate.

The correct procedure to identify a SCC*mec* cassette is to define, in subsequent steps, the *ccr* gene complex, the *mec* gene complex and specific structures in the joining regions. The combination of the different results allows us to assign the isolate to a previously known SCC*mec* type or subtype. Kondo et al. proposed a method based on 5 to 6 different mPCRs that is able to identify the most common staphylococcal cassettes, also allowing to detect new subtypes or variants [25]. In this case, the result should be confirmed by reference laboratories (see www.sccmec.org and Ref. [26]). This system is widely used and extremely reliable and complete; however, its main disadvantages are that it is based on different mPCRs, which may not be feasible for routine purposes, is time-consuming, and requires skilled personnel.

Other molecular approaches may be more practical in their application. As an example, the protocol proposed by Lawung et al. describes a multiplex PCR in two steps that is able to amplify all the eleven SCC*mec* [27].

In our personal opinion, in any setting one should initially use a method able to detect the SCC*mec*, which most frequently circulate in that geographical area. As an example, to analyze isolates from six different Italian hospitals, we chose the multiplex PCR proposed by Zhang et al., which is able to identify seven different SCC*mec* types and also allows us to distinguish six different SCC*mec* IV subtypes [28]. Utilizing this method, we have been able to immediately type the 70% of the isolates (example in Fig. 17.1). The remaining 30% of the strains were instead analyzed using the methods described in Milheiriço et al. [29], because we suspected a spread of the EMRSA-15 strains as previously happened in Portugal [30]. In fact, the application of this protocol on the isolates that were undetermined using the mPCR of Zhang demonstrated that 90% of them harbored the SCC*mec* IVh, that is, they belonged to the clone EMRSA-15 (Fig. 17.2).

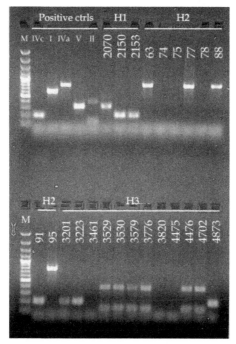

FIG. 17.1 Example of multiplex PCR able to identify the SCC*mec* types and subtypes. H1, H2, and H3 = hospitals 1-2-3. Methods cited in Ref. [28].

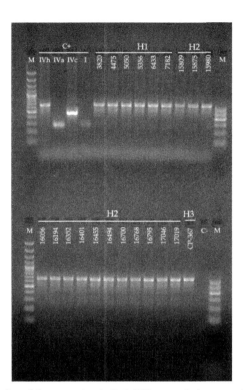

FIG. 17.2 Application of the protocol as in Ref. [29] to type isolates negative with the method of Zhang et al. [28]. The 663 bp band is characteristic of *SCCmec* IVh. H1, H2, and H3 = hospitals 1-2-3.

17.6 COMMUNITY ACQUIRED MRSA AND HEALTHCARE ASSOCIATED MRSA

During the past decade, community acquired MRSA (CA-MRSA) spread worldwide and, probably, replaced healthcare associated MRSA (HA-MRSA) in healthcare facilities. Therefore, nowadays the definition of CA-MRSA can be challenging and is based on clinical definition, rather than on microbiological findings.

The Centers for Disease Control and Prevention (CDC) defines CA-MRSA as MRSA strains isolated in an outpatient setting, or isolated from patients within 48 hours of hospital admission. These patients must have no medical history of MRSA infection or colonization, and no medical history in the past year of any kind of hospitalization. Moreover, the patient should not have any medical devices that pass through the skin. However, genetic markers can also be used to define CA-MRSA, which is characterized by SCCmec type IV, V, or VII and the presence of PVL. Moreover, CA-MRSA can be associated with specific SA lineages [3].

Some authors proposed a mathematical model that allowed them to postulate that CA-MRSA will become the dominant MRSA strain in healthcare facilities, as a result of the documented expanding community reservoir and increasing influx into the hospital of individuals who harbor CA-MRSA [31]. However, this hypothesis is difficult to confirm, because strains harboring SCC*mec* IV now have a worldwide circulation, but PVL is rarely present in MRSA isolated in healthcare settings. Considerable additional data will be needed to prove the underpinnings of this working model.

17.7 OTHER GENES RELATED TO METHICILLIN RESISTANCE

Different genes are involved in the mechanism of the methicillin resistance. As an example, plasmids codifying for staphylococcal β-lactamase are known to prevent spontaneous deletion of the *mecA* gene [2]. Loss or deletion of the *mecA* gene may rarely occur, mainly due to factors affecting the stability of SCC*mec*. Vancomycin may induce deletion of the *mecA* gene in *S. aureus* [32]; there have been reports of cases of implant-associated infection due to methicillin-resistant *Staphylococcus epidermidis* that lost the *mecA* gene after prolonged treatment with glycopeptides [33].

Six auxiliary genes known as *fem* (factor essential for methicillin resistance) exist, namely *femA*, *femB*, *femC*, *femD*, *femE*, and *femF*. They are involved in the synthesis of peptidoglycan and are required for the correct formation of the cell wall. When a mutation of these genes occurs, the bacterial strain shows a progressive reduction of its level of resistance to β-lactams [2].

The accessory gene regulator (*agr*) is a quorum sensing operon that controls staphylococcal virulence factors and other accessory gene functions. Mutations of these genes leading to defective *agr* activities are associated with a substantially reduced virulence in SA. Four different *agr* groups have been described. Inactivation of *agr* locus in a heterogeneous population of SA resulted in a slight decrease in the number of highly resistant cells, although the level of resistance expressed by the single isolates was unaffected [2].

17.8 BORDERLINE OXACILLIN RESISTANT *S. AUREUS*

Borderline resistance is characterized by oxacillin MICs of 4–8 μg/mL. Borderline resistant strains are essentially divided into two categories, based on the presence or absence of the *mecA* gene. The presence of *mecA* associated with borderline resistance can be observed only in heterogeneous populations, where the resistant subpopulation is able to grow at increased drug concentrations [2]. Subcultures of these strains express after few passages of high levels of resistance.

In *mecA* negative strains, modifications of the PBP genes, such as point mutations in the penicillin binding domain, or the overexpression of PBPs (particularly 4) may result in borderline resistance, because more unbound PBPs are present. In this case, the net effect is that cell wall synthesis continues despite some PBPs being bound to a β-lactam. Another possibility is that the strain harbors some *mecA* variant, such as *mecC* [2,8,19].

Overproduction of staphylococcal β-lactamase is another proposed mechanism, although less understood [7].

REFERENCES

[1] Barber M. Methicillin-resistant staphylococci. J Clin Pathol 1961;14:385–93.

[2] Chambers HF. Methicillin resistance in staphylococci: molecular and biochemical basis and clinical implications. Clin Microbiol Rev 1997;10(4):781–91.

[3] Deurenberg RH, Stobberingh EE. The evolution of *Staphylococcus aureus*. Infect Genet Evol 2008;8(6):747–63.

[4] Boucher HW, Corey GR. Epidemiology of methicillin-resistant *Staphylococcus aureus*. Clin Infect Dis 2008;46(Suppl. 5):S344–9.

[5] European Centre for Disease Prevention and Control. Antimicrobial resistance surveillance in Europe 2016. Annual report of the European Antimicrobial Resistance Surveillance Network (EARS-Net), Stockholm: ECDC; 2017.

[6] Llarrull LI, Fisher JF, Mobashery S. Molecular basis and phenotype of methicillin resistance in *Staphylococcus aureus* and insights into new beta-lactams that meet the challenge. Antimicrob Agents Chemother 2009;53(10):4051–63.

[7] McDougal LK, Thornsberry C. The role of beta-lactamase in staphylococcal resistance to penicillinase-resistant penicillins and cephalosporins. J Clin Microbiol 1986;23(5):832–9.

[8] Tomasz A, Drugeon HB, de Lencastre HM, Jabes D, McDougall L, Bille J. New mechanism for methicillin resistance in *Staphylococcus aureus*: clinical isolates that lack the PBP 2a gene and contain normal penicillin-binding proteins with modified penicillin-binding capacity. Antimicrob Agents Chemother 1989;33(11):1869–74.

[9] de Lencastre H, Oliveira D, Tomasz A. Antibiotic resistant *Staphylococcus aureus*: a paradigm of adaptive power. Curr Opin Microbiol 2007;10(5):428–35.

[10] Tsubakishita S, Kuwahara-Arai K, Sasaki T, Hiramatsu K. Origin and molecular evolution of the determinant of methicillin resistance in staphylococci. Antimicrob Agents Chemother 2010;54(10):4352–9.

[11] Ito T, Ma XX, Takeuchi F, Okuma K, Yuzawa H, Hiramatsu K. Novel type V staphylococcal cassette chromosome *mec* driven by a novel cassette chromosome recombinase, *ccrC*. Antimicrob Agents Chemother 2004;48(7):2637–51.

[12] Ito T, Katayama Y, Hiramatsu K. Cloning and nucleotide sequence determination of the entire mec DNA of pre-methicillin-resistant *Staphylococcus aureus* N315. Antimicrob Agents Chemother 1999;43(6):1449–58.

[13] Ito T, Katayama Y, Asada K, Mori N, Tsutsumimoto K, Tiensasitorn C, et al. Structural comparison of three types of staphylococcal cassette chromosome mec integrated in the chromosome in methicillin-resistant *Staphylococcus aureus*. Antimicrob Agents Chemother 2001;45(5):1323–36.

[14] Berglund C, Ito T, Ikeda M, Ma XX, Soderquist B, Hiramatsu K. Novel type of staphylococcal cassette chromosome *mec* in a methicillin-resistant *Staphylococcus aureus* strain isolated in Sweden. Antimicrob Agents Chemother 2008;52(10):3512–6.

[15] Ma XX, Ito T, Tiensasitorn C, Jamklang M, Chongtrakool P, Boyle-Vavra S, et al. Novel type of staphylococcal cassette chromosome *mec* identified in community-acquired methicillin-resistant *Staphylococcus aureus* strains. Antimicrob Agents Chemother 2002;46(4):1147–52.

[16] Oliveira DC, Milheirico C, de Lencastre H. Redefining a structural variant of staphylococcal cassette chromosome *mec*, SCC*mec* type VI. Antimicrob Agents Chemother 2006;50(10):3457–9.

[17] Zhang K, McClure JA, Elsayed S, Conly JM. Novel staphylococcal cassette chromosome *mec* type, tentatively designated type VIII, harboring class A *mec* and type 4 *ccr* gene complexes in a Canadian epidemic strain of methicillin-resistant *Staphylococcus aureus*. Antimicrob Agents Chemother 2009;53(2):531–40.

[18] Li S, Skov RL, Han X, Larsen AR, Larsen J, Sorum M, et al. Novel types of staphylococcal cassette chromosome *mec* elements identified in clonal complex 398 methicillin-resistant *Staphylococcus aureus* strains. Antimicrob Agents Chemother 2011;55(6):3046–50.

[19] Paterson GK, Harrison EM, Holmes MA. The emergence of *mecC* methicillin-resistant *Staphylococcus aureus*. Trends Microbiol 2014;22(1):42–7.

[20] Garcia-Alvarez L, Holden MT, Lindsay H, Webb CR, Brown DF, Curran MD, et al. Meticillin-resistant *Staphylococcus aureus* with a novel *mecA* homologue in human and bovine populations in the UK and Denmark: a descriptive study. Lancet Infect Dis 2011;11(8):595–603.

[21] Shore AC, Deasy EC, Slickers P, Brennan G, O'Connell B, Monecke S, et al. Detection of staphylococcal cassette chromosome mec type XI carrying highly divergent mecAmecI, mecR1blaZ, and ccr genes in human clinical isolates of clonal complex 130 methicillin-resistant *Staphylococcus aureus*. Antimicrob Agents Chemother 2011;55(8):3765–73.

[22] Loncaric I, Kubber-Heiss A, Posautz A, Stalder GL, Hoffmann D, Rosengarten R, et al. Characterization of methicillin-resistant *Staphylococcus* spp. carrying the mecC gene, isolated from wildlife. J Antimicrob Chemother 2013;68(10):2222–5.

[23] Harrison EM, Paterson GK, Holden MT, Morgan FJ, Larsen AR, Petersen A, et al. A *Staphylococcus xylosus* isolate with a new *mecC* allotype. Antimicrob Agents Chemother 2013;57(3):1524–8.

[24] Harrison EM, Paterson GK, Holden MT, Ba X, Rolo J, Morgan FJ, et al. A novel hybrid SCCmec-mecC region in *Staphylococcus sciuri*. J Antimicrob Chemother 2014;69(4):911–8.

[25] Kondo Y, Ito T, Ma XX, Watanabe S, Kreiswirth BN, Etienne J, et al. Combination of multiplex PCRs for staphylococcal cassette chromosome *mec* type assignment: rapid identification system for *mecccr*, and major differences in junkyard regions. Antimicrob Agents Chemother 2007;51(1):264–74.

[26] Ito T, Hiramatsu K, Tomasz A, de Lencastre H, Perreten V, Holden MT, et al. Guidelines for reporting novel *mecA* gene homologues. Antimicrob Agents Chemother 2012;56(10):4997–9.

[27] Lawung R, Chuong LV, Cherdtrakulkiat R, Srisarin A, Prachayasittikul V. Revelation of staphylococcal cassette chromosome *mec* types in methicillin-resistant *Staphylococcus aureus* isolates from Thailand and Vietnam. J Microbiol Methods 2014;107:8–12.

[28] Zhang K, McClure JA, Conly JM. Enhanced multiplex PCR assay for typing of staphylococcal cassette chromosome *mec* types I to V in methicillin-resistant *Staphylococcus aureus*. Mol Cell Probes 2012;26(5):218–21.

[29] Milheirico C, Oliveira DC, de Lencastre H. Multiplex PCR strategy for subtyping the staphylococcal cassette chromosome *mec* type IV in methicillin-resistant Staphylococcus aureus: 'SCCmec IV multiplex'. J Antimicrob Chemother 2007;60(1):42–8.

[30] Amorim ML, Faria NA, Oliveira DC, Vasconcelos C, Cabeda JC, Mendes AC, et al. Changes in the clonal nature and antibiotic resistance profiles of methicillin-resistant *Staphylococcus aureus* isolates associated with spread of the EMRSA-15 clone in a tertiary care Portuguese hospital. J Clin Microbiol 2007;45(9):2881–8.

[31] D'Agata EM, Webb GF, Horn MA, Moellering Jr. RC, Ruan S. Modeling the invasion of community-acquired methicillin-resistant *Staphylococcus aureus* into hospitals. Clin Infect Dis 2009;48(3):274–84.

[32] Wang A, Zhou K, Liu Y, Yang L, Zhang Q, Guan J et al. A potential role of transposon IS *431* in the loss of *mecA* gene. Sci Rep 2017;7:41237.

[33] Sendi P, Graber P, Zimmerli W. Loss of *mecA* gene in *Staphylococcus epidermidis* after prolonged therapy with vancomycin. J Antimicrob Chemother 2005;56(4):794–5.

Chapter 18

In Vivo Resistance Mechanisms: Staphylococcal Biofilms

Barbara Różalska, Beata Sadowska
Department of Immunology and Infectious Biology, Faculty of Biology and Environmental Protection, University of Lodz, Lodz, Poland

18.1 INTRODUCTION

The classic definition of microbial biofilm describes it as "a structured community of cells enclosed in a self-produced polymeric matrix and adherent to an inert or living surface" [1]. This concept has recently been extended with free-floating aggregates of the microorganisms. What was the reason? It was because these aggregates, similar to the biofilm, are covered by slime, which correlates with their increased tolerance to hazardous environments, including host defense mechanisms and antibiotic treatment. Research in vitro conducted by Haaber et al. [2] on *Staphylococcus aureus* model showed that the aggregation already starts in the early exponential growth phase and one of the main structural component of *S. aureus* aggregates is polysaccharide intercellular adhesin (PIA), which until now has been associated with biofilm formation by staphylococci. On the other hand, the results obtained indicate that the properties of aggregated cells differ in some aspects from the cells representing other lifestyles. For example, they displayed higher metabolic activity than corresponding planktonic and biofilm cells [2]. However, regardless of the definition, the biofilm is a predominant form of microbial life occurring practically in every possible environment, including eukaryotic cells and tissues. It is proved that so-called microbiome (human and animal natural microflora) usually takes the form of biofilm, such as dental plaque, intestinal biofilm, and aggregates of bacteria on the mucous membranes of the upper respiratory tract and urogenital tract or microcolonies existing among keratinocyte layers in the skin [3–6]. Because of the numerous advantages for the microorganisms encased in extracellular polymeric substances (EPS), the biofilm is often also created by pathogens [7–9]. Staphylococci belong to the group of bacteria forming biofilms very easily, which is regarded as one of their crucial virulence factors [10–14].

Pet-to-Man Travelling Staphylococci: A World in Progress. https://doi.org/10.1016/B978-0-12-813547-1.00018-2

18.2 MOLECULAR BASE FOR STAPHYLOCOCCAL BIOFILM FORMATION AND ITS STRUCTURE

The process of biofilm formation by staphylococci does not generally differ from the way of its creation by other microorganisms. The main phases include: microbial adhesion and coaggregation, multiplication, microcolonies' formation, EPS production, and finally, biofilm maturation involving phenotypic and functional differentiation of cells within the microcolonies, and release of free-floating cells/aggregates from a complete structure [15–18]. If we consider the biofilm in the classic sense, everything starts from microbial adhesion to the surface. The first phase of staphylococcal adhesion, like in other microorganisms, depends on unspecific forces: electrostatic bonds, hydrophobic interactions, or van der Waals forces. Then, a specific bindings of the receptor-ligand type are involved, gaining the importance for both processes characteristic for biofilm formation: co-adhesion and co-aggregation. Staphylococci possess a wide range of surface adhesins, including MSCRAMMs (microbial surface components recognizing adhesive matrix molecules), which bind host extracellular matrix proteins (ECM), such as fibronectin, fibrinogen, collagen, vitronectin or laminin. In the case of *S. aureus*, the list of these adhesins comprises staphylococcal protein A (SpA), clumping factor A and B (Clf A and B), Bap (biofilm-associated protein), SasG (*S. aureus* surface protein G), Eap (extracellular adhesive protein) and Efb (extracellular fibrinogen binding protein). *Staphylococcus epidermidis* produces Aap (accumulation-associated protein)—the homolog of Sas G, SdrF (*S. epidermidis* surface proteins binding collagen), SdrG (*S. epidermidis* surface proteins binding fibrinogen)—the homolog of Clf and Ebp (extracellular binding protein). All of them can participate in staphylococcal adhesion to inert or living surface and thus are significant for biofilm formation [2,11,18–21]. Nevertheless, the ability to produce intercellular adhesin PIA is the most important determinant of cell-to-cell interaction in staphylococci, leading to the accumulation of bacterial biomass. PIA, in biochemical terms, is a homopolymer of at least 130 units of partially deacetylated β-1,6-linked N-acetyl-glucosamine (PNAG), encoded by intercellular adhesion operon *ica*, containing *icaADBC* genes. In addition to polysaccharides, EPS also contains teichoic acids, proteins, extracellular DNA (eDNA), and presumably host factors too. Interestingly, α-toxin (hemolysin α) produced by most *S. aureus* strains has recently been described as a component forming the skeletal framework stabilizing staphylococcal biofilm by covalent cross-links with eDNA [11,13,17,22–25]. Recent data indicate, however, that the possession of an active *ica* operon and PNAG synthesis are not an absolute condition for the disclosure of a biofilm phenotype in staphylococci. When polysaccharide production is low, there is often a noticeable increase in the importance of proteinaceous and other interconnecting materials. It was revealed that the components of the staphylococcal cell wall such as teichoic acids and previously mentioned adhesion proteins (mainly Bap, SasG and Aap) do not only participate in microbial adhesion, but also replace PIA/PNAG in its role as a mediator of intercellular interactions [22,26,27].

Regardless of the ecosystem in which the biofilm exists, the main stages of its formation are similar, but even at the phase of cell adhesion, the influence of specific characteristics of the microorganism on the final morphology and function of the consortium can be observed. For example, if single cells of the microorganism attach to the surface, the biofilm develops usually as a monolayer. In vitro studies conducted in strictly controlled physio-chemical conditions show that the biofilm in a monolayer form usually occurs when microbial interactions with the surface are more intense than cell-to-cell interactions. However, when microorganisms adhere as cell clusters, the biofilm forms much more complex structures and commonly has many layers. It can be of a flat shape with the layers arranged one upon the other; otherwise, the biofilm can develop in a form of columns ("mushroom-like" structures) with the channels in between for water, ions, and nutrient exchange. Different structures can also be expected for single-species and multispecies biofilms, depending on synergistic or antagonistic interactions between microorganisms remaining in the same community [3,15,16,28]. However, recent data from the studies of the physiology of biofilms present in different ecosystems indicate the need for a more flexible approach to the naming of such structures. Certainly in relation to the biofilm formed in vivo, in association with human and animal tissues, it seems unnecessary for the biofilm to fulfill the condition of firm contact with the solid surface. Indeed, because of the reported similarities, biofilm should be also regarded when the aggregates of microorganisms are loosely submerged in the mucus layer or free-floating in the body fluids [2,7,9,16].

18.2.1 Biofilm Maturation and Disassembly— The Mechanisms and Clinical Implications

Mature microbial biofilm possesses such a complex structure and physiology that it is even compared to a multicellular organism, in which the cells communicate with each other and exhibit a certain division of their function based on metabolic differences in various layers of the biofilm. The cells occupying the outer layers of the biofilm are usually metabolically active and intensively divide to compensate for any damage caused by the impact of the external environment. Besides, their role also involves an acquisition and delivery of nourishment to the deeper layers of the biofilm. Inside the biofilm, availability of oxygen and carbon sources is decreased, which induces phenotypical changes in the microbial cells leading to the formation of their less active counterparts [4,29,30]. Functioning of this small microbial ecosystem requires coordination of the simultaneous activities of many cells so that they may react in a similar mode to different stimuli. The system called "quorum-sensing" (Q-S) is responsible for such a behavior. This is the mechanism of cell-to-cell communication based on the secretion and response to biochemical signal molecules via gene expression. In a biofilm, where a great number of microbial cells are close to each other and the concentration of signal molecules is very high, Q-S operates best. Staphylococci respond to autoinducing peptides (AIP), and their numerous gene regulatory systems (including two the most important: accessory gene regulator—*agr*, and staphylococcal accessory regulator A—*sarA*) are functioning under Q-S restriction [31–33].

Despite unquestioned benefits provided to bacteria through the biofilm mode of growth, this community cannot develop infinitely. External forces and the deterioration of local conditions cause a mechanical or intentional (strictly genetically controlled) detachment of single bacterial cells or their clusters from the biofilm, ensuring dissemination and survival of microbes that can rebuild the biofilm in a new location. The research on *S. aureus* and *S. epidermidis* points to the Q-S system *agr*, which controls the production of small pore-forming toxins called phenol-soluble modulins (PSMs) and enzymes (proteases, DNases) as directly involved in the disassembly process. PSMs playing a detergent-like function (surfactants) and enzymes disrupt and solubilize the components of the biofilm matrix. PSMs have been demonstrated to promote *S. epidermidis* biofilm dissemination both in vitro and in a mouse model of biomaterial-associated infection. In *S. aureus*, the mutations leading to strong upregulation of the extracellular proteases decrease their ability to form biofilm and keep these bacteria in a planktonic state. A similar observation was made for DNases (e.g., thermonuclease produced by *S. aureus*), which can function as an endogenous mediator of biofilm dispersion [22,34]. An opportunity to influence the biofilm dispersion or inhibition of this process has potential therapeutic value. To remove mature *Staphylococcus* sp. biofilm from an inert or living surface, including eukaryotic tissues, the use of selected enzymes, surfactants or Q-S signal particles are considered. The action of the following has been reported: proteinase K, tripsin, bovine DNase I, *S. aureus* thermonuclease, dispersin B produced by *Aggregatibacter actinomycetemcomitans*, lysostaphin secreted by *Staphylococcus simulans*, surfactants derived from *Lactobacillus acidophilus* and many others [22,34–36]. On the other hand, staphylococcal biofilm disassembly can be also a disadvantageous process because of possible medical complications; in fact, bacteria detached from the biofilm localized on medical devices or living tissues can cause severe acute infections such as sepsis, pneumonia, endocarditis (when detached bacteria from the device may reach heart valves through the bloodstream) or devastating embolic events during endocarditis development (when bacteria detach from the biofilm formed on heart tissue, e.g., valves) [22,34].

18.3 BIOFILM RESISTANCE

The high resistance of microbial biofilm/aggregates to environmental stress conditions, including antibiotics and host immune response (even 100- to 1000-fold higher than planktonic forms), is well known. A number of factors that contribute to this effect, based mainly on the very complex structure and unique physiology of biofilms, have been listed. Facilitated gene transfer inside biofilms and the impairment of physiological processes within host tissue, induced by microbial products and host cells secreting proinflammatory cytokines/chemokines, are also indicated [2,37–39]. Special attention has been paid to an extracellular polymer, which may form a physical and chemical diffusion barrier for both the drug/antiseptics and host defense factors. The enzymes degrading antibiotics (e.g., β-lactamases), produced by bacteria residing in microcolonies, can accumulate in this layer and neutralize the medications used before they reach a bacterial cell [30]. However, it is only part of the protective strategy of these microbial communities. Still not fully understood and explained phenomenon of weak activity of neutrophils and macrophages against the biofilm (called "frustrated phagocytosis") can be observed [38,39]. For many years, it was believed that the phagocytes do not penetrate biofilm through the slime layer. Nevertheless, this opinion was verified during the study based on microscopic observations, demonstrating attachment and penetration of matured bacterial biofilm by human leukocytes [38,40]. Our own research points to biofilm as a structure, which is not strictly isolated from the external environment. Despite the fact that bacterial biofilm is covered by EPS, many *S. aureus* virulence factors (peptidoglycan, lipoteichoic acids, hemolysins) leak into the external environment and may play a role as bacterial modulins affecting host cells during infection, as we observed in vitro [41]. A protective function is also associated with varying metabolic activity of the cells in the biofilm. As we mentioned herein, changing conditions in the deeper layers of this structure (restricted oxygen and nutrient availability) generate the formation of less active cell subpopulations, such as starving cells, dormant cells, viable-but-nonculturable cells, and persisters. The last group shows extremely reduced sensitivity to drugs that target normally growing cells. It is one of the reasons why the treatment of infections involving biofilms is so difficult and complete removal of the mature biofilm from the surface seems to be almost impossible [4,30,42].

Another biofilm resistance strategy is the stronger expression of efflux pumps. About 30 open reading frames, encoding proteins possessing homology with multidrug resistance (MDR) efflux pumps, have been found in the genome of *S. aureus* strains. The proteins from major facilitator superfamily predominate, but the members of all five drug transporters' families are also present. Staphylococcal MDR pump proteins called MepA, MdeA, NorA-B-C, SepA, SdrM, and TetK have been described in detail and are associated with reduced susceptibility of these bacteria to such chemically and structurally diverse compounds as antibiotics (e.g., fluoroquinolones, tetracyclines), other biocides (e.g., benzalkonium chloride, chlorhexidine) and some dyes (such as ethidium bromide) [43,44]. The overexpression of MDR efflux pumps may be significant for *S. aureus* infections through demonstrating an unfavorable effect on the therapeutic outcome. It has been shown that NorB facilitates *S. aureus* survival in an animal model of abscess formation. Also, nearly 50% of

S. aureus clinical bloodstream isolates overexpress at least one MDR pump gene [43]. The biofilm mode of growth upregulates efflux pumps expression, which, along with other resistance mechanisms, contributes to its increased tolerance to biocides [45].

18.4 THE CLINICAL IMPACT OF STAPHYLOCOCCAL BIOFILM

Because bacteria in biofilms show much higher resistance to antimicrobial agents than their planktonic counterparts, treatment of biofilm-associated infections is difficult, especially when the causative organism is multidrug resistant. Under antibiotic treatment, metabolically active bacteria from the outer layers of the biofilm are first eliminated, while persisters usually survive. When the concentration of the drugs is going down they turn to metabolically active cells, begin to divide, and very quickly rebuild the biofilm [8,30]. Moreover, bacteria forming the biofilm structure can still affect the host immune system efficiently by the expression of virulence factors, which reduce its ability to eradicate such an infection. That explains why *S. aureus* and some other coagulase negative staphylococci (CNS) cause a diverse range of biofilm-type pathologies, such as various skin and soft tissues infections or systemic infection with a possibility of septic shock development. Many of them are connected with invasive medical procedures [8,46–48]. For example, epidemiological data indicate that staphylococci account for a large proportion of prosthetic device-related infections, called FBRIs (foreign-body-related infections). These comprise all entities with respect to local and bloodstream infections associated with inserted or implanted medical devices, including intravascular catheters, prosthetic valves, orthopaedic and neurosurgical implants, vascular grafts, and many others. The adherence of bacteria to foreign bodies and biofilm formation often result in device dysfunction necessitating its removal. Left in place, it represents instead a potential nidus for either persistent or recurrent infections because pathogens can disperse from the original site of colonization and cause pathology in other niches. This occurs due to the host defence mechanism's inability to clear aggregates/biofilm-type infection. When surgical removal and replacement of the device is not a viable option, patients require intermittent antibiotic therapy; however, chemotherapy is rarely effective [7,48,49]. Numerous clinical observations indicate that FBRIs caused by *S. aureus* are often severe, with local or systemic manifestations, while those caused by CNS are usually milder. It is due to the fact that the virulence factors produced by *S. aureus* are much wider and more "aggressive" than those of CNS [46–48].

Recent research, however, has shown the occurrence of staphylococcal biofilms in many other conditions than FBRIs, including chronic wound infections, chronic bronchopulmonary infections in patients with cystic fibrosis, infective endocarditis (IE), and septic transfusions. Among the local infections with biofilm participation, special attention is paid to the high frequency of surgical site infections, diabetic foot, venous leg ulcers, and cancer-associated chronic wounds with *S. aureus*/*S. epidermidis* etiology [21,37,38,50,51]. What is the reason? Numerous factors from the host allow the establishment of chronic wound and biofilm-type infection. These are, for example, poor perfusion, hyperglycaemia, pressure, trauma, or the presence of foreign body. The efficiency of local defense mechanisms is often compromised, and microorganisms, as well as their extracellular components within the biofilm, will enhance and prolong inflammation, delaying the physiological healing process [42]. Studies on the causes of impaired healing of chronically infected wounds have shown that the cells responsible for repairing damage (fibroblasts, as well as vascular endothelial cells) proliferate and migrate more slowly. Additionally, the cells tend to produce fewer enzymes and growth factors. The activity of keratinocytes, which are extremely important in the process of wound healing, is also deficient—the cells synthesize far fewer cytokines, which impairs their migration and proliferation abilities. Furthermore, in chronically infected wounds, eukaryotic metalloproteases are synthesized in quantities much higher than in normal conditions, while the synthesis of tissue protease inhibitors is weaker [14,42]. It is clear that the type of wounds, their origin, and etiology influence the outcome of the infection, which may have a different clinical course. It largely depends on the competence of the host immune system described herein and the virulence of the strains colonizing a given niche. For example, the prevalence of Gram-positive microorganisms, including multidrug resistant pathogens such as methicillin-resistant *S. aureus* and methicillin-resistant *S. epidermidis*, has become common in patients with cancer, particularly those with underlying neutropenia. These patients may be at a higher risk because of their immunodeficient status caused by preoperative exposure to radiotherapy, chemotherapy, and corticosteroids, frequent exposure to healthcare settings, weight loss related to the neoplastic process, then the need for invasive devices to support systemic treatments and parenteral nutrition [52,53].

Staphylococci are also a significant cause of life-threatening biofilm-related invasive diseases, such as bacteremia, sepsis, osteomyelitis, pneumonia or IE, unrelated or related to FBRIs metastatic infections. IE is a dangerous, often fatal pathology, which develops as a consequence of bacteremia or as a complication after heart surgery. The high-risk group for IE includes intravenous drug users, patients with heart lesions, elderly people with valve sclerosis, patients with intravascular prostheses, persons with inflammation of the oral cavity (mucositis), those exposed to nosocomial bacteremia or hemodialysis. Moreover, IE is observed with increased frequency in young people (under the age of 35) who have not previously been reported with abnormalities in the cardiovascular system [10,54]. The most important and frequent etiological agents of IE are streptococci—mainly *Streptococcus mutans*, *Streptococcus salivarius* and *Streptococcus sanguinis*, enterococci and staphylococci, including *S. aureus* and *S. epidermidis* [54]. This is mostly related to the surface MSCRAMMs adhesins of these microorganisms, which promote bacterial attachment to both

ECM and host cells. The formation of aggregates composed of platelets, fibrin and bacteria within the endocardium, so called vegetation, is a typical manifestation of IE. Several *S. aureus* MSCRAMMs promote adhesion to and activation of platelets by these bacteria. ClfA is involved in the aggregation of staphylococci in the presence of fibrinogen/fibrin, which reduces bacterial phagocytosis. It also participates in the adhesion of these bacteria to the surfaces conditioned by fibrinogen/fibrin, including endothelial cells or platelets. A similar function is played by fibronectin-binding proteins (FnBP), which bind to ECM. Both ClfA and FnBP form "protein bridges" by the fibrinogen, fibrin, and thrombospondin with glycoprotein IIb/IIIa (GPIIb/IIIa) receptor on platelets. It has been also demostrated that ClfA participates in *S. aureus* anchoring to vascular endothelium via the complex formed by von Willebrand binding protein (vWFbp) and von Willebrand factor (vWF). SpA principally operates as antiopsonin, binding Fc portion of IgG and IgM. However, it also interacts with several other host proteins/molecules, including vWF, gC1qR/p33 molecules, and tumor necrosis factor receptor 1. This makes SpA a multipotent factor, affecting the activity of a number of eukaryotic cells and the course of many physiological processes in a host. Interaction between SpA and vWF with gC1qR/p33 seems to be particularly significant for the development of IE. *S. aureus* through the activity of coagulase (Coa) and vWFbp also has an impact on clot formation by prothrombin activation [55,56]. Although, in the context of IE complications, another staphylococcal virulence factor, staphylokinase (SAK), known as well fibrinolysin, should also be mentioned. This protein is one of the specific human immune innate modulators, acting as an antiopsonic agent and an inactivator of ß-defensins. Staphylococcal fibrinolysin belongs to a group of bacterial plasminogen activators, which are the precursors of fibrinolytic protease—plasmin. SAK enhances proteolytic activity of *S. aureus* strains, which seems to play an important role in the release of thromboembolic lesions containing bacteria from heart tissue in the course of IE [10,12].

Regardless of how bacteria get into the bloodstream and whether they are single cells, aggregates or biofilm-derived emboli, in order to survive they must escape a variety of innate immune mechanisms, such as antimicrobial peptides, complement, and phagocytic killing. Immediate and essential host defense against *S. aureus* is provided by neutrophils. However, these bacteria have also evolved an abundant repertoire of the factors aimed to evade the innate immune system, including defense strategies utilized by neutrophils. These may be subverted at many different stages: producing protective capsular polysaccharides, free-floating aggregates or biofilm formation to protect before the phagocytosis; secreting the specific molecules blocking phagocyte receptor function; decreasing the efficiency of antimicrobial mechanisms; or producing toxins that lyse phagocytes. After being ingested, *S. aureus* uses a wide range of effective immune evasion molecules. SOK (surface factor promoting resistance to oxidative killing), catalase, superoxide dismutase (SOD) and staphyloxanthin eliminate ROS (reactive oxygen species), thus avoiding intracellular killing. *S. aureus*, like some other bacterial pathogens, generates also adenosine (using adenosine synthase A—AdsA) to promote its ability to escape phagocytic clearance in the blood. Adenosine is known to inhibit neutrophil degranulation, adhesion to vascular surfaces, and oxidative burst [21]. Recently, the role of secreted staphylococcal factors has been reappraised in IE models. Among these, coagulase, the extracellular matrix binding protein, or the extracellular adhesive protein are the most prominent ones, playing a role in endovascular diseases. Further research on the complex functional role of these "secretable expanded repertoire adhesive molecules" (SERAMMs) may not only help to increase our understanding of *S. aureus* infection pathogenesis, but also specify novel targets for preventive or therapeutic strategies [57].

18.5 NEW APPROACHES FOR TREATING STAPHYLOCOCCAL BIOFILM INFECTIONS

Research into new treatment options effective in combating biofilms involve looking for the substances with various types of activity. These could be the compounds that have not only direct antimicrobial activity, but also exhibit synergistic effect with classic pharmacological agents, restrict the expression of microbial virulence factors, prevent microbial adhesion and aggregation, or activate host immune defense mechanisms designed to combat infections. It is known that nature is a good source of such substances, and the best examples are plant-derived products exhibiting a wide range of biological activity. They usually fall into the class of phytoalexins—secondary metabolites synthesized to prevent insects, fungal and microbial infections and to repair tissue damage. Among them, polyphenols (with flavonoids as the biggest and the best known group) possess strong biological activity of interest. The highly differentiated chemical structures of polyphenolic compounds reflect their varied chemical and physical properties, and their multidirectional activity, such as antioxidant, antiinflammatory, antiallergic, antitumor, and finally, antimicrobial [58–60]. Other interesting plant-derived compounds are essential oils—volatile and chemically complex products characterized by a strong aroma and antimicrobial effect. When used in order to combat colonization/infection, their hydrophobic constituents directly contact the phospholipid bilayer of the microbial cell membrane, leading to an increase in ion permeability, leakage of vital intracellular constituents and/or impairment of bacterial enzyme systems. Essential oils also cause other biological effects, mostly depending on their concentration: antiinflammatory, locally anesthetic, antioxidant or prooxidant, cytotoxic (membranes damage, cell lysis), proapoptotic, antimutagenic, or carcinogenic [58,61–63].

Some plant products exhibit direct biostatic/biocidal effects close to those of antibiotics. For example, it has been found that the extract from *Myrtus communis* leaves inhibited the growth of more than 99% of the studied clinical isolates of *S. aureus*, similar to the widely used pharmacological complex trimethoprim-sulfamethoxazole [64]. The essential oils of *Salvia officinalis* (sage), *Salvia triloba* (sage shrub), *Origanum minutiflorum* (wild oregano), *Origanum onites* (oregano), *Thymbra spicata* (black thyme), *Satureja cuneifolia* (wild savory) and their components as carvacrol limited the growth of *S. aureus* efficiently [58]. Moreover, it was demonstrated that some of these compounds are also active against the biofilm form of microorganisms, often by interfering at the stage of their initial adhesion and aggregation. Chaieb et al. [65] described thymoquinone, an active principle of *Nigella sativa*, as a potent antibiofilm factor, which induced prevention of 90% of biofilm formation by *S. aureus*, *S. epidermidis*, and *Enterococcus faecalis*, but not by *Pseudomonas aeruginosa*. Our group showed that both plant-derived and microbial-derived natural products, such as diterpenoids salvipisone and aethiopinone from *Salvia sclarea* hairy roots, taxodione-derivative isolated from *Salvia austriaca* hairy roots, essential oils of *Lavendula angustifolia*, *Melaleuca alternifolia*, *Melissa officinalis* and their major constituents (linalool, linalyl acetate, α-terpineol, terpinen-4-ol), preparations obtained from *Humulus lupulus* L., even surfactants produced by *L. acidophilus* and *Saccharomyces cerevisiae* display encouraging activity against *Staphylococcus* biofilms. These products influenced staphylococcal adhesion and biofilm development, and were capable of partial destruction and eradication of mature 24-h-old biofilms [35,66–69]. Phytocompounds can also enhance the effects of other biocides (antibiotics) by blocking the specific mechanisms of drug resistance. They can deactivate the enzymes that degrade antibiotics, increase the permeability of bacterial cell envelopes, impair MDR efflux pumps, or inhibit the synthesis of selected cell wall proteins of resistant bacteria [60,70].

Since Q-S systems started to be considered essential for staphylococcal virulence genes expression and biofilm maturation, a search for new therapeutic agents among Q-S signal molecules has been carried out. Halogenated furanones from Australian red algae *Delisea pulchra* were the first described products of the signal-mimic type. Many other plants, fungi, and bacteria are rich sources of the autoinducer analogues exhibiting inhibitory or stimulatory activity for Q-S systems [71]. However, in the context of biofilm formation, the possibility of using so called Q-S-quenching phenomenon is not so simple. Theoretically, *agr*-inhibiting substances, such as AIP produced by staphylococci belonging to other *agr* group (e.g., known Q-S antagonism of *S. aureus* and *S. epidermidis*), might be beneficial in the treatment of acute staphylococcal infections by inhibiting the production of extracellular toxins and enzymes. On the other hand, the loss of *agr* function may enhance the expression of staphylococcal adhesins, the long-term survival in the host and contribute to persistent, often biofilm-associated infections [33,72].

The interest in bacteriophage therapy, especially against multidrug-resistant microorganisms and these forming biofilms has recently increased. As indicated by the literature, the most important advantages of bacteriophages include safety for the host organism, activity against both Gram-positive and Gram-negative bacteria, also against microbial biofilm independently of its maturation stage, and the presence of an EPS barrier. Moreover, phages produce polysaccharide depolymerases, which allow them to better penetrate the biofilm biomass [73,74]. Even with a lack of nutrients and during the slower metabolism of bacteria (as in the deeper layers of the biofilm), bacteriophages are able to perform their lytic cycle. However, the use of bacteriophage therapy is limited by high specificity of bacteriophages, which makes it necessary to implement previous in vitro research on the susceptibility of the etiological agent of infection and to apply a phage cocktail against multispecies biofilms. The potential to cause systemic inflammatory response to massively released components from lysed bacterial cells should also be taken into account [73,75,76].

18.6 CONCLUSIONS

Staphylococci are perfectly predisposed to the biofilm lifestyle, which largely contributes to their success as pathogens. The most significant problem related to the treatment of biofilm-type infections is the resistance to antibiotics resulting not only from the increasing number of drug-resistant microorganisms, but first of all from unique structure and physiology of biofilms as well as the facilitated gene transfer within. Ineffective antibiotic treatment of the biofilm leads to both health consequences for individual patients, and global economic losses related to prolonged hospitalizations and the necessity of applying additional medical procedures. In the face of the current limitations of therapeutic strategies, it is urgently needed to better understand the pathogenic mechanisms of *S. aureus* and other staphylococci as a new targets for the drugs. Moreover, the emphasis should be placed on the necessity to search for an efficient treatment of infections caused by microbial biofilms.

REFERENCES

[1] Costerton JW, Stewart PS, Greenberg EP. Bacterial biofilms: a common cause of persistent infections. Science 1999;284(5418):1318–22.

[2] Haaber J, Cohn MT, Frees D, Andersen TJ, Ingmer H. Planktonic aggregates of *Staphylococcus aureus* protect against common antibiotics. PLoS One 2012;7(7):e41075.

[3] Elias S, Banin E. Multispecies biofilms: living with friendly neighbours. FEMS Microbiol Rev 2012;36:990–1004.

[4] Flemming H-C, Wingender J, Szewzyk U, Steinberg P, Rice SA, Kjelleberg S. Biofilms: an emergent form of bacterial life. Nat Rev Microbiol 2016;14:563–75.

[5] Rickard AH, Gilbert P, High NJ, Kolenbrander PE, Handley PS. Bacterial coaggregation: an integral process in the development of multi-species biofilms. Trends Microbiol 2003;11(2):94–100.

[6] Vlassova N, Han A, Zenilman JM, James G, Lazarus GS. New horizons for cutaneous microbiology: the role of biofilms in dermatological disease. Br J Dermatol 2011;165:751–9.

[7] Costerton JW, Post JC, Ehrlich GD, Hu FZ, Kreft R, Nistico L, et al. New methods for the detection of orthopedic and other biofilm infections. FEMS Immunol Med Microbiol 2011;61(2):133–40.

[8]   Høiby N, Ciofu O, Johansen HK, Song ZJ, Moser C, Jensen PØ, et al. The clinical impact of bacterial biofilms. Int J Oral Sci 2011;3(2):55–65.

[9]   Leid JG, Cope E. Population level virulence in polymicrobial communities associated with chronic disease. Front Biol 2011;6(6):435–45.

[10]   Fowler VG, Miro JM, Hoen B, Cabell CH, Abrutyn E, Rubinstein E, et al. *Staphylococcus aureus* endocarditis. A consequence of medical progress. JAMA 2005;293(24):3012–21.

[11]   Frankenberger RH, Zähringer U, Mack D. Structure, function and contribution of polysaccharide intercellular adhesin (PIA) to *Staphylococcus epidermidis* biofilm formation and pathogenesis of biomaterial-associated infections. Eur J Cell Biol 2010;89:103–11.

[12]   Nethercott C, Mabbett AN, Totsika M, Peters P, Ortiz JV, Graeme R, et al. Molecular characterization of endocarditis-associated *Staphylococcus aureus*. J Clin Microbiol 2013;51(7):2131–8.

[13]   Rohde H, Frankerberger RH, Zähringer U, Mack D. Structure, function and contribution of polysaccharide intercellular adhesin (PIA) to *Staphylococcus epidermidis* biofilm formation and pathogenesis of biomaterial-associated infections. Eur J Cell Biol 2010;89(1):103–11.

[14]   Schierle CF, De la Garza M, Mustoe TA, Galiano RD. Staphylococcal biofilms impair wound healing by delaying reepithelialization in a murine cutaneous wound model. Wound Repair Regen 2009;17:354–9.

[15]   Donlan RM. Biofilms: microbial life on surfaces. Emerg Infect Dis 2002;8(9):881–90.

[16]   McCormick DW, Stevens MRE, Boles BR, Rickard AH. Does it take two to tango? The importance of coaggregation in multi-species biofilms. Culture 2011;32(2):1–5.

[17]   Merino N, Toledo-Arana A, Vergara-Irigaray M, Valle J, Solano C, Calvo E, et al. Protein A-mediated multicellular behaviour in *Staphylococcus aureus*. J Bacteriol 2009;191(3):832–43.

[18]   Otto M. Staphylococcal biofilms. Curr Top Microbiol Immunol 2008;322:207–28.

[19]   Cucarella C, Tormo MA, Knecht E. Expression of the biofilm-associated protein interferes with host protein receptors of *Staphylococcus aureus* and alters the infective process. Infect Immun 2002;70(6):3180–6.

[20]   Hussain M, von Eiff C, Sinha B, Joost I, Herrmann M, Peters G, et al. *Eap* gene as a novel target for specific identification of *Staphylococcus aureus*. J Clin Microbiol 2008;46(2):470–6.

[21]   Kim HK, Thammavongsa V, Schneewind O, Missiakas D. Recurrent infections and immune evasion strategies of *Staphylococcus aureus*. Curr Opin Microbiol 2012;15(1):92–9.

[22]   Boles BR, Horswill AR. Staphylococcal biofilm disassembly. Trends Microbiol 2011;19(9):449–55.

[23]   Diemond-Hernández B, Solórzano-Santos F, Leaños-Miranda B, Peregrino-Bejarano L, Miranda-Novales G. Production of *icaADBC*-encoded polysaccharide intercellular adhesin and therapeutic failure in pediatric patients with staphylococcal device-related infections. BMC Infect Dis 2010;10:68–74.

[24]   Nasr RA, AbuShady HM, Hussein HS. Biofilm formation and presence of *icaAD* gene in clinical isolates of staphylococci. Egypt J Med Hum Genet 2012;13:269–74.

[25]   O'Neill E, Pozzi C, Houston P, Humphreys H, Robinson DA, Loughman A, et al. A novel *Staphylococcus aureus* biofilm phenotype mediated by the fibronectin-binding proteins, FnBPA and FnBPB. J Bacteriol 2008;190:3835–50.

[26]   Cucarella C, Solano C, Valle J, Amorena B, Lasa I, Penades JR. Bap, a *Staphylococcus aureus* surface protein involved in biofilm formation. J Bacteriol 2001;183:2888–96.

[27]   Kogan G, Sadovskaya I, Chaignon P, Chokr A, Jabbouri S. Biofilms of clinical strains of *Staphylococcus aureus* that do not contain polysaccharide intercellular adhesin. FEMS Microbiol Lett 2006;255:11–6.

[28]   Dalton T, Dowd SE, Wolcott RD, Sun Y, Watters C, Griswold JA, et al. An *in vivo* polymicrobial biofilm wound infection model to study interspecies interactions. PLoS One 2011;6(11):e27317.

[29] Monds RD, O'Toole GA. The developmental model of microbial biofilms: ten years of a paradigm up for review. Trends Microbiol 2009;17:73–87.

[30] Schoenfelder SMK, Lange C, Eckart M, Eckart M, Hennig S, Kozytska S, et al. Success through diversity—how *Staphylococcus epidermidis* establishes as a nosocomial pathogen. Int J Med Microbiol 2010;300(6):380–6.

[31] Cheung AL, Bayer AS, Zhang G, Gresham H, Xiong Y-Q. Regulation of virulence determinants *in vitro* and *in vivo* in *Staphylococcus aureus*. FEMS Immunol Med Microbiol 2004;40:1–9.

[32] Irie Y, Parsek MR. Quorum sensing and microbial biofilms. Curr Top Microbiol Immunol 2008;322:67–84.

[33] Yarwood JM, Schlievert PM. Quorum sensing in *Staphylococcus* infections. J Clin Invest 2003;112:1620–5.

[34] Kaplan JB. Biofilm dispersal: mechanisms, clinical implications, and potential therapeutic uses. J Dent Res 2010;89(3):205–18.

[35] Walencka E, Różalska S, Sadowska B, Różalska B. The influence of *Lactobacillus acidophilus*-derived surfactants on staphylococcal adhesion and biofilm formation. Folia Microbiol 2008;53(1):61–6.

[36] Walencka E, Sadowska B, Różalska S, Hryniewicz W, Różalska B. Lysostaphin as a potential therapeutic agent for staphylococcal biofilm eradication. Pol J Microbiol 2005;54(3):191–200.

[37] Bjarnsholt T, Kirketerp-Møller K, Jensen PØ, Madsen KG, Phipps R, Krogfelt K, et al. Why chronic wounds will not heal: a novel hypothesis. Wound Repair Regen 2008;16(1):2–10.

[38] Leid JG. Bacterial biofilms resist key host defenses. Microbe 2009;4(2):66–70.

[39] Thurlow LR, Hanke ML, Fritz T, Angle A, Aldrich A, Williams SH, et al. *Staphylococcus aureus* biofilms prevent macrophage phagocytosis and attenuate inflammation *in vivo*. J Immunol 2011;186(11):6585–96.

[40] Leid JG, Shirtliff ME, Costerton JW, Stoodley P. Human leukocytes adhere to, penetrate, and respond to *Staphylococcus aureus* biofilms. Infect Immun 2012;70(11):6339–45.

[41] Sadowska B, Więckowska-Szakiel M, Paszkiewicz M, Różalska B. The immunomodulatory activity of *Staphylococcus aureus* products derived from biofilm and planktonic cultures. Arch Immunol Ther Exp 2013;61(5):413–20.

[42] Percival SL, Cutting KF. Biofilms: possible strategies for suppression in chronic wounds. Nurs Stand 2009;23(32):64–72.

[43] Frempong-Manso E, Raygada JL, DeMarco CE, Seo SM, Kaatz GW. Inability of a reserpine-based screen to identify strains overexpressing efflux pump genes in clinical isolates of *Staphylococcus aureus*. Int J Antimicrob Agents 2009;33(4):360–3.

[44] Stavri M, Piddock LJV, Gibbons S. Bacterial efflux pump inhibitors from natural sources. J Antimicrob Chemother 2007;59(6):1247–60.

[45] Høiby N, Bjarnsholt T, Givskov M, Molin S, Ciofu O. Antibiotic resistance of bacterial biofilms. Int J Antimicrob Agents 2010;35(4):322–32.

[46] Pozzi C, Waters EM, Rudkin JK, Schaeffer CR, Lohan AJ, Tong P, et al. Methicillin resistance alters the biofilm phenotype and attenuates virulence in *Staphylococus aureus* device-associated infections. PLoS Pathog 2012;8(4):e1002626.

[47] Stevens NT, Greene CM, O'Gara JP, Humphreys H. Biofilm characteristics of *Staphylococcus epidermidis* isolates associated with device-related meningitis. J Med Microbiol 2009;58:855–62.

[48] von Eiff C, Jansen B, Kohnen W, Becker K. Infections associated with medical devices. Drugs 2005;65(2):179–214.

[49] Arciola CR, Montanaro L, Costerton JW. New trends in diagnosis and control strategies for implant infections. Int J Artif Organs 2011;34(9):729–36.

[50] Owens CD, Stoessel K. Surgical site infections: epidemiology, microbiology and prevention. J Hosp Infect 2008;70(Suppl. 2):3–10.

[51] Savini V, Catavitello C, Astolfi D, Balbinot A, Masciarelli G, Pompilio A, et al. Bacterial contamination of platelets and septic transfusion: review of the literature and discussion on recent patents about biofilm treatment. Recent Pat Antiinfect Drug Discov 2010;5:168–76.

[52] Chemaly RF, Hachem RY, Husni RN, Bahna B, Abou Rjaili G, Waked A, et al. Characteristics and outcomes of methicillin-resistant *Staphylococcus aureus* surgical-site infections in patients with cancer: a case-control study. Ann Surg Oncol 2010;17:1499–506.

[53] Rolston K, Mihu C, Tarrand J. Current microbiology of surgical site infections associated with breast cancer surgery. Wounds 2010;22(5):132–5.

[54] Widmer E, Que Y-A, Entenza JM, Moreillon P. New concepts in the pathophysiology of infective endocarditis. Curr Infect Dis Rep 2006;8:271–9.

[55] Claes J, Liesenborghs L, Peetermans M, Veloso TR, Missiakas D, Schneewind O, et al. Clumping factor A, von Willebrand factor-binding protein and von Willebrand factor anchor *Staphylococcus aureus* to the vessel wall. J Thromb Haemost 2017;15:1009–19.

[56] McAdow M, Missiakas DM, Schneewind O. *Staphylococcus aureus* secretes coagulase and von Willebrand factor binding protein to modify the coagulation cascade and establish host infections. J Innate Immun 2012;4:141–8.

[57] Chavakis T, Wiechmann K, Preissner KT, Herrmann M. *Staphylococcus aureus* interactions with the endothelium: the role of bacterial "secretable expanded repertoire adhesive molecules" (SERAM) in disturbing host defense systems. Thromb Haemost 2005;94(2):278–85.

[58] Alviano DS, Alviano CS. Plant extracts: search for new alternatives to treat microbial diseases. Curr Pharm Biotechnol 2009;10:106–21.

[59] Fraga CG, Galleano M, Verstraeten SV, Oteiza PI. Basic biochemical mechanisms behind the health benefits of polyphenols. Mol Asp Med 2010;31:435–45.

[60] Tegos G, Stermitz FR, Lomovskaya O, Lewis K. Multidrug pump inhibitors uncover remarkable activity of plant antimicrobials. Antimicrob Agents Chemother 2002;46(10):3133–41.

[61] Bakkali F, Averbeck S, Averbeck D, Idaomar M. Biological effects of essential oils—a review. Food Chem Toxicol 2008;46:446–75.

[62] Budzyńska A, Sadowska B, Lipowczan G, Maciąg A, Kalemba D, Różalska B. Activity of selected essential oils against *Candida* spp. strains. Evaluation a new aspects of their specific pharmacological properties, with special reference to Lemon balm. Adv Microbiol 2013;3:317–25.

[63] Martos J, Luque CMF, González-Rodriguez MP, Arias-Moliz MT, Baca P. Antimicrobial activity of essential oils and chloroform alone and combined with cetrimide against *Enterococcus faecalis* biofilm. Eur J Microbiol Immunol 2013;3(1):44–8.

[64] Gholamhoseinian N, Mansouri S, Rahighi S. Effects of sub-inhibitory concentration of *Myrtus communis* leave extracts on the induction of free radicals in *Staphylococcus aureus*: a possible mechanism for the antibacterial action. Asian J Plant Sci 2009;8:551–6.

[65] Chaieb K, Kouidhi B, Jrah H, Mahdouani K, Bakhrouf A. Antibacterial activity of thymoquinone, an active principle of *Nigella sativa* and its potency to prevent bacterial biofilm formation. BMC Complement Altern Med 2011;11(29):1–6.

[66] Budzyńska A, Więckowska-Szakiel M, Sadowska B, Kalemba D, Różalska B. Antibiofilm activity of selected plant essential oils and their major components. Pol J Microbiol 2011;60(1):35–41.

[67] Kuźma Ł, Wysokińska H, Różalski M, Budzyńska A, Więckowska-Szakiel M, Sadowska B, et al. Antimicrobial and anti-biofilm properties of new taxidione derivative from hairy roots of *Salvia austriaca*. Phytomedicine 2012;19(14):1285–7.

[68] Różalski M, Micota B, Sadowska B, Stochmal A, Jędrejek D, Więckowska-Szakiel M, et al. Antiadherent and antibiofilm activity of *Humulus lupulus* L. derived products: new pharmacological properties. Biomed Res Int 2013;1–7. https://doi.org/10.1155/2013/101089.

[69] Walencka E, Różalska S, Wysokińska H, Różalski M, Kuźma L, Różalska B. Salvipisone and aethiopinone from *Salvia sclarea* hairy roots modulate staphylococcal antibiotic resistance and express anti-biofilm activity. Planta Med 2007;73(6):545–51.

[70] Nguyen HM, Graber CJ. Limitations of antibiotic options for invasive infections caused by methicillin-resistant *Staphylococcus aureus*: is combination therapy the answer? J Antimicrob Chemother 2010;65:24–36.

[71] Rasmussen TB, Givskov M. *Quorum-sensing* inhibitors as anti-pathogenic drugs. Int J Med Microbiol 2006;296:149–61.

[72] Otto M. *Quorum-sensing* control in staphylococci—a target for antimicrobial drug therapy? FEMS Microbiol Lett 2004;241:135–41.

[73] Gutiérrez D, Rodríguez-Rubio L, Martínez B, Rodríguez A, García P. Bacteriophages as weapons against bacterial biofilms in the food industry. Front Microbiol 2016;7:1–15.

[74] Verbeken G, Pirnay J-P, Lavigne R, Jennes S, De Vos D, Casteels M, et al. Call for a dedicated european legal framework for bacteriophage therapy. Arch Immunol Ther Exp 2014;62:117–29.

[75] Fu W, Forster T, Mayer O, Curtin JJ, Lehman SM, Donlan RM. Bacteriophage cocktail for the prevention of biofilm formation by *Pseudomonas aeruginosa* on catheters in an *in vitro* model system. Antimicrob Agents Chemother 2010;54:397–404.

[76] Górski A, Międzybrodzki R, Borysowski J, Weber-Dabrowska B, Lobocka M, Fortuna W, et al. Bacteriophage therapy for the treatment of infections. Curr Opin Investig Drugs 2009;10:766–77.

Chapter 19

Autovaccines in Individual Therapy of Staphylococcal Infections

Stefania Giedrys-Kalemba*, Danuta Czernomysy-Furowicz†,
Karol Fijałkowski†, Joanna Jursa-Kulesza*
*Department of Microbiology and Immunology, Pomeranian Medical University, Szczecin, Poland
†Department of Immunology, Microbiology and Physiological Chemistry, Faculty of Biotechnology
and Animal Husbandry, West Pomeranian University of Technology, Szczecin, Poland

19.1 INTRODUCTION

Autogenous (autologous) vaccine, or autovaccine, is a therapeutic vaccine prepared for an individual patient suffering from chronic or recurrent infections. Autovaccines are also commonly used in veterinary medicine to treat infectious diseases occurring in domestic and farm animals (stable or herd specific vaccines) [1–3]. However, it should be emphasized that autovaccines are not useful for treatment of acute diseases, as well as viral infections and those by spirochetes.

On the basis of our own experiences, the most prominent indications for tailoring of autogenous vaccines in humans are currently chronic infections caused by *Staphylococcus aureus* such as furunculosis, diabetic food, surgical site infection, osteomyelitis, sinusitis, otitis, bronchitis, and otherwise noninfectious diseases that may be complicated nevertheless by staphylococci colonization (e.g., atopic dermatitis) [4–7]. Autovaccines are also prepared in the event of recurrent urinary tract infections (caused by *Escherichia coli* or other Gram-negative rods) [8–10], acne (*Propionibacterium* spp.) [11,12] and vaginal candidiasis [5,13]. Sometimes polyvalent autogenous vaccines, for example, consisting of *S. aureus* and *Pseudomonas aeruginosa* are manufactured in mixed infections such as osteomyelitis and diabetic foot [5,7]. Polyvalent vaccines prepared in allergic-infectious processes of the respiratory tract often consist of pathogenic strains (e.g., *S. aureus, Haemophilus influenzae, Streptococcus pneumoniae*) and microorganisms physiologically colonizing the throat (such as *Neisseria* spp. and oral streptococci) [14–16].

Pet-to-Man Travelling Staphylococci: A World in Progress. https://doi.org/10.1016/B978-0-12-813547-1.00019-4

An important characteristic of autogenous vaccines is their specificity. The used microorganisms are in fact obtained from a lesion or an infection site of the patient for whom the autogenous vaccine is being prepared. It is therefore a patient- and strain-specific therapeutic. The intention of vaccine treatment is to stimulate or modulate innate and specific immune response against agents of infections. The patient's immune system, particularly, should be able to prevent recurrent episodes or stop the chronic inflammatory process.

19.2 HISTORY OF AUTOGENOUS VACCINES

So-called vaccine therapy was introduced by Sir Almroth Edward Wright at the beginning of the 20th century (between 1900 and 1904). At that time, he was notable for developing a system of antityphoid fever inoculations and a method of measuring protective substances (opsonins) in human blood [17]. His first three patients had boils on their faces or necks for months. They insisted that Wright vaccinate them against boils as he had vaccinated normal individuals against typhoid fever. Wright isolated staphylococci from each patient and prepared heat-killed "vaccines." At the same time, he observed that subjects with recurrent staphylococcal infections had a reduced ability to phagocytize their own staphylococci, and their opsonic index varied from 0.1 to 0.88 of normal value, which he designated as 1. As the doses of vaccine were administered, the rate of phagocytosis increased and a complete cure of recurrent infections was achieved [18,19].

Low opsonin levels as the cause of recurring infections was never accepted by Wright's generation, because many agglutinins, precipitins, and even staphylococcal antitoxins were shown by other scientists in the sera of patients who had never had boils. However, there was a period of rapid increase in the use of autogenous vaccines in England, the United States, and other countries. At the beginning of 1907, a number of case reports, extending indications and improving protocols for autovaccine preparation were published. The autologous vaccines were used in adults, children, and infants in the treatment of different chronic infections of the skin, colon, lung, and upper respiratory tree, the urinary tract, the female genital tract, as well as in cases of septicemia, bronchial asthma, ozaena, gonorrhea and gonorrheal arthritis, brucellosis, osteomyelitis, cerebrospinal meningitis, chronic dysentery, ulcerous blepharitis, and candidiasis. Until now, more than 650 publications voted to autogenous vaccines have been recorded, with the peak (up to 30 papers yearly) in the 1930's. A list of references concerning autogenous vaccines therapy is available on the website www.autovaccine.de, edited by Dr. Oliver Nolte.

The efficacy of autogenous vaccine therapy is variable. In many patients, results were satisfactory as they had no relapses through many years, whereas in others, there seemed to be no effects, and in some patients, harmful local and systemic reaction occurred, particularly if doses were too heavy. In 1917, intracutaneous skin tests for detecting the degree of hypersensitivity had been introduced by Walker in Boston [20]. A number of investigators concluded that an excess of allergy is present in patients with chronic and recurrent infections, and autogenous vaccines provide hyposensitizing effects. Properly selected doses can suppress the tuberculin-type hypersensitivity and release the inhibition of phagocytosis while simultaneously stimulating a rapid increase of specific antibodies.

The use of bacterial vaccines, then the importance of their role as therapeutics, decreased together with the introduction of antibiotics in medicine. However, a lack of satisfactory therapeutic effects after long-lasting antibiotic courses for chronic infections, additional clinical complications (e.g., fungal infections), and the constant selection of bacterial resistance traits led to the resumption of vaccination based on autogenous vaccines. Today, autovaccination, mainly in eastern countries of Europe, and less in central and western zones, is used to replace antimicrobial drugs in the treatment of chronic and recurrent diseases in humans, as well as in veterinary medicine.

Therapy with autovaccines provides the best clinical effects in the case of chronic staphylococcal infections, even by MRSA (methicillin-resistant *S. aureus*) [21]. Unfortunately, exact mechanisms of autovaccine action are indeed still unknown.

19.3 PREPARATION AND ADMINISTRATION OF AUTOVACCINES IN HUMANS

The autovaccines are made from a culture of pathogenic microorganisms that are isolated from a site of infection (i.e., pus from furuncle or abscess, bone fistula, sputum, urine, vaginal discharge, etc.). However, the recommended protocols of preparation and administration of autovaccines differ more or less in details depending on the manufacturer. In general, the cultured microorganisms are suspended in sterile 0.9% NaCl solution and inactivated, either by chemicals (0.4% formalin solution) or by heat. The inactivated whole cell suspension is bottled in a series of vials (density in our laboratory is as follows: 5×10^8 bacteria/mL; 1×10^9 bacteria/mL; 2.5×10^9 bacteria/mL) based on the McFarland quantification. For preservation, 0.1 mg/mL thiomersal or 0.5% phenol is added. Finally, sterility of suspensions is tested by cultivation in liquid enrichment medium. Manufacturing of autovaccines is a time consuming process, requiring up to 3 to 4 weeks. Before the beginning of the treatment, an intradermal skin test is performed to exclude a patient's hypersensitivity to any of the vaccine components. An autovaccine is then applied subcutaneously in intervals of several days or weekly over 3, 6, or 9 months. Usually, each injection introduces an increased dose and intervals are gradually prolonged. Sometimes therapy is individualized by physician according to observed clinical effects. Oral administration of autovaccines is also used, especially in bronchial asthma and children [22].

Autologous vaccines are generally considered safe. However, other vaccines can cause a mild reaction at the site of injection (redness, swelling) and, very rarely, systemic reactions (fever, headache, malaise, sore throat, increased blood pressure, and breath frequency). Following the application, the patient should be under observation for an additional hour.

19.4 AUTOVACCINES IN THE TREATMENT OF STAPHYLOCOCCAL INFECTIONS IN ANIMALS

The autovaccines are commonly used for treatment of various animal infections such as purulent inflammation of skin (dermatitis pyogenes), upper respiratory tract infections (sinusitis, pharyngitis, laryngitis), otitis externa and mastitis in cows and goats. All these pictures can be caused by diverse Gram-positive bacteria, mainly *Staphylococcus* spp., *Erysipelothrix* spp. and *Actinomyces* spp., as well as Gram-negative microorganisms, including *Pseudomonas* spp., *Escherichia* spp. *Bordetella* spp., *Pasteurella* spp., *Yersinia* spp., *Moraxella* spp. and even dermatophytes and yeasts.

It is generally accepted that indications for treatment with autovaccine are

(a) resistance of pathogenic microorganisms to antibiotics used in the treatment
(b) recurrence of infection despite therapy
(c) insufficient innate immune response
(d) lack of appropriate commercial vaccines

The interest in antistaphylococcal autovaccines results from the widespread dissemination of asymptomatic carriers of different staphylococci, which may behave as opportunistic pathogens as soon as an impairment of the immune system occurs. Such autovaccines prepared for animals generally contain one pathogenic strain. In large farm herds, especially under adverse environmental conditions, an infectious agent may rapidly disseminate. In this case, after isolation of one or several pathogens, the autovaccine can be prepared for the entire herd: it is the so-called herd-specific autovaccine, and contains one or more strains of the same species [3].

The autovaccines are made for single individuals, usually for companion animals such as dogs, cats, rabbits, or horses. Conversely, herd autovaccines may apply to herds of cows, horses, goats, chinchillas, minks, rabbits, pigs, poultry, and pigeons. Autovaccines for animals are always prepared from whole microorganisms. Preparation, particularly, does not differ in principle from production of a standard vaccine. The autovaccines are medicaments used to treat individual animals with chronic diseases when standard treatments do not provide the desired results; otherwise they can be used for all animals in the herd, where spread of a certain disease has occurred (i.e., *S. aureus* mastitis). Conventional vaccines are administered in order to promote a humoral response while application of autovaccines is also recommended to stimulate cell-mediated immunity.

A thorough review of the available literature surrounding the issues discussed in this chapter revealed that published material is poor. Therefore, critical discussion of the reported cases must be limited and presented results and discussion are mostly based on our experience due to many years of work with autovaccines.

19.4.1   Autovaccine for Dogs

The autovaccines are commonly prepared for dogs with pyoderma and in-flammation of the outer and middle ear, whose etiological agents include both coagulase-positive staphylococci (CoPS) such as *S. aureus, Staphylococcus intermedius, Staphylococcus pseudintermedius, Staphylococcus hyicus* (co-agulase of the latter is variably expressed, as it is strain-dependent) and coagulase-negative staphylococci (CoNS), for example, *Staphylococcus xylosus*. Particularly, autovaccines are usually given to animals that were previously and unsuccessfully treated with antibiotics. Moreover, besides staphylococci (usually *S. aureus, S. intermedius* and *S. pseudintermedius*), fungi are used, including *Malassesia pachydermatis*. Therefore, the autovaccine is often made up of two components; that is, a bacterium and a yeast (bivalent autovaccine).

However, it was reported that administration of an autovaccine to treat dog pyoderma by staphylococci may result in variable degrees of success; 43.7% of dogs were free of symptoms following autovaccination, and 42.5% did not recover completely [23].

19.4.2   Autovaccine for Cats and Rabbits

Rabbits reared on farms, and those kept at home may suffer from subcutaneous abscesses, caused by *S. aureus*. The disease is associated with the formation of encysted abscesses that are difficult to treat and require surgery, which is difficult to perform with rabbit mass diseases. The autovaccine is therefore an option and is made by using the microorganisms isolated from the abscesses and administered twice. On the basis of our observations, the treatment frequently results in reversal of abscesses within ~2 weeks. Notably, in rabbits kept in farms, infection of an animal quickly spreads to others. In this case, preparation of a herd autovaccine and its administration to all rabbits of the herd is always recommended. This way, the infected rabbits receive an autovaccine for a thera-peutic purpose, while healthy animals get it as prophylaxis.

Some reports also show therapeutic and protective effect of autovaccines against highly virulent *S. aureus* isolates, which are the main cause of subcuta-neous abscesses in rabbits. This autovaccine is manufactured with the specific *S. aureus* strain and its administration to sick animals results in decrease of abscesses diameter; nevertheless, it is not able to prevent formation of new ab-scesses in infected rabbits [24].

19.4.3   Autovaccine for Cats

Autovaccines for cats are made using staphylococci that are isolated mainly from the leakage of purulent lesions of nose and eyes. Bacteria that cause such clinical pictures are usually *S. aureus, S. intermedius, S. pseudintermedius, S. hyicus, S. cohni,* and *S. xylosus* [25–27].

19.4.4 Autovaccine for Horses

Antistaphylococcal autovaccines are rarely made for horses, and exclusively when staphylococci are the only etiological agent of a disease. Horses are often asymptomatic carriers of CoNS and CoPS [28]. When the immune system is impaired, staphylococci inhabiting a horse's upper respiratory tract can cause inflammation of the lining of the nose and sinuses, resulting in a purulent discharge from the nostrils [29,30]. In these cases, an autovaccine results in the relief of clinical symptoms and even elimination of the microorganisms from the upper airways. In the autumn and winter, when horses reside longer in the stable, purulent exudate from their nostrils, and a cough is often observed. If antibiotic treatment achieves unsatisfactory results, and additional staphylococci are isolated from nose purulent exudate or throat cultures, the veterinarian may consider whether there is an indication to perform autovaccine. However, their limitation, when working with horses, is represented by simultaneous isolation of molds belonging to the genus *Rhizopus* [31]. Spores of these fungi float in the air in the stable and may irritate mucous membranes, thus causing allergy and cough [32]. In such cases, the use of antistaphylococcal autovaccines is ineffective as symptoms are caused primarily by the fungus and staphylococcal infections occurs a second time, as a complication.

19.4.5 Autovaccine for Pigs

Autovaccines for pigs are most frequently prepared to treat exudative infection of the skin of piglets (*dermatitis exudativa procellorum*) whose etiologic agent is *S. hyicus* [33,34]. Organisms such as *S. aureus* [35], *S. chromogenes* [33], and *S. sciuri* [36] can produce exfoliative toxins and have been isolated, although rarely, from cases of this disease, as well. This type of autovaccine is particularly useful in the case of rapid development of the infection and is administrated to all piglets (with and without clinical symptoms). Autovaccination of pregnant sows 3 weeks before delivery is also possible and, in this case, IgG antibodies produced by sows are transferred via the colostrum to pigs, thus protecting them against the disease.

19.4.6 Autovaccine for American Mink

Staphylococci that inhabit the oral cavity of American minks (*Neovison vison*) are often responsible for infection as a result of cross-bites between these animals. Minks are common carriers of staphylococci due to the type of food fed to them (raw meat) [37]. In this case, an important role in the spread of different species of *Staphylococcus* is played by the method of preserving meat which comprises the addition of staphylococci. As a result of bites, in some animals, staphylococcal abscesses develop and some bitten minks die due to systemic staphylococcal infection. Prophylactic administration of the herd autovaccines helps to reduce the rate of deaths. However, in large farms, for technical reasons, this type of therapy has not gained great interest.

Staphylococci also cause mastitis and neonatal food poisoning in minks [38,39]. In these cases, satisfactory results have been achieved through administration of autovaccines to sick, nursing females for 3 weeks before delivery.

19.4.7 Autovaccine for Pigeons

The gastrointestinal tract of pigeons is naturally colonized by several species of staphylococci (including *S. aureus*) [40,41]. These microorganisms cause diarrhea in adult birds, while they often lead young pigeons to death. Based on our observations, administration of antistaphylococcal autovaccine to infected animals leads to complete recovery of birds.

19.4.8 Autovaccine for Cows

Dairy cows and goats often suffer from mastitis involving one or more mammary glands. The most common agents are staphylococci, including both CoPS and CoNS [42–44]. In cows, autovaccines are only administered to treat subclinical disease [3] and, as far as etiological agents are CoNS, this kind of therapy is highly effective (up to 100%), elimination of the microorganisms from the mammary gland occurring approximately after 10 days. The potential of autovaccines against CoNS diseases has been gaining increasing importance due to the rapid selection of antibiotic resistance in this group of microorganisms [45,46]. In *S. aureus* mastitis, nevertheless, effectiveness of autovaccines is lower (about 50%–60%), but similar to that shown by antibiotics. However, administration of an anti-*S. aureus* autovaccine combined with proper, parenteral antimicrobial may remove *S. aureus* and lead to prolonged udder protection against its invasion [3].

19.5 MODE OF AUTOLOGOUS VACCINE ACTION

The mechanism of autovaccine action is not fully understood, however, it is considered to be associated with activation of both nonspecific and specific immune response. Following subcutaneous administration of autovaccines, swelling at the injection site is observed that results from activation of the innate immune system (the so-called acute phase reaction). The antigen injection sites are reached by large amounts of phagocytic cells, including macrophages, which, after ingestion and digestion of killed staphylococci, can present antigens to Th1 cells (adaptive immune system). IFN-γ produced by Th1 cells activate macrophages and neutrophils to enhance the oxygen dependent and independent mechanisms of engulfed staphylococci killing. Cytokines produced by Th1 and Th2 lymphocytes help B cells to produce immunoglobulins, primarily IgG2. IFN-γ also stimulates the formation of Fc receptors for IgG on the surface of neutrophils and macrophages. The antigens present in autovaccines are also processed by dendritic cells. It should be emphasized, particularly, that surface antigens in the autovaccine are the same as the antigens of staphylococci that cause the disease.

The mode of action of staphylococcal autovaccines in humans is similar. Our earlier publications confirmed Wright's results; that is, that administration of autovaccines to patients with chronic staphylococcal ostitis and carbunculosis leads to an intensification of phagocytic activity, particularly concerning the phagocytic index and strain-specific agglutinating antibodies, but also increases the level and activation of T cells [4,47]. Similarly, clinical improvements of acne after vaccination with killed *Propionibacterium acnes* are accompanied by the synthesis of specific antibodies against bacterial structures [12]. Finally, studies using two-dimensional immunoblotting and flow-cytometry showed that although patients with furunculosis possess many different antibodies versus a broad range of staphylococcal antigens before autovaccine treatment, vaccination moderately enhances the immunoglobulin response to some surface proteins (e.g., ClfA, ClfB, SdrD, and SdrE), but in different degrees in diverse individuals [48]. Higher humoral and also cellular immune response to homologous strains rather than heterologous staphylococci was also showed in experiments with mice and rabbits vaccinated with different *S. aureus* strains [49–54]. Other authors reported that autovaccine downregulates Th1 cell function and reduces delayed types of hypersensitivity reactions [55]. Further studies analyzed the levels of circulating cytokines produced by Th cells (IFN-γ, IL-4, IL-10, IL-17A) and by monocytes/macrophages (TNF-α, IL-1β, IL-12) in patients with chronic *S. aureus* infections thus indicating that, when induced by a pathogen, Th17/Th1 cells are mainly responsible for promotion of chronic inflammatory response (higher levels of IFN-γ and IL-17A were noted) [56]. After autovaccine treatment, a significant increase in the levels of IFN-γ, TNF-α, and IL-12 was found [57].

19.6 HERD SPECIFIC AUTOVACCINES

Prior to herd specific autovaccine preparation, bacteria were collected from several animals, only to determine the etiological agents of diseases [3]. Given that bacterial strains from different animals inhabiting the same herd can produce different antigens, it is necessary to determine the antigenic differences among diverse isolates. Animals in large herds can be subjected to infections from a variety of sources. Therefore staphylococci that, in herd animals, are responsible for the same clinical pictures can be isolated. In such a case, a polyvalent autovaccine is prepared where the percentage of microorganisms corresponds to their involvement in the infection.

19.7 DOSES AND METHODS OF ADMINISTRATION OF AUTOVACCINE TO ANIMALS

The efficacy of autovaccines depends on the dose volume, the number of doses, the administration site and the density of the bacterial suspension. In contrast to vaccines, autovaccines are not combined with the adjuvants. However, in order to intensify the nonspecific cellular immune response, immunostimulation with the preparations containing bacteria (e.g., *Bacillus subtilis* and *P. acnes*) preceding autovaccine therapy can be carried out. In order to activate macrophages, the immunostimulator is administered to animals for 3–5 days before the first dose of autovaccine. Activated macrophages more effectively phagocyte bacteria from the autovaccine and also begin to digest bacteria that cause the disease. The immunomodulator is administered to animals subcutaneously and only once.

Autovaccine, depending on animal species, can be administered in different ways, for example, in the wing membrane to birds, intramuscularly to pigs, intramuscularly or into the lymph node region to mastitis ruminants, and usually subcutaneously to other animals. Anti-*Staphylococcus* autovaccines are usually administered three times. As exceptions, cows and goats with subclinical mastitis generally receive one only dose of autovaccine in the upper udder lymph node region. In this case, the site selection is related to the anatomy of the mammary gland and the strategic role played by the upper udder lymph nodes in protection of this organ. Administration of the autovaccines to cows and goats in the upper udder lymph node region is intended to increase the flow of IgG from the supramammary lymph node to the mammary gland and to activate neutrophils in it.

The autovaccine dose is associated with the animal's body weight. Particularly, two methods of administration are generally recommended; that is, first, three doses of the same volume, with each subsequent dose of ~50% increase in density; second, three doses of the same density, the first one having the largest volume while the other two behave as booster doses. Antigen concentration in the autovaccine ranges from 3.0 to 4.0 McFarland standard. In animals infected with staphylococci, very promising results are achieved when autovaccine therapy is combined with antibiotic treatment. Antibiotics destroy bacteria, while the autovaccine increases phagocytic activity. What's more, after the application of both therapies, relapses are not usually observed.

19.8 CONCLUSION

Autovaccines are commonly used in human and veterinary medicine to treat chronic infectious diseases. They provide a number of advantages over conventional treatments, for example, they are strain-specific, thus enabling treatment of diseases for which no traditional preventive vaccine is yet available or there are no alternative and effective treatments. To date, it is demonstrated that autogenous vaccination against various *Staphylococcus* species might be advantageous, reducing the rate and severity of recurrent episodes of infection. However, full clinical studies concerning the efficacy of autovaccines are still lacking and scientific reports trying to explain their exact mode of action are limited. It has been suggested that autovaccination may induce specific humoral and cellular response and activate parts of the innate immune system. However, further studies are needed to demonstrate this hypothesis.

REFERENCES

[1] Nolte O, Morscher J, Weiss HE, Sonntaga HG. Autovaccination of dairy cows to treat post-partum metritis caused by *Actionomyces pyogenes*. Vaccine 2001;19:3146–53.

[2] Nawrotek P, Czernomysy-Furowicz D, Borkowski J, Fijałkowski K, Pobucewicz A. The effect of auto-vaccination therapy on the phenotypic variation of one clonal type of *Staphylococcus aureus* isolated from cows with mastitis. Vet Microbiol 2012;155:434–7.

[3] Czernomysy-Furowicz D, Fijałkowski K, Silecka A, et al. Herd specific autovaccine and antibiotic treatment in elimination of *Staphylococcus aureus* mastitis in dairy cattle. Turk J Vet Anim Sci 2014;38:496–500.

[4] Giedrys-Galant S. Effect of autovaccine on various immunologic indices in chronic staphylococcal infections. Ann Acad Med Stetin 1977;23:279–303.

[5] Giedrys-Galant S, Hałasa J. Activities of the autovaccine laboratory of the szczecin health service group after 2 years. Pol Tyg Lek 1981;36(37):1421–4.

[6] Giedrys-Galant S, Hałasa J, Podkowińska I. Persistence of *Staphylococcus aureus* in the pharynx of children with chronic tonsillitis treated with autovaccine. Pediatr Pol 1985;60(11–12):761–7.

[7] Hałasa J, Podkowińska I, Giedrys-Galant S. Autovaccine in the treatment of chronic inflammation of the bones and periosseous tissues. Pol Przegl Chir 1990;29(6):704–9.

[8] Georgescu C, Boboc F, Iancu L, et al. Immunotherapy with autovaccines in urinary tract infections caused by Gram-negative bacteria. Rev Ig Bacteriol Virusol Parazitol Epidemiol Pneumoftiziol 1982;2792:109–19.

[9] Dacco L, Bonati PA, Cucinotta D, Butturini L, Girardello R. Possibility of oral vaccine therapy in elderly patients with recurrent infections of the urinary tract, wearers of permanent bladder catheters. G Clin Med 1985;66(7–8):269–80.

[10] Nolte O, Bindewald A, Weiss HE, Weiss H, Sonntag HG. Eradication of multi-resistant *E. coli* from a patient with urinary tract infection using autovaccination therapy. Proceedings of the 3rd annual conference on vaccine research; 2000 April 30-May 2; Washington, DC, USA. Available from http://www.olivernolte.de/poster/ACVR_2000%20UTI%20autovaccine.pdf.

[11] Rubisz-Brzezińska J, Wilk-Czyż R, Brzezińska-Wcislo L, Kasprowicz A. Clinical evaluation of serious forms of acne treated with autovaccine. Med Dośw Mikrobiol 1994;46(1–2):35–42.

[12] Załuga E. Skin reactions to antigens of *Propionibacterium acnes* in patients with acne vulgaris treated with autovaccine. Ann Acad Med Stetin 1998;44:65–85.

[13] Rusch K, Schwiertz A. *Candida* autovaccination in the treatment of vulvovaginal *Candida* infections. Int J Gynaecol Obstet 2007;96(2):130.

[14] Hałasa J, Hałasa M, Wojciechowska W, Podkowińska I, Kucharska E. Clinical efficacy of autovaccine in the treatment of infectious nonatopic asthma and COPD—double blind placebo controlled trial. Alergia Astma Immunol 2001;6(2):109–13.

[15] Hałasa J, Millo B. Treatment with an autovaccine of a patient with catarrhal rhinitis, nasal polyps and infectious bronchial asthma. Pneumonol Alergol Pol 1992;60(2):145–6.

[16] Kollarova K, Sonak R. Treatment of allergic diseases with autovaccine. Bratisl Lek Listy 1960;42:36–43.

[17] Wright AE, Semple D. Remarks on vaccination against typhoid fever. Br Med J 1897;1:256–9.

[18] Wright AE. On therapeutic inoculations of bacterial vaccines, and their practical exploitation in the treatment of disease. Br Med J 1903;1:1069–74.

[19] Wright AE, Douglas SR. On the action exerted upon the *Staphylococcus pyrogenes* by human body fluids and an elaboration of protective elements in the human organism in response to inoculation of a *Staphylococcus* vaccine. Proc R Soc Lond 1904;74:147–59.

[20] Walker IC. Studies on the sensitization of patients with bronchial asthma to bacterial protein as demonstrated by the skin test reaction. J Med Res 1917;35:487–96.

[21] Rizzo C, Brancaccio G, De Vito D, Rizzo G. Efficacy of autovaccination therapy on post-coronary artery bypass grafting methicillin-resistant *Staphylococcus aureus* mediastinitis. Interact Cardiovasc Thorac Surg 2007;6(2):228–9.

[22] Chachaj W, Suchnicka R. Oral administration of autovaccine in bronchial asthma of bacterial origin. Pol Tyg Lek 1960;15:1263–6.

[23] Mayr A, Selmair J, Schels H. Erfahrungen mit einer autovakzine-therapie bei der staphylokokken-pyodermie des hundes. Tierarztl Umsch 1987;42(2):112–8.

[24] Meulemans G, Hermans K, Lipinska U, Duchateau L, Haesebrouck F. In: Possible protective effect of an autovaccine against high virulence *Staphylococcus aureus* in a rabbit skin infection model. Proceedings of the 9th World Rabbit Congress; 2008 June 10-13; Verona, Italy, Pathol. Hyg; 2008. p. 1019–23.

[25] Ozaki K, Yamagami T, Nomura K, Haritani M, Tsutsumi Y, Narama I. Abscess-forming inflammatory granulation tissue with Gram-positive cocci and prominent eosinophil infiltration in cats: possible infection of methicillin-resistant *Staphylococcus*. Vet Pathol 2003;40:283–7.

[26] Casanova C, Iselin L, von Steiger N, Droz S, Sendi P. *Staphylococcus hyicus* bacteremia in a farmer. J Clin Microbiol 2011;49(12):4377–8.

[27] Wang N, Neilan AM, Klompas M. *Staphylococcus intermedius* infections: case report and literature review. Infect Dis Rep 2013;5(1):6–11.

[28] Karakulska J, Fijałkowski K, Nawrotek P, Pobucewicz A, Poszumski F, Czernomysy-Furowicz D. Identification and methicillin resistance of coagulase-negative staphylococci isolated from nasal cavity of healthy horses. J Microbiol 2012;50(3):444–51.

[29] Weese JS, Archambault M, Willey BM, et al. Methicillin-resistant *Staphylococcus aureus* in horses and horse personnel, 2000-2002. Emerg Infect Dis 2006;11(3):430–5.

[30] Weese JS, Rousseau J, Willey BM, Archambault M, McGeer A, Low DE. Methicillin-resistant *Staphylococcus aureus* in horses at a veterinary teaching hospital: frequency, characterization, and association with clinical disease. J Vet Intern Med 2006;20(1):182–6.

[31] Carrasco L, Tarradas MC, Gomez-Villamandosa JC, Arenas LA, Méndeza A. Equine pulmonary mycosis due to *Aspergillus niger* and *Rhizopus stolonifer*. J Comp Pathol 1997;117(3):191–9.

[32] Cafarchiaa C, Figueredob LA, Otrantoa D. Fungal diseases of horses. Vet Microbiol 2013;167(1–2):215–34.

[33] Andresen LO, Ahrens P, Daugaard L, Bille-Hansen V. Exudative epidermitis in pigs caused by toxigenic *Staphylococcus chromogenes*. Vet Microbiol 2005;105:291–300.

[34] Victor I, Akwuobu CA, Akinleye OA, Tyagher JA, Buba E. Management of exudative epidermitis (greasy pig disease) in 4 week old piglets. J Vet Med Anim Health 2013;5(7):180–5.

[35] Van Duijkeren E, Ikawaty R, Broekhuizen-Stins MJ, et al. Transmission of methicillin-resistant *Staphylococcus aureus* strains between different kinds of pig farms. Vet Microbiol 2008;126:383–9.

[36] Chen S, Wang Y, Chen F, Yang H, Gan M, Zheng SJ. A highly pathogenic strain of *Staphylococcus sciuri* caused fatal exudative epidermitis in piglets. PLoS One 2007;2:147.

[37] Juokslahti T, Lindroth S, Niskanen A. Pathogenic, enterotoxin-producing staphylococci in mink feed and mink feed raw materials. Acta Vet Scand 1980;21(4):516–22.

[38] Ryan MJ, O'Connor DJ, Nielsen SW. *Staphylococcus aureus* mastitis in nursing mink affected with Aleutian disease. J Wildl Dis 1979;15(4):533–5.

[39] Trautwein GW, Helmboldt CF. Mastitis in mink due to *Staphylococcus aureus* and *Escherichia coli*. J Am Vet Med Assoc 1966;149(7):924–8.

[40] Losito P, Vergara A, Muscariello T. Antimicrobial susceptibility of environmental *Staphylococcus aureus* strains isolated from a pigeon slaughterhouse in Italy. Poult Sci 2005;84(11):1802–7.

[41] Futagawa-Saito K, Ba-Thein W, Sakurai N, Fukuyasu T. Prevalence of virulence factors in *Staphylococcus intermedius* isolates from dogs and pigeons. BMC Vet Res 2006;26:2–4.

[42] Wang SC, Wu CM, Xia SC, Qi YH, Xia LN, Shen JZ. Distribution of superantigenic toxin genes in *Staphylococcus aureus* isolates from milk samples of bovine subclinical mastitis cases in two major diary. Vet Microbiol 2009;137(3–4):276–81.

[43] Fijałkowski K, Masiuk H, Czernomysy-Furowicz D, et al. Superantigen gene profiles, genetic relatedness and biological activity of exosecretions of *Staphylococcus aureus* isolates obtained from milk of cows with clinical mastitis. Microbiol Immunol 2013;57:674–83.

[44] Fijałkowski K, Struk M, Karakulska J, et al. Comparative analysis of superantigen genes in *Staphylococcus xylosus* and *Staphylococcus aureus* isolates collected from a single mammary quarter of cows with mastitis. J Microbiol 2014;52(5):366–72.

[45] Jarløv JO. Phenotypic characteristics of coagulase-negative staphylococci: typing and antibiotic susceptibility. Acta Pathol Microbiol Scand 1999;91:1–42.

[46] John JF, Harvin AM. History and evolution of antibiotic resistance in coagulase-negative staphylococci: susceptibility profiles of new anti-staphylococcal agents. J Ther Clin Risk Manag 2007;3(6):1143–52.

[47] Hałasa J, Giedrys-Galant S, Podkowińska I, Braun J, Strzelecka G, Dąbrowski W. Evaluation of certain immunological parameters in the course of autovaccine treatment in patients with chronic ostitis and carbunculosis. Arch Immunol Ther Exp 1978;26(1–6):589–93.

[48] Holtfreter S, Jursa-Kulesza J, Masiuk H, et al. Antibody responses in furunculosis patients vaccinated with autologous formalin-killed *Staphylococcus aureus*. Eur J Clin Microbiol Infect Dis 2011;30:707–17.

[49] Giedrys-Galant S, Podkowińska I, Hałasa J, Bohatyrewicz R, Daszko J. Evaluation of immunological parameters in rabbits inoculated with *Staphylococcus aureus* in view of specific and non-specific stimulation. Arch Immunol Ther Exp 1978;26(1/6):583–7.

[50] Giedrys-Galant S, Mikłaszewicz A, Hałasa J. In: Jeljaszewicz J, editor. The cellular response in mice and rabbits immunized by Staphylococcus aureus Smith. The staphylococci: proceedings of V international symposium on staphylococci and staphylococcal infections, Warszawa, June 26-30, 1984; Stuttgart, Zbl. Bakt. Suppl. 14. 1985. p. 425–8.

[51] Bohatyrewicz R, Giedrys-Galant S, Hałasa J. In: Jeljaszewicz J, editor. Induction of chemotactic factors in the immune serum of rabbits by *Staphylococcus aureus* suspension. The staphylococci: proceedings of V international symposium on staphylococci and staphylococcal infections, Warszawa, June 26-30, 1984; Stuttgart, Zbl. Bakt. Suppl. 14. 1985. p. 457–8.

[52] Giedrys-Galant S. Some aspects of specific immunological activity of staphylococcal vaccine favoring the use of autologous vaccines. I. Evaluation of phagocytosis and intracellular killing of staphylococci by rabbit neutrophiles. Med Dośw Mikrobiol 1987;39(2):65–82.

[53] Giedrys-Galant S. Some aspects of specific immunological activity of staphylococcal vaccine favoring the use of autologous vaccines. II. Influence of control and immune rabbit serum on staphylococcal growth. Med Dośw Mikrobiol 1987;39(2):83–93.

[54] Giedrys-Galant S. Some aspects of specific immunological activity of staphylococcal vaccine favoring the use of autologous vaccines. III. Evaluation of chemotactic of staphylococci *in vivo*. Med Dośw Mikrobiol 1987;39(4):213–22.

[55] Rusch V, Ottendorfer D, Zimmermann K. Results of an open and non-placebo controlled pilot study investigating the immunomodulatory potencial of autovaccine. Arzneimittelforschung 2001;51(8):690–7.

[56] Szkaradkiewicz A, Karpiński TM, Zeidler A, Szkaradkiewicz AK, Masiuk H, Giedrys-Kalemba S. Cytokine response in patients with chronic infections caused by *Staphylococcus aureus* strains and diversification of their Agr system classes. Eur J Clin Microbiol Infect Dis 2012;31(10):2809–15.

[57] Szkaradkiewicz A, Karpiński TM, Goślińska-Pawłowska O, Szkaradkiewicz AK, Giedrys-Kalemba S. Cytokine response in autocvaccine-treated patients with chronic *Staphylococcus aureus* infections. Eur J Inflamm 2013;11(1):103–10.

Chapter 20

Experimental Animal Models in Evaluation of Staphylococcal Pathogenicity

Jacek Międzobrodzki, Maja Kosecka-Strojek
Department of Microbiology, Faculty of Biochemistry, Biophysics and Biotechnology, Jagiellonian University, Krakow, Poland

20.1 INTRODUCTION

20.1.1 The Aims of Research Using the Animal Experimental Model

Animal models are used in the research of pathogenetic mechanisms of infections, with the goal of disease therapy, limitation, and prevention. Nowadays the flood of publications causes difficulties in interpretation of presented results of investigations because the used models are very diverse and frequently not properly chosen. Moreover, it happens that a model is used as it is easy to obtain and available at a low cost, and not because it properly reflects the infection process.

The use of experimental models dates back to the beginning of the history of medicine. Curiosity about the human body moved also toward research of the mysteries of animal organisms. Though the science of laboratory animals emerged as a separate field after the Second World War, knowledge about experimental animals had been collected previously. In the field of bacteriology, the first documented experiments relying on the use of animals were done by Delafond in 1856 and Koch in 1876 [1,2].

The research based on animal experimental models is necessary to study staphylococci, particularly certain strains of *Staphylococcus aureus* (while it is exceptional among other and opportunistic microbes) because of the great amounts of virulence factors they express and the exceptional ability of adaptations to changing environmental conditions (in vivo and in vitro); particularly, the latter may result in the emergence of new strains with a stronger pathogenicity and a higher resistance to antibiotics. In light of this, staphylococci are in the focus of research in bacteriology laboratories [3–5].

Pet-to-Man Travelling Staphylococci: A World in Progress. https://doi.org/10.1016/B978-0-12-813547-1.00020-0

20.1.2 Koch's Postulates

The deliberations about the causes of infections and identification of etiological agents was started thanks to Koch's research which, in an unusual and astute way, observed nature, thus formulating conclusions that were far ahead of his epoch. Koch's postulates were published in 1884 and they are still valid, nowadays. Currently, conventional methods of microbial identification are supported by genetic methods, as diverse phenotypic features are today investigated, such as virulence determinants, resistance to antibiotics, tissue tropism, and the ability to synthetize substances in general [6,7]. Hence, research using animal models of infections is not only desirable, but irreplaceable [8].

Of crucial importance in this area are Koch's and Loeffler's discoveries, dating back to 1884, focusing on the requirements to be fulfilled aiming to prove the true pathogenic role of a microorganism in a specific case of infection. Koch and Loeffler formulated postulates to establish a causal relationship between a microbe and a disease, starting with tuberculosis and cholera, as well as other infectious conditions. Koch's postulates are:

(1) The microorganism must be found in abundance in all organisms suffering from the disease, but should not be found in healthy organisms [9,10].
(2) The microorganism must be isolated from a diseased organism and grown in a pure culture [9,10].
(3) The cultured microorganism should cause disease when introduced into a healthy organism [9,10].
(4) The microorganism must be reisolated from the inoculated diseased experimental host and identified as being identical to the original specific causative agent [9,10].

There are, however, some factors limiting the validity of Koch's postulates; first, the natural physiological flora, including potential opportunistic pathogens, like staphylococci; second, immunosuppression.

Also, staphylococci, like other bacteria, may show phenotype variations, due to diverse gene expression under a modified environment and/or host conditions, that can affect the reliability of conventional phenotypic assays [11,12].

Finally, the proper experimental animal model must be selected, [8,13], and this choice is preceded by meticulous analyses and discussions, and based both on scientific publications, and the researchers' experience, should be ruled by defined criteria.

20.1.3 The Criteria of Choice of Animal Host Species

The animals used as models in scientific research should be characterized by the following features:

(1) small body size, breeding that does not require large laboratory space, relatively simple and cheap maintenance [14];

(2) high fertility and a large number of individuals, allowing reliable statistical analysis of obtained results [14];

(3) short life cycle, enabling the observations of patterns inherited in next generations [14];

(4) availability of information and proper investigative techniques for particular species [14].

In the case of experimental infections with staphylococci, the variety of animal models allows researchers to identify and characterize specific virulence factors. Unfortunately, diversity of models described in the specialized literature has frequently caused discrepant results and erroneous conclusions, as the influence of individual factors differs between models. Moreover, not all animal models may be dedicated to research of all organisms' pathogenicity; therefore, the choice of a proper investigative model has a critical impact on the results of experiments, also due to the fact that virulence factors can be specific to certain hosts, and not necessarily universal [15].

Several experimental animal models were proposed for research of staphylococcal infections. From the early 1960's, rabbits, rats and mice were used. Though rats remain the preferred endocarditis models [16,17], and rabbits those for toxic shock syndrome (TSS) [18], the most used animals for many bacterial pathological conditions are mice.

20.2 CHARACTERISTIC OF SELECTED EXPERIMENTAL ANIMAL MODELS OF STAPHYLOCOCCAL INFECTION WITH REFERENCE TO PATHOLOGICAL CHANGES

20.2.1 Model of Domestic Mouse (*Mus musculus*)

The domestic mouse is often used in biomedical research, most frequently in two areas; that is, in experimental oncology [19] and experimental research of infectious disorders (i.e., wound and bloodstream infections [20,21], infections of internal organs and joints [21], mastitis [22], otitis media, pneumonia, nephritis and [23], vaginitis [24]). Again, the mouse is a convenient model in the research of colonization of nose mucosa [25], in comparative research of virulence of different staphylococcal strains of staphylococci [26] and in research of skin infections [27], hypodermic and intradermic abscesses [28], and staphylococcal scalded skin syndrome (SSS). Concerning SSS, particularly, experimental models are preferably based on mice litters [29]. Mouse skin (without fur) is similar to human skin, and experiences similar changes as those caused in man by staphylococcal epidermolytic toxins [30,31]. Again, the mouse model discloses new opportunities for studying mechanisms of histopathologic and immunological progression to tissue destruction in chronic arthritis and bone infections [32]. In serological studies, additionally, the mouse model enables the research of specific staphylococcal virulence determinants (i.e., secreted proteinases) based on investigation of sera from patients and infected mice [33,34].

20.2.2 Rat Model (*Rattus rattus*)

Rats are naturally relatively resistant to staphylococcal infections. So, they are not used for research of mechanisms of wounds infections, although they remain preferred models for research in the field of endocarditis [16,17], and convenient models for that of staphylococcal adhesion to plastic surfaces (i.e., catheters) [35]. The rat is, again, a good model for studying experimental infections of bones (osteomyelitis), following both placement of artificial limbs [36] and bacteremic dissemination [37], as well as for investigating pathogenesis and therapy of vaginitis [24] and prevention and treatment of biofilm-associated staphylococcal communities [38].

20.2.3 Hamster Model (*Mesocricetus auratus*)

Cricetine includes small rodents, among which the most popular laboratory animal is the hamster. Two features make it particularly useful for scientific research: a high tolerance to alcohol (relatively to body mass), which enables its use in development of medicaments soluble in alcohol, as well as medicaments addressed to patients addicted to alcohol, and people of high susceptibility to infections, which makes it useful in the study of *Mycobacterium tuberculosis* and *Brucella* infections, as well as flu, rabies, foot-and-mouth disease, and tularemia [39]; and finally, the hamster may be used to investigate mechanisms of tumor diseases [40]. The natural resistance of the hamster to infections is well recognized, and it is known that normal (untreated) hamsters are highly resistant to colonization and infection [41]. Most investigators have utilized 6- to 8-week-old male Syrian (golden) hamsters weighing about 60–100 g with the aim of studying colonization and infections of gastrointestinal tract, less frequently of other organs, as well for research of chemotherapy compounds. Experimental contamination is performed by introducing bacteria into the muzzle, or through the nozzles. Subsequently, the pathological tissue changes develop after 12 weeks. In the field of staphylococci, hamsters are used, particularly, for the study of drugs facing encysting strains in the experimental model of pneumonia [42], in the research of human-hamster transmission of methicillin-resistant *S. aureus* (MRSA) [43], as well as that of inflammation processes and hemorrhagic reactions [44].

20.2.4 Model of Mongolian Gerbil (*Meriones unguiculatus*)

Gerbils from inbred husbandry are used as experimental models in otitis media research [45], as well as for studies of drug farmacokinetic [46,47], postantibiotic effects [48], immunological response in infectious diseases, and efficiency of immunoglobulins when administered in infections with bacterial damage of gastric mucosa due to the infection [49].

20.2.5 Model of Bird—Chicken Embryo (Domestic Hen—*Gallus gallus*)

The model of the chicken embryo is an animal model that is placed between models of vertebrata and invertebrata and is used for research in the field of sepsis. Some researchers, also, treat this model as a particular type of tissue culture, because of its location inside an egg shell and the lack of immunological system expression. This simple model is used with success in research of staphylococcal virulence [50,51], as well as that expressed by other Gram-positive and Gram-negative bacteria [52,53] and *Candida albicans* [54,55]. A crucial difference in discriminating mammal and bird models is the host immunological system. Generally, in fact, sepsis is studied in immunocompetent animals, whereas the adaptive immunological system of a chicken embryo is not activated until a late stage of embryonal development (namely the 17th day of chicken embryo development). Bird embryos, and particularly chicken embryos, are very useful not only in bacteriology, but in cell biology, virology, general biology, tumor biology, and neurology [56,57]. Main features of this embryo are the high similarity to the human embryo, on the molecular, cellular, and anatomical levels, along with fast development (21 days), the availability for visualization and experimental manipulation and, finally, the relatively big size and flat structure it shows in the early stage of development. Furthermore, parameters for optimal chicken embryo incubation are the same as those of most staphylococci. This is the reason why, among several animals used as models of bacterial virulence, the chicken embryo is the best known [58,59]. It offers a wide spectrum of microbe delivery, that is, into yolk, into egg white, into the amniotic-allantois fluid, onto or under the chorion-amniotic membrane, and into an air sack [60]. Last but not least, the chicken embryo model has been gaining novel research potential over the years owing to complete hen genome mapping, and there are nowadays specialized internet databases gathering the knowledge of chicken embryos *G. gallus*, among the other GEISHA (Gallus Expression in Hybridization Analysis—www.geisha.arizona.edu) database and Chicken Genome Resources database supervised by the NCBI (National Center for Biotechnology Information—www.ncbi.nlm.nih.gov).

20.2.6 Model of the Domestic Rabbit (*Oryctolagus cuniculus*)

The rabbit still remains a very suitable model in immunology studies as it produces immunoglobulins under chronic antigen administration. In microbiology, it is a well-known model for infection by *Treponema pallidum* (syphilis). Bacteria *T. pallidum* do not grow in artificial culture media, and require special conditions of growth in rabbit epithelial cells tissue culture. Until 1980, bacteria multiplication coursed under in vivo condition in rabbits, and the in vitro multiplication system was first described by Fieldsteel et al. in 1981 [61]. In the ambit of staphylococcal infections, the rabbit is a suitable model for endocarditis [27,62]. The inoculums is delivered by a catheter, and, for *S. aureus* and *S. epidermidis*, the dose range is 103–104 CFUs (colony forming units) and ≥108 CFUs, respectively [63]. This model also allows studies on methicillin-resistant *S. epidermidis* (MRSE) [64], experimental staphylococcal osteomyelitis [65,66], as well as experimental abscesses and wound infections [67], and investigation of staphylococcal virulence determinants in the context of skin and soft tissue disease [68]. The rabbit is a convenient model for experimental otitis media, and chronic infections of peritoneum and hypodermis connected with polystyrene or steel mesh containers located either in intraperitoneally or hypodermically. This kind of research also allows researchers, furthermore, to gain knowledge about the in vivo antibiotic killing versus *S. aureus* and other bacteria and microbial morphology during infection and therapy [69]. The rabbit is also used to study staphylococcal meningitis [70], endophthalmitis [71], vaginitis, enterotoxin-related diarrhea, and for the evaluation of diet [72].

20.2.7 Guinea Pig Model (*Cavia porcellus*)

Guinea pigs are covered with thick fur, but possess a delicate (after cautious depilation) skin that is suitable for delivering suspensions of staphylococci for studies on abscesses [73] and wound infections by *S. aureus* and *S. epidermidis* [74]. The guinea pig is also a good model for study of otitis media and keratitis, [75]. The species is particularly susceptible to *M. tuberculosis* and *Mycobacterium bovis* infections, therefore it serves nowadays to investigate antituberculosis medicaments. It is a very good model for studies on colitis as well as endotoxemia caused by various aerobic or anaerobic bacteria. Particularly, while postantibiotic dysentery develops in mice and rats, the guinea pig is resistant to toxins, even those released by *Clostridium difficile*; again, this model is useful to study staphylococcal infections that are known to affect all among humans, chinchillas, monkeys, and dogs [41,76].

20.2.8 Domestic Pig Model (*Sus scrofa domestica*)

The skin of the domestic pig, particularly of piglets, optimally reflects the infectious processes occurring in human skin, therefore, it is a useful model for studies on this topic, including local dermal changes, inflammatory reaction, and fibroblast response observed after direct infection of open, nonbacteremic wounds [77]. Also, this model also allows quantitative evaluation of bacterial reproduction in the dermal tissue [78]. The domestic pig is suitable for investigation of different stages of sepsis and sepsis antibiotic therapy [79], as well as for comparative studies of cardio-vascular system experimental infections, particularly those related to vascular prosthesis placement [80].

20.2.9 Dog Model (*Canis familiaris*)

The dog is a useful model for studies on chronic staphylococcal infections. Adult mongrels are mostly used, after antiworm therapy and vaccination against rabies and pacifying (*dis temper*). The dog allows researchers, particularly, to investigate experimental otitis media, pneumonia, endocarditis meningitis [81], and vascular grafts (performed hypodermically or by anastomosis to the infrarenal artery or to the carotid artery). Studies with *S. aureus* and *S. epidermidis* mostly emphasized the important role of their adhesion as a key step in development of vascular and bacteremic infection [82,83].

20.2.10 Nematode Model (*Caenorhabditis elegans*)

The *Caenorhabditis elegans*, because of the transparency and small body size, as well as short life cycle, has become a popular model organism in research of pathogenetic mechanisms of various bacterial species, including *S. aureus* [84–86]. This nematode is a free-living, about 1-mm-long worm, which colonizes soils in moderate climate regions. The use of the *C. elegans* for molecular biology research, begun by Brenner, dates back to 1965 [87,88]. Since then, the *C. elegans* has become a model organism for studying development processes, embryogenesis, morphogenesis, and aging, as well as pathogenic potential and difference in virulence of various *S. aureus* strains, as well as of other bacterial species [89,90].

20.2.11 Insect Model (*Insecta*)

Among insects, the most used model organism for studies on *S. aureus* virulence are fruit-flies (*Drosophila melanogaster*), larvas of the wax moth (*Galleria mellonella*), and silkworms (*Bombyx mori*) [91–93]. However, these models are far from mammalian models, even from chicken embryos and nematodes, concerning physiology. *D. melanogaster* is a very small host, easy and cheap to breed, enabling simple experimental systems for research of host-pathogen interaction and studies on the influence of antibiotic activity mechanisms versus *S. aureus* [94]. This model enables research in the field of MRSA virulence and response to antimicrobials [95].

20.2.12 Fish Model—Zebrafish (*Danio rerio*)

The zebrafish (*Danio rerio*) serves as a model organism in research on development of vertebrata, due to its easy reproduction and short life-cycle; again, it is employed for investigations on mechanisms of bacterial infections. Additionally, as this fish body is transparent in early periods of development (until about the 8th day), it allows observation of internal organs. However, when used with microbes, including *S. aureus*, the optimal temperature for fish, namely 26–28°C, is not optimal, unfortunately, for bacterial growth (that requires 37°C) [96,97].

20.2.13 Other Animal Models

Other animal species that have not been mentioned herein are more rarely used in experimental infections by staphylococci. For instance, for experimental meningitis, dogs, cats, monkeys, and goats are used [98]. Chinchillas and cats find employment in the research of otitis media [99,100], as well as rhesus monkeys and squirrels [101]. The *rhesus* model is also useful in experimental infections following eye surgery [102]. Monkeys are, again, good models for studying gastric poisoning by staphylococcal enterotoxins (after vagotomy) [103]. The sheep model may support investigations in the ambit of osteotomy, that is, at risk of staphylococcal infections (mostly skin species), and in studies on biofilm formation on implant surfaces, as well as on the efficiency of anti-*S. aureus* therapies [104].

20.3 LABORATORY SAFETY AND ETHICAL CONCERNS

Research on bacterial virulence based on animal models is long lasting, slow, and sometimes even boring, but always dangerous, not only due to the manipulation of infectious material (for which the experimenter is prepared), but also owing to additional factors, such as live animals, physiologically active, living under stress, animals suffering pain or other discomforts. Hence, working with animal models requires a special researcher's predisposition, greater than that needed when manipulating microorganisms through in vitro methods. The experimenter should have mental predispositions for work with animals, should be very patient, should like animals, and be aware, finally, that he/she performs manipulation on a living organism, which feels and sometimes suffers. The experimenter also must expect various extreme reactions of animals, be prepared for unexpected behaviors, forecast them in order to prevent risks such as damage of the animal body, self-contamination, and exposure to aggressive animal behavior. Despite the application of analgesic drugs, which is not always possible, and in spite of legal regulations followed in treatment of animals, risk always exists [104,105]. Research in vivo is necessary, and it is conducted in case of unconditional necessity, when study with in vitro or in silico models cannot provide necessary scientific knowledge. Studies with animal models provide the highest scientific results, and are supported by a multidisciplinary approach including pathology, immunology, cytology, histology, biochemistry, clinical laboratory diagnostics, and others. It is an ethical duty for the scientist who works with animal models to use, as well as possible, a full set of all obtained results.

20.4 CONCLUSIONS

Models of mammal infections, especially mice models, are considered by many researchers as the gold standard in studies of host-pathogen interactions. However, use of mammal models is limited by a range of factors, including ethical ones, as well as costs and specialized requirements. Under these conditions, alternative models of hosts, namely evolutionary lower organisms or on early phases of ontogenesis, as invertebrates or bird's embryos, can represent important and alternative instruments for research on staphylococcal virulence [55].

In the field of *S. aureus*, a variety of animal models allows identification and characterization of specific virulence factors responsible for host colonization and infection. Nevertheless, the variety of animal models described in specialized scientific papers led to contradictory experimental results and erroneous conclusions, as the influence of diverse virulence factors differs from one model to another [87,106,107]. Currently, it appears that no animal model for virulence research is suitable for getting conclusions that may be directly applied to humans. Nonetheless, specific aspects of staphylococcal physiology require specific, dedicated research models to be studied. As an example, the rabbit model allowed studies on the Panton-Valentine leukocidin (PVL), while mice permit investigation on skin diseases; in fact, mouse neutrophils are poorly sensitive to PVL if compared with rabbits or man [68,108]. Additionally, it is known that mice show a naturally low susceptibility to bacterial toxins [109] and super-antigens, such as toxic shock syndrome toxin-1 (TSST-1), [110]; also, mouse susceptibility can rise, but only after early endotoxin application [111]. It was also proven that not only a careful choice of animal models, but even that of specific staphylococcal strains is important.

A uniform level of virulence expressed by various staphylococcal strains may be observed with a specific animal model, while virulence variations occurring with other models may reflect the host-pathogen co-evolution, and it has been shown that efficient staphylococcal colonization of and pathogenicity versus a given host relies on specific virulence factors [87,112,113].

Research based on animal models for experimental staphylococcal diseases is crucial and represents a key stage in studies of infection mechanisms and for evaluation of drug efficiency, as well as antibiotic toxicity for the host. Therefore, in vivo research is of paramount importance, and establishes a connection between in vitro studies on bacterial behavior under antimicrobial exposure, and clinical studies in vivo. Animal models of experimental infections are more advanced than ex vivo and mono-parametric models, as they exactly reproduce the infection, allowing researchers to gain knowledge about all among bacterial pathogenicity, the influence of drugs on bacterial pathogens, and host-pathogen/host-drug interactions [2,114]. Animal models are, and will remain, essential for research finalized to fight infectious diseases, and improve mankind's well-being and health.

CONFLICT OF INTEREST

None declared.

ACKNOWLEDGMENTS

This work was supported in part by the grant no. N N401 017740 from the National Science Centre, Poland.

REFERENCES

[1] Zak O. Scope and limitations of experimental chemotherapy. Experientia 1980;36(1):479–83.

[2] Sandle MA. Animal models in the evaluation of antimicrobial agents. Infect Dis 1981;4(11):4–20.

[3] Chang S, Sievert DM, Hageman JC, et al. Infection with vancomycin-resistant *Staphylococcus aureus* containing the *vanA* resistance gene. N Engl J Med 2003;348(14):1342–7.

[4] Ito T, Okuma K, Ma XX, et al. Insights on antibiotic resistance of *Staphylococcus aureus* from its whole genome: genomic island SCC. Drug Resist Updat 2003;6(1):41–52.

[5] Krzyszton-Russjan J, Gniadkowski M, Polowniak-Pracka H, et al. The first *Staphylococcus aureus* isolates with reduced susceptibility to vancomycin in Poland. J Antimicrob Chemother 2002;50(6):1065–9.

[6] Fredericks DN, Relman DA. Sequence-based identification of microbial pathogens: a reconsideration of Koch's postulates. Clin Microbiol Rev 1996;9(1):18–33.

[7] West SA, Griffin AS, Gardner A, Diggle SP. Social evolution theory for microorganisms. Nat Rev Microbiol 2006;4(8):597–607.

[8] Binek M. A glance at Koch's postulates after 100 years of his death. Postepy Mikrobiol 2010;49(3):157–64.

[9] Koch R. Die Aetiologie der Tuberkulose. Mitt Kaiser Gesundh 1884;1(1):1–88.

[10] Inglis TJ. Principia aetiologica: taking causality beyond Koch's postulates. J Med Microbiol 2007;56(1):1419–22.

[11] Miedzobrodzki J, Malachowa N, Markiewski T, et al. Differentiation of *Staphylococcus aureus* isolates based on phenotypical characters. Postepy Hig Med Dosw 2008;62(1):322–7.

[12] Savini V, Carretto E, Polilli E, et al. Small colony variant of methicillin-resistant *Staphylococcus pseudintermedius* ST71 presenting as a sticky phenotype. J Clin Microbiol 2014;52(4):1225–7.

[13] Falkov S. Molecular Koch's postulates applied to bacterial pathogenicity—a personal recollection 15 years later. Nat Rev Microbiol 2004;2(1):67–72.

[14] Garcia-Lara J, Needham AJ, Foster SJ. Invertebrates as animal models for *Staphylococcus aureus* pathogenesis: a window into host-pathogen interaction. FEMS Immunol Med Microbiol 2005;43(3):311–23.

[15] Sung JM, Lloyd DH, Lindsey JA. *Staphylococcus aureus* host specificity: comparative genomics of human versus animal isolates by multi-strain microarray. Microbiology 2008;154(7):1949–59.

[16] Santoro J, Levison ME. Rat model of experimental *endocarditis*. Infect Immun 1978;19(3):915–8.

[17] Malachowa N, Kohler PL, Schlievert PM, et al. Characterization of *Staphylococcus aureus* surface virulence factor that promotes resistance to oxidative killing and infectious *endocarditis*. Infect Immun 2011;79(1):342–52.

[18] McCormick JK, Yarwood JM, Schlievert PM. Toxic shock syndrome and bacterial superantigens: an update. Annu Rev Microbiol 2001;55(55):77–104.

[19] Grabacka M, Placha W, Urbanska K, et al. PPAR gamma- regulates MITF and beta-catenin expression and promotes a differentiated phenotype in mouse melanoma S91. Pigment Cell Melanoma Res 2008;21(1):388–96.

[20] Bunce C, Wheeler L, Reed G, et al. Murine model of cutaneous infection with Gram-positive cocci. Infect Immun 1992;60(7):2636–40.

[21] Tarkowski A, Collins LV, Gjertsson I, et al. Model systems: modeling human staphylococcal *arthritis* and *sepsis* in the mouse. Trends Microbiol 2001;9(7):321–6.

[22] Brouillette E, Grondin G, Lefebvre C, et al. Mouse *mastitis* model of infection for antimicrobial compound efficacy studies against intracellular and extracellular forms of *Staphylococcus aureus*. Vet Microbiol 2004;101(4):253–62.

[23] Weiss WJ, Lenoy E, Murphy T, et al. Effect of *srtA* and *srt*B gene expression on the virulence of *Staphylococcus aureus* in animal models of infection. J Antimicrob Chemother 2004;53(3):480–6.

[24] Polak A. Experimental *vaginitis*. In: Zak O, Sande MA, editors. Experimental models in antimicrobial chemotherapy. vol. 1. London: Academic Press Inc.; 1986. p. 21–42.

[25] Kiser KB, Cantey-Kiser JM, Lee JC. Development and characterization of a *Staphylococcus aureus* nasal colonization model in mice. Infect Immun 1999;67(10):5001–6.

[26] Miedzobrodzki J, Tadeusiewicz R, Heczko PB. Virulence of coagulase-negative staphylococci for chick embryos. Zbl Bakt Int J Med Microbiol Suppl 1985;14:477–80.

[27] Malachowa N, Kobayashi SD, Braughton KR, et al. Mouse model of *Staphylococcus aureus* skin infection. Methods Mol Biol 2013;1031:109–16.

[28] Ford CW, Hamel JC, Stapert D, et al. Antibiotic therapy of an experimental *S. epidermidis* subcuteneus abscess in mice. In: Pulverer G, Quie PG, Peters G, editors. Pathogenicity and clinical significance of coagulase-negative staphylococci. Stuttgart, New York: Gustav F; 1987. p. 247–58.

[29] Melisch ME, Glasgow LA. The *staphylococcal* scalded skin syndrome—development of an experimental model. N Engl J Med 1970;282(20):1114–9.

[30] Kinsman OS. Models for staphylococcal colonization and skin infections. In: Zak O, Sande MA, editors. Experimental models in antimicrobial chemotherapy. vol. 2. London: Academic Press Inc.; 1986. p. 135–46.

[31] Prabhakara R, Foreman O, DePascalis R, et al. Epicutaneous model of community-acquired *Staphylococcus aureus* skin infections. Infect Immun 2013;81(4):1306–15.

[32] Bremell T, Lange S, Yacoub A, et al. Experimental *Staphylococcus aureus arthritis* in mice. Infect Immun 1991;59(1):2615–23.

[33] Naidu AS, Schalen C, Flock JI, et al. Serological variation in the fibronectin binding to protein-A-deficient mutants of *Staphylococcus aureus*. In: Wadstrom T, Eliasson I, Holder I, Ljungh A, editors. Pathogenesis of wound and biomaterial-associated infections. London: Springer Verlag; 1990. p. 353–60.

[34] Zdzalik M, Karim AY, Wolski K, et al. Prevalence of genes encoding extracellular proteases in *Staphylococcus aureus*—import ant target triggering immune response in vivo. FEMS Immunol Med Microbiol 2012;66(2):220–9.

[35] Rupp ME, Ulphani JS, Fey PD, et al. Characterization of *Staphylococcus epidermidis* polysaccharide adhesion/hemagglutinin in the pathogenesis of intravascular catheter-associated infection in rat. Infect Immun 1999;67(5):2656–9.

[36] Soe NH, Jensen NV, Nurnberg BM, et al. A novel knee prosthesis model of implant-related *osteomyelitis* in rats. Acta Orthop 2013;84(1):92–7.

[37] Veloso TR, Chaouch A, Roger T, et al. Use of human-like low-grade bacteremia model of experimental *endocarditis* to study the role of *Staphylococcus aureus* adhesins and platelet aggregation in early *endocarditis*. Infect Immun 2013;81(3):697–703.

[38] Simonetti O, Cirioni O, Mocchegiani F, et al. The efficacy of the quorum sensing inhibitor FS8 and tigecycline in preventing prosthesis in an animal model of staphylococcal infection. Int J Mol Sci 2013;14(8):16321–32.

[39] Gibson SV, Brady AG. Syrian hamsters: bacterial and mycotic diseases. Laboratory Animal Medicine Science. Series II. Washington: University of Washington; 2000.

[40] Urbanska K, Romanowska-Dixon B, Elas M, et al. Experimental ruthenium plaque of amelanotic and melanotic melanomas in the hamster eye. Melanoma Res 2000;10(1):26–35.

[41] Fekety R. Animal models of antibiotic-induced colitis. In: Zak O, Sande MA, editors. Experimental models in antimicrobial chemotherapy. vol. 2. London: Academic Press Inc.; 1986. p. 61–72.

[42] Verghese A, Haire C, Franzus B, et al. LY146032 in a hamster model of *Staphylococcus aureus* pneumonia—effect on in vitro clearance and mortality and in vitro opsonophagocytic killing. Chemotherapy 1988;34(6):497–503.

[43] Ferreira JP, Fowler VG, Correra MT, et al. Transmission of methicillin-resistant *Staphylococcus aureus* between human and hamster. J Clin Microbiol 2011;49(4):1679–80.

[44] Gustafson G. Hemorragic reactions in the hamster produced by interaction of protein A from *Staphylococcus aureus* human gamma-globulin and endotoxin. Acta Pathol Microbiol Immunol Scand 1968;74(1):127–38.

[45] Worthington JM, Fulghum RS. Cecal and fecal bacterial flora of the Mongolian gerbil and the chinchilla. Appl Environ Microbiol 1988;54(5):1210–5.

[46] Girard AE, Girard D, English AR, et al. Pharmacokinetic and in vivo studies with azithromycin (CP-62,993), a new macrolide with an external half-life and excellent tissue distribution. Antimicrob Agents Chemother 1987;31(12):1948–54.

[47] Doyle WJ. Animal models of *otitis media*: other pathogens. Pediatr Infect Dis J 1989;8(1):48–94.

[48] MacGowan A. Pharmacokinetic and pharmacodynamic profile of linezolid in healthy volunteers and patients with Gram-positive infections. J Antimicrob Chemother 2003;51(2):7–25.

[49] Shin JH, Yang M, Nam SW, et al. Use of egg yolk-derived immunoglobulin as an alternative to antibiotic treatment for control of *Helicobacter pylori* infection. Clin Vaccine Immunol 2002;9(5):1061–6.

[50] McCabe WR. Studies of staphylococcal infections. I. Virulence of *staphylococci* and characteristics of infections in embryonated eggs. J Clin Invest 1964;43(1):2146–57.

[51] Miedzobrodzki J, Tadeusiewicz R. Estimation of staphylococcal pathogenicity for chick embryos. Acta Biol Cracov Ser Bot 1988;1(1):1–8.

[52] Aly R, Maibach HI, Shinefield HR, et al. Protection of chicken embryos by viridans streptococci against the lethal effect of *Staphylococcus aureus*. Infect Immun 1974;9(3):559–63.

[53] Nix EB, Cheung KKM, Wang D, et al. Virulence of *Francisella* spp. in chicken embryos. Infect Immun 2006;74(8):4809–16.

[54] Norris RF, Shorey WK, Bongiovanni AM. Lesions produced in chick embryos by *Candida (Monilia) albicans*. Arch Pathol Lab Med 1948;45(4):506–12.

[55] Jacobsen ID, Grosse K, Berndt A, et al. Pathogenesis of *Candida albicans* infections in the alternative chorio-allantoic membrane chicken embryo model resembles systemic murine infections. PLoS ONE 2011;6(5):e19741.

[56] Endo Y. Chick embryo culture and electroporation. Curr Protoc Cell Biol 2012;19(1):15–9.

[57] Mangioris G, Chiodini F, Dosso A. New strategy to study corneal endothelial cell transplantation: the chick cornea model. Cornea 2011;30(12):1461–4.

[58] Fergelot P, Bernhard JC, Soulet F, et al. The experimental renal cell carcinoma model in the chick embryo. Angiogenesis 2013;16(1):181–94.

[59] Ribatti D. Chicken chorioallantoic membrane angiogenesis model. Methods Mol Biol 2012;843(1):47–57.

[60] Vergara MN, Canto-Soler MV. Rediscovering the chick embryo as a model to study retinal development. Neural Dev 2012;7(1):22–5.

[61] Fieldsteel AH, Cox DL, Moeckli RA. Further studies on replication of virulent *Treponema pallidum* in tissue cultures of Sf1Ep cells. Infect Immun 1982;35(2):449–55.

[62] Perlman BB, Freeman LR. Experimental *endocarditis*. II. Staphylococcal infection of the aortic valve following placement of a polyethylene catheter in the left side of the heart. Yale J Biol Med 1971;44(2):206–13.

[63] Freedman LR, Valone JR. Experimental infective *endocarditis*. Prog Cardiovasc Dis 1979;12(1):169–80.

[64] Drake A, Sande MA. Experimental *endocarditis*. In: Zak O, Sande MA, editors. Experimental models in antimicrobial chemotherapy. vol. 1. London: Academic Press Inc.; 1986. p. 257–84.

[65] Mader JT, Brown GL, Guckian JC, et al. A mechanism for the amelioration by hyperbaric oxygen of experimental staphylococcal *osteomyelitis* in rabbits. J Infect Dis 1980;142(6):915–22.

[66] Smeltzer MS, Thomas JR, Hickmon SG, et al. Characterization of rabbit model of staphylococcal *osteomyelitis*. J Orthop Res 1997;15(1):414–21.

[67] Wills QF, Kerrigan CS, Soothill JS. Experimental bacteriophage protection against *Staphylococcus aureus* abscesses in a rabbit model. Antimicrob Agents Chemother 2005;49(3):1220–1.

[68] Kobayashi SD, Malachowa N, Whitney AR, et al. Comparative analysis of USA300 virulence determinants in a rabbit model of skin and soft tissue infection. J Infect Dis 2011;204(6):937–41.

[69] Bruhin H. Principles of animal care. In: Zak O, Sande MA, editors. Experimental models in antimicrobial chemotherapy. vol. 1. London: Academic Press Inc.; 1986. p. 7–18.

[70] Scheld WM. Experimental animal models of bacterial *Meningitis*. In: Zak O, Sande ME, editors. Experimental models in antimicrobial chemotherapy. vol. 1. London: Academic Press; 1986. p. 139–86.

[71] Barza M, Kane A, Baum J. Ocular penetration of subgonjunctival oxacillin, methicillin, and cefazolin in rabbit with staphylococcal *endophtalmitis*. J Infect Dis 1982;145:899–903.

[72] Hayama T, Sugiyama H. Diarrhea in cecectomized rabbits induced by staphylococcal enterotoxin. Proc Soc Exp Biol Med 1964;117(1):115–8.

[73] Miedzobrodzki J, Tadeusiewicz R. Staphylococcal dermonecrotic reactions in ginea pigs. Int J Biomed Comput 1987;21:67–74.

[74] Kernodle DS, Gates H, Kaiser AB. Prophylactic anti-infected activity of poly-[1-6]-beta-D-glucopyranosyl[1-3]-beta-D-glucopyranose glucan in a guinea pig model of staphylococcal wound infection. Antimicrob Agents Chemother 1998;42(3):545–9.

[75] Vijay AK, Saukaridurg P, Zhu H, et al. Guini pigs models of acute *keratitis* responses. Cornea 2009;28(10):1153–9.

[76] Baldoni D, Furustrand Tafin U, Aeppli S, Angevaare E, Oliva A, Haschke M, et al. Activity of dalbavancin, alone and in combination with rifampicin, against meticillin-resistant *Staphylococcus aureus* in a foreign-body infection model. Int J Antimicrob Agents 2013;42(3):220–5.

[77] Svedman P, Ljungh A, Rausing A, et al. Staphylococcal wound infection in the pig. Part I. Course. Ann Plast Surg 1989;23(3):212–8.

[78] Sandén G, Ljungh A, Wadström T, et al. Staphylococcal wound infection in the pig. Part II. Inoculation, quantification of bacteria, and reproducibility. Ann Plast Surg 1989;23(3):219–23.

[79] Sauer M, Altrichter J, Mencke T, et al. Role of different replacement fluids during extracorporeal treatment in a pig model of sepsis. Ther Apher Dial 2013;17(1):84–92.

[80] Johnson JJ, Jacocks AM, Gauthier SC, et al. Establishing a swine model to compare vascular prostheses in a contamineted field. J Surg Res 2013;181(2):355–8.

[81] Pennington JE. Use of animal models to evaluate antimicrobial therapy for bacterial pneumonia. In: Zak O, Sande ME, editors. Experimental models in antimicrobial chemotherapy. vol. 1. London: Academic Press Inc.; 1986. p. 237–56.

[82] Moore WS, Chvapil M, Seiffert G, et al. Development of an infection-resistant vascular *prosthersis*. Arch Surg 1981;116(1):1403–7.

[83] Rutladge R, Burnham SJ, Johnson GR. The use of animal models in studying vacular graft infection. In: Zak O, Sande ME, editors. Academic experimental models in antimicrobial chemotherapy. vol. 1. London: Academic Press; 1986. p. 285–93.

[84] Bae T, Banger AK, Wallace A, et al. *Staphylococcus aureus* virulence genes identified by bursa aurealis mutagenesis and nematode killing. PNAS 2004;101(33):12312–7.

[85] Garsin DA, Sifri CD, Mylonakis E, et al. A simple model host for identifying Gram-positive virulence factors. PNAS 2001;98(19):10892–7.

[86] Irazoqui JE, Troemel ER, Feinbaum RL, et al. Distinct pathogenesis and host responses during infection of *C. elegans* by *P. aeruginosa* and *S. aureus*. PLoS Pathog 2010;6(1):e1000982.

[87] Polakowska K, Lis MW, Helbin WM, et al. The virulence of *Staphylococcus aureus* correlates with strain genotype in a chicken embryo model but not a nematode model. Microbes Infect 2012;14(14):1352–62.

[88] Brenner S. The genetics of *Caenorhabditis elegans*. Genetics 1974;77(1):71–94.

[89] Sifri CD, Baresch-Bernal A, Calderwood SB, et al. Virulence of *Staphylococcus aureus* small colony variants in the *Caenorhabditis elegans* infection model. Infect Immun 2006;74(2):1091–6.

[90] Bogaerts A, Beets I, Temmerman L, et al. Proteome changes of *Caenorhabditis elegans* upon a *Staphylococcus aureus* infection. Biol Direct 2010;5(1):11–6.

[91] Desbois AP, Coote PJ. Wax moth larva (*Galleria mellonella*): an in vivo model for assessing the efficacy of antistaphylococcal agents. J Antimicrob Chemother 2011;66(8):1785–90.

[92] Kaito C, Akimitsu N, Watanabe H, et al. Silkworm larvae as an animal model of bacterial infection pathogenic to humans. Microb Pathog 2002;32(4):183–90.

[93] Needham AJ, Kibart M, Crossley H, et al. *Drosophila melanogaster* as a model host for *Staphylococcus aureus* infection. Microbiology 2004;150(7):2347–55.

[94] Diaz L, Kontoyiannis DP, Panesso D, et al. Dissecting the mechanisms of linezolid resistance in a *Drosophila melanogaster* infection model of *Staphylococcus aureus*. J Infect Dis 2013;208(1):83–91.

[95] Ben-Ami R, Watson CC, Lewic RE, et al. *Drosophila melanogaster* as a model to explore the effects of MRSA strain type on virulence and response to linezolid treatment. Microb Pathog 2013;55:16–20.

[96] Li Y, Hu B. Establishment of multi-site infection model in zebrafish larvae for studying *Staphylococcus aureus* infectious disease. J Genet Genomics 2012;39(9):521–34.

[97] Prajsnar TK, Hamilton R, Garcia-Lara J, et al. A privileged intraphagocyte niche is responsible for disseminated infection of *Staphylococcus aureus* in a zebrafish model. Cell Microbiol 2012;14(10):1600–19.

[98] Scheld WM. Experimental animal models of bacterial *Meningitis*. In: Zak O, Sande ME, editors. Experimental models in antimicrobial chemotherapy. vol. 1. London: Academic Press; 1986. p. 139–86.

[99] Spector BC, Reinisch L, Smith D, et al. Noninvasive-fluorescent identification of bacteria causing acute *otitis media* in a chinchilla model. Laryngoscope 2000;110(7):1119–23.

[100] Taylor WM, Grady AW. Catheter-tract infections in rhesus macaques (*Macaca mulatta*) with indwelling intravenous catheters. Lab Anim Sci 1998;48(5):448–54.

[101] Goycoolea MV, Paparella MW, Carpenter AM, et al. A longitudinal study of cellular changes in experimental *otitis media*. Otolaryngol Head Neck Surg 1979;87(5):685–700.

[102] Bever TL, Vogler G, Sharma D, et al. Protective barrier effect of the posterior lens capsule in exogenous bacterial *endophtalmitis*—an experimental primate study. Invest Opthalmol Vis Sci 1984;25(1):108–12.

[103] Sugiyama H, Hayama T. Comparative resistance of vagotomized monkeys to intravenous vs. intragastric staphylococcal enterotoxin challenges. Proc Soc Exp Biol Med 1964;115:243–6.

[104] Stewart S, Barr S, Engiles J, et al. Vancomycin modified implant surface inhibits biofilm formation and supports bone-healing in an infected osteotomy model in sheep: a prof-of-concept study. J Bone Joint Surg Am 2012;94(15):1406–15.

[105] European Convention for the Protection of Vertebrate Animals used for Experimental and Other Scientific Purposes, European Treaty Series No. 123, Council of Europe, 1991.

[106] Darouiche RO, Landon GC, Patti JM, et al. Role of *Staphylococcus aureus* surface adhesins in orthopaedic device infections: are results model-dependent? J Med Microbiol 1997;46(1):75–9.

[107] Polakowska K. Studies of genetic and phenotypic determinants of host-specific virulence of *Staphylococcus aureus* strains using in vivo models. PhD Thesis, Kraków: Jagiellonian University; 2013.

[108] Lipinska U, Hermans K, Meulemans L, et al. Panton-valentine leukocidin does play a role in the early stage of *Staphylococcus aureus* skin infections: a rabbit model. PLoS ONE 2011;6(8):e22864.

[109] Peavy DL, Adler WH, Smith RT. The mitogenic effects of endotoxin and staphylococcal enterotoxin B on mouse spleen cells and human peripheral lymphocytes. J Immunol 1970;105(6):1453–8.

[110] Herbert S, Ziebandt A-K, Ohlsen K, et al. Repair of global regulators in *Staphylococcus aureus* 8325 and comparative analysis with other clinical isolates. Infect Immun 2010;78(6):2877–89.

[111] Stiles BG, Campbell YG, Castle RM, et al. Correlation of temperature and toxicity in murine studies of staphylococcal enterotoxins and toxic shock syndrome toxin 1. Infect Immun 1999;67(3):1521–5.

[112] Sung JM, Lloyd DH, Lindsey JA. *Staphylococcus aureus* host specificity: comparative genomics of human versus animal isolates by multi-strain microarray. Microbiology 2008;154(7):1949–59.

[113] Herron-Olson L, Fitzgerald JR, Musser JM, et al. Molecular correlates of host specialization in *Staphylococcus aureus*. PLoS ONE 2007;2(10):e1120.

[114] Zak O. Scope and limitations of experimental chemotherapy. Experientia 1980;36(1):479–83.

Chapter 21

Application of Staphylococci in the Food Industry and Biotechnology

Benedykt Władyka, Emilia Bonar

Department of Analytical Biochemistry, Faculty of Biochemistry, Biophysics and Biotechnology, Jagiellonian University, Krakow, Poland

21.1 INTRODUCTION

Bacteria belonging to the genus *Staphylococcus* are generally considered as human and animal-associated commensals as well as opportunistic pathogens. In particular, much attention has been given to their behavior as agents of disease due to increasing antibiotic resistance, the reported cases of host jumps, and the selection and global spread of highly virulent strains. Diverse aspects supporting the ecological success of these bacteria as well as their efficiency in host colonization and infection are described throughout chapters of this volume. However, it should be emphasized that association of staphylococci with both humans and animals, along with the vast inter- and intra-species diversity of strains, have led to the application of these organisms in certain areas of food industry and modern biotechnology. It is striking that different staphylococcal strains may act as agents of severe infections or harmless microorganisms involved in the development of a pleasant aroma in sausages or cheeses. Moreover, some proteins primarily considered as virulence factors, such as lipase, nuclease, protease, and protein A, have found applications in modern biotechnology as biocatalysts and tools for molecular biology. Finally, *Staphylococcus carnosus* is used as a host for expression of recombinant proteins, especially in a powerful cell display system.

21.2 STAPHYLOCOCCI IN THE FOOD INDUSTRY

The genus *Staphylococcus* comprises more than 50 validly described species that are grouped traditionally into coagulase-positive staphylococci (CoPS) and coagulase-negative staphylococci (CoNS). Generally, CoPS are considered to be pathogenic and/or toxigenic organisms causing infections and food intoxications,

Pet-to-Man Travelling Staphylococci: A World in Progress. https://doi.org/10.1016/B978-0-12-813547-1.00021-2

whereas some CoNS (*Staphylococcus epidermidis, Staphylococcus haemolyticus, Staphylococcus saprophyticus,* and *Staphylococcus warneri*) are regarded as opportunistic pathogens; some others, again, such as *Staphylococcus xylosus* and *S. carnosus* are described as completely apathogenic, and even regarded as safe (GRAS status; Generally Regarded as Safe). The list of staphylococcal species isolated from food products is long and contains CoPS and CoNS species [1–4]. Nevertheless, this chapter only focuses on those involved in food production, while spoilage and intoxication are outside the scope of this section.

Production of fermented products is one of the oldest and best known food processes for humans. Although nowadays fermentation has become an industrial process controlled by the use of defined starter cultures, still, some traditional fermentation processes remain in use. These mainly apply to traditional dry fermented sausages and some cheeses that are produced without the addition of selected starter cultures. Fermentation of these products only relies on the indigenous microflora whose composition is variable and depends in part on contamination of raw materials as well as environmental conditions modulating the bacterial charge during the process. However, bacteria that have gained fundamental biotechnological interest belong to two groups, that is, lactic acid bacteria (LAB) and CoNS. LAB are responsible for pH decrease (mainly through lactic acid production), that ensures the stability of fermented products by preventing growth of pathogens. CoNS contribute to the development and stability of the product color and, due to their antioxidant activities, avoid rancidity. Moreover, they enhance the flavor mainly through amino and fatty acids degradation. Although CoNS are not routinely identified to the species level, dedicated studies on microflora of dry fermented sausages and cheese revealed a huge diversity of staphylococcal species and strains associated with these products. Generally, the cheese staphylococcal microflora is more diverse than that related to fermented meat. Cotton et al., by using an array of molecular biology methods, have identified 15 species in one hundred samples isolated from French cheese, mostly *Staphylococcus equorum* (22%), *S. xylosus* (20%), *S. saprophyticus* (14%), *S. epidermidis* (11%) and *Staphylococcus lentus* (9%). Other isolates, comprising only a minority of samples, were identified as *Staphylococcus succinus, S. haemolyticus, S. warneri, Staphylococcus vitulinus, Staphylococcus hominis, Staphylococcus sciuri, Staphylococcus simulans, Staphylococcus chromogenes, Staphylococcus auricularis,* and *Staphylococcus capitis.* For dry sausage samples (128 isolates), 8 species were identified, with *S. xylosus* (41.4%), *S. equorum* (35.2%) and *S. saprophiticus* (10.9%) being the most prevalent. Conversely, *S. epidermidis, S. succinus, S. carnosus, S. sciuri,* and *S. warneri* were much less abundant [5]. However, the strains and counts of these microorganisms (staphylococci) in meat products depend on the raw materials used, the process conditions, in particular pH and water activity, and also correlate with geographical regions. For example, in Italian sausages, the microflora is mainly composed of *S. xylosus, S. saprophyticus,* and *S. equorum* [6]. Spanish low-acid chorizos are dominated, conversely, by three species, that is, *S. xylosus*, *S. carnosus*, and *S. epidermidis* [7]. However, in dry-cured ham, more than 70% of isolates were identified as *S. equorum* [8]. In spontaneously fermented Swiss meat products, *S. equorum* was prevalent in frequency and cell counts during maturation and in the final products, followed by *S. warneri*, *S. saprophyticus, S. epidermidis*, and *S. xylosus*, whereas in Slovak traditional sausages, *S. xylosus* and *S. carnosus* were identified as the predominant species [9,10]. This diversity in staphylococcal microflora of traditionally fermented meat is often responsible for the uniqueness of the products. However, the expectations for products with repeatable hygienic and organoleptic qualities, as well as the need to shorten the production time, support the use of highly defined starter cultures, such as the *S. carnosus* strain TM300.

The biodiversity of food-associated CoNS is investigated to identify strains and related properties that may serve as defined components of starter cultures in controlled industrial production of fermented meat products. The strains are screened, among others, for nitrate reductase, catalase, proteolitic and lipolytic activities, antibiotic susceptibility, and biogenic amine production. These activities are important in the development of the aroma, color, flavor, and texture, as well as in ensuring the safety of the final fermented products. The high nitrate reductase activity is desirable because the enzyme catalyzes the reduction of nitrate to nitrite, which is important for the formation of nitrosylmyoglobin, which is responsible for the red color in the fermented meat products. Furthermore, nitrate reduction produces nitrite that can limit lipid oxidation. The catalase decomposes hydrogen peroxide that otherwise could oxidize unsaturated fatty acids. The activity of protease and lipase during the processing of dry fermented sausages contributes to the generation of peptides, free amino acids, and free fatty acids. These compounds are known to be involved in the overall flavor of cured meat products by their taste and also as precursors of other volatile aroma compounds. On the other hand, resistance to antibiotics is considered as a potential threat because the resistance genes may be further transmitted to human-associated strains via the food chain. Studies investigating food-associated staphylococcal strains' susceptibility to drugs revealed the occurrence of susceptible as well as mono- and multi-drug resistance strains. Marty and co-workers reported that nearly half of 132 isolates from spontaneously fermented meat products showed resistance to at least one antibiotic. Resistance to penicillins was most frequently detected, followed by tetracycline. All isolates were susceptible to clinically important antibiotics such as vancomycin, gentamycin, kanamycin, chloramphenicol, rifampicin, and mupirocin. Interestingly, although all six commercial *S. carnosus* starter strains tested were susceptible to antibiotics, six of ten commercial *S. xylosus* starter strains were resistant to tetracycline and two to ampicilin [11]. A similar incidence was also reported in other studies [12]. Antibiotic resistance among staphylococcal strains collected at different stages of cheese manufacture was slightly higher (almost 60%), half of them being multidrug-resistant with each of *S. equorum* and *S. saprophyticus* strains showing resistance to up to six antibiotics [13]. An even higher incidence of antibiotic resistances have been reported among CoNS strains from sausage (83%), and hard and soft cheese (87%) [14]. The potential production of biogenic amines (BA) is among the food safety hazards associated with CoNS. BA are generated in various fermented products through amino acid decarboxylation. When consumed in sufficient amount, some BA lead to food poisoning. BA production by CoNS in fermented products is poorly documented. It has never been described in dairy products, and only few studies reported BA production by CoNS from meat products [15,16]. Even and co-workers reported that only 5 out of 129 strains were able to produce detectable amounts of BA, indicating that BA production is not widespread among CoNS, but should not be neglected [12]. This agrees with guidelines concerning starters safety introduced by the European Food Safety Authority for microorganisms deliberately applied to the food chain [17]. The major concerns that had been highlighted in the concept of Qualified Presumption of Safety is the presence of transferable antibiotic resistance genes in the starter genomes, as well as the production of toxins or biogenic amines in the end-products.

The growing number of bacterial genome sequences indicate that substantial attention is given to virulent and antibiotic-resistant staphylococci; however, some food-associated staphylococcal genomes have also been published [18–20]. The genome of the widely used meat starter culture bacterium *S. carnosus* TM300 comprises 2.57 Mb, with predicted 2462 coding sequences, and thus belongs to the smallest sequenced staphylococcal genomes. The comparison of its genome sequence with those of pathogenic staphylococcal species immediately shows remarkable differences in the content of mobile genetic elements (MGEs). *S. carnosus* shows a conspicuous lack of MGEs carrying only remnants of a prophage and a genomic island in its genome. Moreover, the genome carries a number of mutationally inactivated genes, including those of the global regulatory systems *agr* and *sae*. This, most probably, is responsible for the lack of pathogenicity despite the presence of various intact genes encoding proteins that are similar to virulence factors such as lipases, proteases, and nucleases. In general, *S. carnosus* shows no significant differences if compared with other staphylococci concerning the encoded metabolic pathways. However, it is distinguished by a comparably high content of osmoprotective factors, which is consistent with its usage in starter cultures during the ripening of raw sausage under high salt conditions [21].

21.3 THE APPLICATION OF STAPHYLOCOCCAL PROTEINS IN MODERN BIOTECHNOLOGY

The undoubtful ecological success of staphylococci, among others, relies on the unique ability to produce a range of extracellular factors that interact with the host and other bacteria. Interestingly, lipases, nucleases, proteases, and protein A, although primarily recognized as virulence factors, found plenty of applications in modern biotechnology as biocatalysts and tools for molecular biology. Moreover, staphylococcal bacteriocins are considered as potential food-protecting agents against food-borne pathogens.

The extracellular staphylococcal lipases (EC 3.1.1.3) are important proteins for modern biotechnology [22] that hydrolyze emulsions of lipids with long-chain fatty acids [23]. From a chemical point of view, these enzymes are able to catalyze reactions including alcoholysis, transesterification, esterification, and interesterification, even in nonaqueous media, if immobilized. A further advantage of staphylococcal lipases is their activity in acidic pH. On the industrial scale, staphylococcal lipases are employed in production of pharmaceuticals, detergents, food, antioxidants, esters, biopolymers, agrochemicals, cosmetics, and flavor compounds [24]. For example, they are applied to esters' synthesis, such as that of isoamyl acetate and ethyl valerate exhibiting banana and green apple flavor, respectively. Other esters (e.g., tyrosol acetate and propyl gallate), act as antioxidants and are widely used in processed food, cosmetics, and food packing materials. Lipases are applied in the production of biodegradable compounds such as butyl oleate, which acts as biodiesel fuel, and starch esters used as biopolymeric, fully biodegradable thermoplastic material [24].

Staphylococcal nuclease (SNase), primarily recognized as a virulence factor that participates in *S. aureus* dissemination in the infected host, found its application in molecular biology and modern microbial diagnostics. SNase is a secreted nucleic acid degrading enzyme that is used for the digestion of chromatin with the aim to identify the location of nucleosomes [25]. Moreover, the enzyme is employed in the large-scale production of recombinant proteins to clear cellular extracts and improve protein purification through the hydrolysis of chromosomal and plasmid DNA and RNA. Cooke and colleagues used a modified *Escherichia coli* strain carrying a staphylococcal nuclease expression cassette. The recombinant nuclease was translocated to the periplasm and, when released during cell lysis, it hydrolyzed host nucleic acids present in the lysate. This resulted in the decreased lysate viscosity. Then, lower lysate viscosity improved the product yield and avoided the addition of exogenous nuclease, thus lowering costs of large-scale recombinant proteins production [26]. Finally, the SNase gene may also act as a marker in the PCR-based procedures for *S. aureus* detection in food samples [27].

Staphylococcal proteases also play a relevant role in the field of biotechnology. The most commonly employed is the exoprotease, referred to as glutamylendopeptidase, otherwise known as V8 protease, GluV8, or SspA. V8 protease (EC 3.4.21.19) belongs to the class of serine proteases, subfamily of chymotrypsin-like proteolytic enzymes. It specifically cleaves the peptide bonds after the negatively charged residues of glutamic acid (Glu) and/or aspartic acid (Asp), although kinetic studies revealed a higher preference for the former [28]. Because glutamylendopeptidase possesses a very narrow substrate specificity and is resistant to detergents, it is widely used for proteome analysis and protein identification with mass spectrometry. However, the protease is also useful as a catalyst in reactions of peptide synthesis [29,30].

Other staphylococcal enzymes of potential biotechnological applications are a debittering enzyme, naringinase, that is used in the industrial processing of citrus fruit juice [31,32], *S. aureus* sortase A (SrtA), a transpeptidase that naturally covalently anchors cell surface proteins to the cell wall, and may be applied to specific protein modification and immobilization in sortase-mediated protein ligation [33].

Staphylococcal protein A (SPA) is a surface, cell-wall protein expressed by the majority of *S. aureus* strains. SPA, unlike the previously mentioned proteins, is not an enzyme, but an affinity protein capable of interacting with immunoglobulin gamma (IgG) molecules. SPA binds immunoglobulins specifically, with very high affinity and without interrupting their antigen-binding ability. This attribute immediately found application in modern biotechnology. SPA has been used for development of many immunological tools for easy, quick, and inexpensive immunoglobulin purification and analysis. In its immobilized form, covalently bounded to a stationary phase chromatography resin, it is used as a ligand in affinity chromatography [34,35], which constitutes the main step in the purification of antibodies for diagnostic and therapeutic applications [36].

Overuse of antibiotics and the subsequent increased incidence of multi-drug resistant bacteria turned attention to alternative antimicrobial agents. Staphylococcal bacteriocins (staphylococcins) are among compounds showing potential utility in the pharmaceutical and food industries. Bacteriocins produced by staphylococci are mainly lantibiotics (e.g., Pep, epidermin, epilancin K7, epicidin 280, staphylococcin C55/BacR1), that are heat-stable, unmodified peptides (e.g., aureocins A70 and 53), as well as lysostaphin, a cell-wall lytic enzyme used in many microbiological laboratories for staphylococcal DNA isolations [37]. Staphylococcins show a wide spectrum of bactericidal activity, so they have potential biotechnological applications as natural (alternative to chemical) food preservatives. Indeed, aureocins A70 and A53 produced by *S. aureus* or micrococcin P1 produced by *S. equorum* exhibit activity against *Listeria monocytogenes*, the most virulent food-borne pathogen [38]. Lantibiotic RB4 inhibits the growth of *Alicyclobacillus acidoterrestris*, a spore-forming bacteria associated with the fruit juice canning industry. Moreover, staphylococcins inhibit several human and animal pathogens, and therefore focus the attention of the pharmaceutical industry as therapeutic agents in prevention and treatment of bacterial diseases [37].

21.4 STAPHYLOCOCCAL SURFACE DISPLAY

Combinatorial protein engineering for a generation of new affinity proteins is a growing branch of modern biotechnology. Traditionally, different technologies for engineering function and structure of proteins have been broadly classified into rational and evolutionary approaches. The former is considerably complicated because proteins are large molecules built up from combinations of up to 20 different amino acids, all with distinct chemical and physical properties.

Moreover, experimental verification of the predicted changes is relatively laborious because every individual mutant requires subcloning, expression, and purification prior to characterization. In contrast to rational design, directed evolution methods are based on construction of large protein pools (i.e., combinatorial protein libraries) containing up to several billions of different variants; among them, proteins with desired traits are isolated using high-throughput screenings or selections. The cell display technique ideally fulfills requirements for such a system because the cell itself provides the physical linkage between the membrane-encapsulated DNA and the encoded recombinant protein displayed on the cell surface [39].

S. carnosus, already used for more than 40 years in the meat industry, has also found application in surface display techniques. Unlike other staphylococci, it appears predominantly as single cells or pairs, which is an important property for successful sorting of the bacteria in a flow cytometer used for selecting cells expressing proteins of interest. Moreover, unlike *E. coli*, the *S. carnosus* cell membrane is surrounded by a thick cell wall to which recombinant proteins are typically C-terminally anchored with a free N-terminal protruding from the wall itself, resulting in a high tolerance for large fusions. Furthermore, the bacterium is practically devoid of extracellular proteolytic activity, then it has been successfully employed for production of secreted recombinant proteins [40–42]. The story of staphylococcal surface display began in 1995 when Stahl and coworkers described a novel vector for display of recombinant proteins on the surface of *S. carnosus* [43]. The display vector contained a promoter and secretion signal from a lipase that is naturally expressed on surface of the related *S. hyicus*. Moreover, it utilizes a conserved cell wall anchoring region, originating from SPA, for covalent surface anchoring. The vector also contains a staphylococcal origin of replication and a chloramphenicol resistance gene, as well as a cassette for maintenance of the plasmid in *E. coli*, which greatly facilitates genetic manipulations. A few years later, the *S. carnosus* vector was optimized by transferring the expression and anchoring cassette to a more genetically stable staphylococcal vector, previously used for display of recombinant proteins on the related *S. xylosus* [44]. However, the real breakthrough of staphylococcal display for library applications came in 2007 when Lofblom et al. reported a huge improvement (over 10,000-fold to around 10^6 transformants per electroporation) of the previously comparatively low transformation efficiency, sufficient for construction of high-complexity combinatorial libraries [45]. In addition to studies on directed evolution of protein function, the staphylococcal display system has been used in several other areas. The efficient surface display of heterologous proteins (around 10,000 recombinant protein molecules per cell), nonpathogenic properties, and GRAS status of *S. carnosus* opened up for in vivo applications of this bacterium as a platform for display of antigens in the vaccine field [46,47]. The staphylococcal display system can also be exploited for various other biotechnological applications [48]. For example, it has been reported for display of a cellulose-binding domain for directed surface immobilization on cellulose materials and for display of functional enzymes on the bacterial cell surface [49,50]. Surface expression of metal-binding peptides has been intended for use as bioadsorbents in purification of industrial wastewater. Moreover, the system has been applied for display of different affinity reagents, such as functional single-chain antibody fragments and Affibody molecules (small affinity proteins) in order to use the bacteria as whole-cell diagnostic devices [51,52]. Recently, by using a *S. carnosus* surface display system, the high-affinity Affibodies interacting with human tumor necrosis factor alpha (TNF-α) and amyloid peptide beta have been generated and optimized, respectively, thus reaching a potential therapeutic effect against rheumatoid arthritis and Alzheimer's disease [53,54].

21.5 CONCLUSION

Although staphylococci are mainly considered as human and animal opportunistic pathogens, with the documented ability to host jumps, they are also used in the food industry and modern biotechnology. Many CoNS are microbial agents involved in production of fermented meats and cheese. Moreover, enzymes and proteins primarily regarded as virulence factors are used to produce chemicals and facilitate production and purification of recombinant proteins. Finally, genetically adopted strains of *S. carnosus* are applied in modern biotechnology for evolutionary protein engineering.

ACKNOWLEDGMENTS

This work was supported in part by the grants N303 813340 from the Polish Ministry of Science and Higher Education and UMO-2012/07/D/NZ2/04282 from National Science Centre (NCN, Poland) (to BW).

REFERENCES

[1] Marino M, Frigo F, Bartolomeoli I, Maifreni M. Safety-related properties of staphylococci isolated from food and food environments. J Appl Microbiol 2011;110:550–61.

[2] Martins PD, de Almeida TT, Basso AP, et al. Coagulase-positive staphylococci isolated from chicken meat: pathogenic potential and vancomycin resistance. Foodborne Pathog Dis 2013;10:771–6.

[3] Place RB, Hiestand D, Gallmann HR, Teuber M. *Staphylococcus equorum* subsp. *linens*, subsp. nov., a starter culture component for surface ripened semi-hard cheeses. Syst Appl Microbiol 2003;26:30–7.

[4] Zell C, Resch M, Rosenstein R, et al. Characterization of toxin production of coagulase-negative staphylococci isolated from food and starter cultures. Int J Food Microbiol 2008;127:246–51.

[5] Coton E, Desmonts MH, Leroy S, et al. Biodiversity of coagulase-negative Staphylococci in French cheeses, dry fermented sausages, processing environments and clinical samples. Int J Food Microbiol 2010;137:221–9.

[6] Coppola S, Mauriello G, Aponte M, et al. Microbial succession during ripening of Naples-type salami, a southern Italian fermented sausage. Meat Sci 2000;56:321–9.

[7] Aymerich T, Martin B, Garriga M, Hugas M. Microbial quality and direct PCR identification of lactic acid bacteria and nonpathogenic Staphylococci from artisanal low-acid sausages. Appl Environ Microbiol 2003;69:4583–94.

[8] Landeta G, Curiel JA, Carrascosa AV, et al. Characterization of coagulase-negative staphylococci isolated from Spanish dry cured meat products. Meat Sci 2013;93:387–96.

[9] Simonova M, Strompfova V, Marcinakova M, et al. Characterization of *Staphylococcus xylosus* and *Staphylococcus carnosus* isolated from Slovak meat products. Meat Sci 2006;73:559–64.

[10] Marty E, Buchs J, Eugster-Meier E, et al. Identification of staphylococci and dominant lactic acid bacteria in spontaneously fermented Swiss meat products using PCR-RFLP. Food Microbiol 2012;29:157–66.

[11] Marty E, Bodenmann C, Buchs J, et al. Prevalence of antibiotic resistance in coagulase-negative staphylococci from spontaneously fermented meat products and safety assessment for new starters. Int J Food Microbiol 2012;159:74–83.

[12] Even S, Leroy S, Charlier C, et al. Low occurrence of safety hazards in coagulase negative staphylococci isolated from fermented foodstuffs. Int J Food Microbiol 2010;139:87–95.

[13] Soares JC, Marques MR, Tavaria FK, et al. Biodiversity and characterization of *Staphylococcus* species isolated from a small manufacturing dairy plant in Portugal. Int J Food Microbiol 2011;146:123–9.

[14] Resch M, Nagel V, Hertel C. Antibiotic resistance of coagulase-negative staphylococci associated with food and used in starter cultures. Int J Food Microbiol 2008;127:99–104.

[15] Martuscelli M, Crudele MA, Gardini F, Suzzi G. Biogenic amine formation and oxidation by *Staphylococcus xylosus* strains from artisanal fermented sausages. Lett Appl Microbiol 2000;31:228–32.

[16] Martin B, Garriga M, Hugas M, et al. Molecular, technological and safety characterization of Gram-positive catalase-positive cocci from slightly fermented sausages. Int J Food Microbiol 2006;107:148–58.

[17] EFSA. Scientific opinion on the maintenance of the list of QPS biological agents intentionally added to food and feed. EFSA J 2010;8:1–56.

[18] Rosenstein R, Nerz C, Biswas L, et al. Genome analysis of the meat starter culture bacterium *Staphylococcus carnosus TM300*. Appl Environ Microbiol 2009;75:811–22.

[19] Irlinger F, Loux V, Bento P, et al. Genome sequence of *Staphylococcus equorum* subsp. *equorum Mu2*, isolated from a French smear-ripened cheese. J Bacteriol 2012;194:5141–2.

[20] Sung JS, Chun J, Choi S, Park W. Genome sequence of the halotolerant *Staphylococcus* sp. *strain OJ82*, isolated from Korean traditional salt-fermented seafood. J Bacteriol 2012;194:6353–4.

[21] Rosenstein R, Gotz F. Genomic differences between the food-grade *Staphylococcus carnosus* and pathogenic staphylococcal species. Int J Med Microbiol 2010;300:104–8.

[22] Rosenstein R, Gotz F. Staphylococcal lipases: biochemical and molecular characterization. Biochimie 2000;82:1005–14.

[23] Jaeger KE, Ransac S, Dijkstra BW, et al. Bacterial lipases. FEMS Microbiol Rev 1994;15:29–63.

[24] Horchani H, Aissa I, Ouertani S, et al. Staphylococcal lipases: biotechnological applications. J Mol Catal B Enzym 2012;76:125–32.

[25] Chung HR, Dunkel I, Heise F, et al. The effect of micrococcal nuclease digestion on nucleosome positioning data. PLoS ONE 2010;5:e15754.

[26] Cooke GD, Cranenburgh RM, Hanak JA, Ward JM. A modified *Escherichia coli* protein production strain expressing staphylococcal nuclease, capable of auto-hydrolysing host nucleic acid. J Biotechnol 2003;101:229–39.

[27] Alarcon B, Vicedo B, Aznar R. PCR-based procedures for detection and quantification of *Staphylococcus aureus* and their application in food. J Appl Microbiol 2006;100:352–64.

[28] Nemoto TK, Ohara-Nemoto Y, Ono T, et al. Characterization of the glutamyl endopeptidase from *Staphylococcus aureus expressed* in *Escherichia coli*. FEBS J 2008;275:573–87.

[29] Murai M, Sekiguchi K, Nishioka T, Miyoshi H. Characterization of the inhibitor binding site in mitochondrial NADH-ubiquinone oxidoreductase by photoaffinity labeling using a quinazoline-type inhibitor. Biochemistry 2009;48:688–98.

[30] Yehezkel G, Hadad N, Zaid H, et al. Nucleotide-binding sites in the voltage-dependent anion channel: characterization and localization. J Biol Chem 2006;281:5938–46.

[31] Puri M, Banerjee UC. Production, purification, and characterization of the debittering enzyme naringinase. Biotechnol Adv 2000;18:207–17.

[32] Puri M, Kaur A, Barrow CJ, Singh RS. Citrus peel influences the production of an extracellular naringinase by *Staphylococcus xylosus MAK2* in a stirred tank reactor. Appl Microbiol Biotechnol 2011;89:715–22.

[33] Proft T. Sortase-mediated protein ligation: an emerging biotechnology tool for protein modification and immobilisation. Biotechnol Lett 2010;32:1–10.

[34] Goding JW. Use of staphylococcal protein A as an immunological reagent. J Immunol Methods 1978;20:241–53.

[35] Ayyar BV, Arora S, Murphy C, O'Kennedy R. Affinity chromatography as a tool for antibody purification. Methods 2012;56:116–29.

[36] Jungbauer A, Hahn R. Engineering protein A affinity chromatography. Curr Opin Drug Discovery Dev 2004;7:248–56.

[37] Bastos MC, Ceotto H, Coelho ML, Nascimento JS. Staphylococcal antimicrobial peptides: relevant properties and potential biotechnological applications. Curr Pharm Biotechnol 2009;10:38–61.

[38] Carnio MC, Holtzel A, Rudolf M, et al. The macrocyclic peptide antibiotic micrococcin P(1) is secreted by the food-borne bacterium *Staphylococcus equorum* WS 2733 and inhibits *Listeria monocytogenes* on soft cheese. Appl Environ Microbiol 2000;66:2378–84.

[39] Kronqvist N, Malm M, Rockberg J, et al. Staphylococcal surface display in combinatorial protein engineering and epitope mapping of antibodies. Recent Pat Biotechnol 2010;4:171–82.

[40] Gotz F. *Staphylococcus carnosus*: a new host organism for gene cloning and protein production. Soc Appl Bacteriol Symp Ser 1990;19:49S–53S.

[41] Dilsen S, Paul W, Sandgathe A, et al. Fed-batch production of recombinant human calcitonin precursor fusion protein using *Staphylococcus carnosus* as an expression-secretion system. Appl Microbiol Biotechnol 2000;54:361–9.

[42] Hansson M, Samuelson P, Nguyen TN, Stahl S. General expression vectors for *Staphylococcus carnosus* enabled efficient production of the outer membrane protein A of *Klebsiella pneumoniae*. FEMS Microbiol Lett 2002;210:263–70.

[43] Samuelson P, Hansson M, Ahlborg N, et al. Cell surface display of recombinant proteins on *Staphylococcus carnosus*. J Bacteriol 1995;177:1470–6.

[44] Wernerus H, Stahl S. Vector engineering to improve a staphylococcal surface display system. FEMS Microbiol Lett 2002;212:47–54.

[45] Lofblom J, Kronqvist N, Uhlen M, et al. Optimization of electroporation-mediated transformation: *Staphylococcus carnosus* as model organism. J Appl Microbiol 2007;102:736–47.

[46] Liljeqvist S, Stahl S. Production of recombinant subunit vaccines: protein immunogens, live delivery systems and nucleic acid vaccines. J Biotechnol 1999;73:1–33.

[47] Cano F, Liljeqvist S, Nguyen TN, et al. A surface-displayed cholera toxin B peptide improves antibody responses using food-grade staphylococci for mucosal subunit vaccine delivery. FEMS Immunol Med Microbiol 1999;25:289–98.

[48] Wernerus H, Stahl S. Biotechnological applications for surface-engineered bacteria. Biotechnol Appl Biochem 2004;40:209–28.

[49] Lehtio J, Wernerus H, Samuelson P, et al. Directed immobilization of recombinant staphylococci on cotton fibers by functional display of a fungal cellulose-binding domain. FEMS Microbiol Lett 2001;195:197–204.

[50] Strauss A, Gotz F. In vivo immobilization of enzymatically active polypeptides on the cell surface of *Staphylococcus carnosus*. Mol Microbiol 1996;21:491–500.

[51] Lofblom J, Feldwisch J, Tolmachev V, et al. Affibody molecules: engineered proteins for therapeutic, diagnostic and biotechnological applications. FEBS Lett 2010;584:2670–80.

[52] Gunneriusson E, Samuelson P, Ringdahl J, et al. Staphylococcal surface display of immunoglobulin A (IgA)- and IgE-specific in vitro-selected binding proteins (affibodies) based on *Staphylococcus aureus* protein A. Appl Environ Microbiol 1999;65:4134–40.

[53] Lindberg H, Johansson A, Hard T, et al. Staphylococcal display for combinatorial protein engineering of a head-to-tail affibody dimer binding the Alzheimer amyloid-beta peptide. Biotechnol J 2013;8:139–45.

[54] Kronqvist N, Lofblom J, Severa D, et al. Simplified characterization through site-specific protease-mediated release of affinity proteins selected by staphylococcal display. FEMS Microbiol Lett 2008;278:128–36.

Index

Note: Page numbers followed by *f* indicate figures, and *t* indicate tables.

Printed in the United States
By Bookmasters